Essential Developmental Biology

The Practical Approach Series

SERIES EDITORS

D. RICKWOOD
Department of Biology, University of Essex
Wivenhoe Park, Colchester, Essex CO4 3SQ, UK

B. D. HAMES
Department of Biochemistry and Molecular Biology, University of Leeds
Leeds LS2 9JT, UK

Affinity Chromatography

Anaerobic Microbiology

Animal Cell Culture
(2nd Edition)

Animal Virus Pathogenesis

Antibodies I and II

Behavioural Neuroscience

Biochemical Toxicology

Biological Data Analysis

Biological Membranes

Biomechanics—Materials

Biomechanics—Structures
and Systems

Biosensors

Carbohydrate Analysis

Cell–Cell Interactions

The Cell Cycle

Cell Growth and Division

Cellular Calcium

Cellular Interactions in
Development

Cellular Neurobiology

Centrifugation (2nd Edition)

Clinical Immunology

Computers in Microbiology

Crystallization of Nucleic
Acids and Proteins

Cytokines

The Cytoskeleton

Diagnostic Molecular
Pathology I and II

Directed Mutagenesis

DNA Cloning I, II,
and III

Drosophila

Electron Microscopy in
Biology

Electron Microscopy in
Molecular Biology

Electrophysiology

Enzyme Assays

Essential Molecular Biology I and II

Essential Developmental Biology

Eukaryotic Gene Transcription

Experimental Neuroanatomy

Fermentation

Flow Cytometry

Gel Electrophoresis of Nucleic Acids (2nd Edition)

Gel Electrophoresis of Proteins (2nd Edition)

Gene Targeting

Gene Transcription

Genome Analysis

Glycobiology

Growth Factors

Haemopoiesis

Histocompatibility Testing

HPLC of Macromolecules

HPLC of Small Molecules

Human Cytogenetics I and II (2nd Edition)

Human Genetic Disease Analysis

Immobilised Cells and Enzymes

Immunocytochemistry

In Situ Hybridization

Iodinated Density Gradient Media

Light Microscopy in Biology

Lipid Analysis

Lipid Modification of Proteins

Lipoprotein Analysis

Liposomes

Lymphocytes

Mammalian Cell Biotechnology

Mammalian Development

Medical Bacteriology

Medical Mycology

Microcomputers in Biochemistry

Microcomputers in Biology

Microcomputers in Physiology

Mitochondria

Molecular Genetic Analysis of Populations

Molecular Imaging in Neuroscience

Molecular Neurobiology

Essential Developmental Biology

A Practical Approach

Edited by

CLAUDIO D. STERN

Department of Human Anatomy
South Parks Road
Oxford OX1 3QX

and

PETER W. H. HOLLAND

Department of Zoology
South Parks Road
Oxford OX1 3PS

OXFORD UNIVERSITY PRESS
Oxford New York Tokyo

Oxford University Press, Walton Street, Oxford OX2 6DP
Oxford New York Toronto
Delhi Bombay Calcutta Madras Karachi
Kuala Lumpur Singapore Hong Kong Tokyo
Nairobi Dar es Salaam Cape Town
Melbourne Auckland Madrid
and associated companies in
Berlin Ibadan

Oxford is a trade mark of Oxford University Press

A Practical Approach 🔵 *is a registered trade mark*
of the Chancellor, Masters, and Scholars of the University of Oxford
trading as Oxford University Press

Published in the United States
by Oxford University Press Inc., New York

A catalogue record for this book is available from the British Library

Library of Congress Cataloging in Publication Data
(Data available)
ISBN 0–19–9634238 (h/b)
ISBN 0–19–963422–X (p/b)

Typeset by Cambrian Typesetters, Frimley, Surrey
Printed in Great Britain by
Information Press Ltd, Eynsham, Oxford

Preface

SINCE the last decade, developmental biology has entered a state of profound and exciting intellectual renewal. The mysterious mechanisms which transform the egg's single cell into a complex organism have always been a central issue of biology, but their study has seen somewhat different fortunes over the years. Under the leadership of Hans Spemann and his followers, experimental embryology assumed a commanding position during the first half of this century. The remarkable successes of cell and molecular biology then turned the emphasis to cellular, subcellular, and molecular levels, drawing developmental biologists towards a more reductionist approach. Experimental embryology was at the time draped with the ambiguous epithet of 'classical' science. Fortunately the efforts of experimental embryologists to submit animal development to causal analysis and to discover its rules were not fruitless and, although left unheeded for certain decades, their acquisitions have not fallen into oblivion. The extraordinary revival that we are witness to today is in fact a return to the old issues, but with the powerful new techniques of cell and molecular biology and of immunology. During the past few years a re-evaluation of old experiments and concepts has been undertaken. The solutions of problems that evaded us for nearly a century have been given.

That 'classical' embryology has been rejuvenated is attested by many books showing the major progress accomplished during recent years. The present one, conceived and edited by Claudio Stern and Peter Holland is the most obvious sign of the remarkable dynamism exhibited by the field. Stern and Holland organized a practical course for developmental biologists at Oxford in 1991. This book has been written especially for those who did not have the chance to attend the course, in an attempt to deliver, as efficiently as possible, the most important messages conveyed to the participants. Representatives of some of the most active laboratories in the field describe the experimental methods that they currently use in their work, in a detailed and highly didactic manner. The book is an authoritative report cataloguing the know-how of experimental methods applied to the study of developmental problems, ranging from *in vivo* manipulations of embryos, and tissue culture techniques for whole embryos and embryonic cells, to the most modern applications of molecular biology, including methodology for targeting mutations by homologous recombination in the mouse.

A striking character of the field, displayed in this book, is the variety and richness of the contemporary approach. Developmental biology is far from being confined to a particular group of organisms or to any system within an organism. It deals with the study of developing systems at all possible scales, from molecules to cells, tissues, organs, and organisms, as they evolve as a

ix

function of time. The choice of models is wide. Some are particularly appropriate for answering certain questions or applying certain technical approaches. *Drosophila*, *Caenorhabditis elegans*, *Xenopus*, mouse, and avian embryos are all favoured by contemporary researchers, but others such as fish, echinoderms, ascidians, and amphioxus are emerging, or will emerge, as interesting systems for future developmental, genetic, or evolutionary studies. One can find here how to handle any of these systems and even more importantly, one will find help in selecting the most suitable model.

This book should be for a long while to come the favourite benchwork companion of many developmental biologists.

Nicole Le Douarin

Contents

3. Embryos and larvae of invertebrate deuterostomes   21

Nicholas D. Holland and Linda Z. Holland

Contents

Contents

15. Laser ablation of cells

*Judith Eisen, Rachel M. Warga, Lois G. Edgar, and
William B. Wood*

III CELLULAR TECHNIQUES

16. Studying cell movements *in vivo* by time-lapse video microscopy

Peter V. Thorogood and Dee Amanze

Contents

20. Introduction of genes using retroviral vectors   179

Jack Price

IV MOLECULAR TECHNIQUES

21. Immunocytochemistry of embryonic material   193

Claudio D. Stern

25. Cloning genes using the polymerase chain reaction 243

Peter W. H. Holland

26. *In situ* hybridization 257

David G. Wilkinson

Appendix 1. Stage tables   277

Appendix 2. Suppliers of specialist items   321

Index   325

Contributors

DEE AMANZE
School of Biochemistry, University of Birmingham, Birmingham B15 2TT, UK.

ELISABETH DUPIN
CNRS–Collège de France, Institut d'Embryologie Cellulaire et Moléculaire, 49bis, avenue de la Belle Gabrielle, 94130 Nogent sur Marne, France.

LOIS G. EDGAR
Department of Molecular, Cellular and Developmental Biology, Porter Biosciences Building, Campus Box 347, Boulder, CO 80309–0347, USA.

JUDITH EISEN
Institute of Neuroscience, Huestis Hall, University of Oregon, Eugene, OR 97403, USA.

RONALD M. EVANS
Howard Hughes Medical Institute, Gene Expression Laboratory, The Salk Institute for Biological Studies, PO Box 85800, San Diego, CA 92186–5800, USA.

ZANDY FORBES
ICRF Developmental Biology Unit, Department of Zoology, South Parks Road, Oxford OX1 3PS, UK.

JOHN B. GURDON
Wellcome/CRC Institute of Cancer and Developmental Biology, Tennis Court Road, Cambridge CB2 1QR, UK.

LINDA Z. HOLLAND
Marine Biology Research Division, Scripps Institution of Oceanography, University of California, San Diego, CA 92093, USA.

NICHOLAS D. HOLLAND
Marine Biology Research Division, Scripps Institution of Oceanography, University of California, San Diego, CA 92093, USA.

PETER W. H. HOLLAND
Department of Zoology, South Parks Road, Oxford OX1 3PS, UK.

PHILIP INGHAM
ICRF Developmental Biology Unit, Department of Zoology, South Parks Road, Oxford OX1 3PS, UK.

Contributors

AKIRA KAKIZUKA
Gene Expression Laboratory, The Salk Institute for Biological Studies, PO Box 85800, San Diego, CA 92186–5800, USA.

PETER KOOPMAN
Centre for Molecular Biology and Biotechnology, The University of Queensland, Brisbane 4072, Australia.

NICOLE M. LE DOUARIN
CNRS–Collège de France, Institut d'Embryologie Cellulaire et Moléculaire, 49bis, avenue de la Belle Gabrielle, 94130 Nogent sur Marne, France.

GAIL R. MARTIN
1334 Sciences, Department of Anatomy, School of Medicine, University of California–San Francisco, San Francisco, CA 94143–0452, USA.

GILLIAN M. MORRISS-KAY
Department of Human Anatomy, South Parks Road, Oxford OX1 3QX, UK.

JACK PRICE
National Institute for Medical Research, The Ridgeway, Mill Hill, London NW7 1AA, UK.

ELIZABETH J. ROBERTSON
The Biological Laboratories, Harvard University, 16 Divinity Avenue, Cambridge, MA 02138, USA.

ARIEL RUIZ i ALTABA
Center for Neurobiology and Behavior, College of Physicians and Surgeons of Columbia University, 722 West 168th Street, New York 10032, USA.

CLAUDIO D. STERN
Department of Human Anatomy, South Parks Road, Oxford OX1 3QX, UK.

PETER V. THOROGOOD
Developmental Biology Unit, Division of Cell and Molecular Biology, Institute of Child Health, 30 Guilford Street, London WC1N 1EH, UK.

CHERYLL TICKLE
Departments of Anatomy and Developmental Biology, University College and Middlesex School of Medicine, Windeyer Building, Cleveland Street, London W1P 6DB, UK.

KAZUHIKO UMESONO
Howard Hughes Medical Institute and Gene Expression Laboratory, The Salk Institute for Biological Studies, PO Box 85800, San Diego, CA 92186–5800, USA.

Contributors

RACHEL M. WARGA
Max-Planck Institut für Entwicklungsbiologie, Spemannstrasse 35, D-7400 Tübingen, Germany.

DAVID G. WILKINSON
National Institute for Medical Research, The Ridgeway, Mill Hill, London NW7 1AA, UK.

WILLIAM B. WOOD
Department of Molecular, Cellular and Developmental Biology, Porter Biosciences Building, Campus Box 347, Boulder, CO 80309–0347, USA.

RUTH T. YU
Gene Expression Laboratory, The Salk Institute for Biological Studies, PO Box 85800, San Diego, California 92186–5800, USA.

Acknowledgements

The concept of this book grew out of a practical course on 'Modern Techniques in Developmental Biology', sponsored by the British Society for Developmental Biology, the Company of Biologists, DuPont (UK) Ltd, the Nuffield Foundation, the Universities Funding Council, and the University of Oxford. The instructors and participants must also be thanked for their good humour and stamina.

The techniques covered in this book have been greatly expanded and updated, in an attempt to produce as complete a practical guide as possible to the essential techniques used in modern experimental biology. As an important appendix, we have included developmental stage tables for zebrafish, *Xenopus*, and the chick; these are printed with the kind permission of Drs Monte Westerfield, Pieter Nieuwkoop, Hefzibah Eyal-Giladi, and Viktor Hamburger and of Elsevier Science Publishers, Academic Press, and John Wiley and Sons. We are also grateful to Pat Elliott for her help at various stages of the editing process and to Terry Richards for producing the cover illustration.

Abbreviations

AC	alternating current
bp	base pairs
BSA	bovine serum albumin
cDNA	complementary DNA
c.p.m.	counts per minute
CHAPS	3-[(3-cholamidopropyl) dimethylammonio]-1-propane sulphonate
Ci	curie
CMF	calcium- and magnesium-free
CNS	central nervous system
c.p.s.	counts per second
d.p.m.	disintegrations per minute
DABCO	1,4-diazobicyclo (2,2,2)-octane
DAB	diaminobenzidine
DC	direct current
DEAE	diethylaminoethyl
DEPC	diethylpyrocarbonate
DiI	1,1′-dioctadecyl-3,3,3′,3′-tetramethyl indocarbocyanine perchlorate
DiO	3,3′-dioctadecyloxacarbocyanine perchlorate
DIC	differential interference contrast
DMEM	Dulbecco's modified Eagle's medium
DMSO	dimethyl sulphoxide
DNAase	deoxyribonuclease
d.p.c.	days post-coitum
DTT	dithiothreitol
EDTA	ethylenediaminetetraacetic acid
EGTA	ethylene glycol-bis(β-aminoethyl ether) N,N,N'-tetraacetic acid
ELISA	enzyme-linked immunosorbent assay
ES cells	embryonic stem cells
FCS	fetal calf serum
FIAU	1(2-deoxy-2-fluoro-β-D-arabinofuranosyl-5-iodouracil
FITC	fluorescein isothiocyanate
GANC	Gancyclovir
Hepes	N-2-hydroxyethyl piperazine-N'-2-ethanesulphonic acid
HPLC	high performance liquid chromatography
HRP	horseradish peroxidase
i.d.	internal diameter
IPTG	isopropyl β-D-thiogalactopyranoside
kb	kilobases
kd	kilodaltons
LTR	long terminal repeat
MEM	minimum essential medium
NC	neural crest

NP-40	Nonidet P-40
o.d.	outer diameter
p.c.	post-coitum
p.f.u.	plaque-forming units
PAGE	polyacrylamide gel electrophoresis
PBS	phosphate-buffered saline
PCR	polymerase chain reaction
PEG	polyethylene glycol
pI	isoelectric point
PIPES	piperazine-N,N'-bis(2-ethane sulphonic acid)
PMSF	phenylmethylsulphonylfluoride
PNS	peripheral nervous system
RACE	rapid amplification of cDNA ends
r.p.m.	revolutions per minute
RNAase	ribonuclease
SDS	sodium dodecyl sulphate
SEM	scanning electron microscopy
TBS	Tris-buffered saline
TEM	transmission electron microscopy
TLC	thin-layer chromatography
Tris	tris (hydroxymethyl)-aminomethane
Tween-20	polyoxyethylenesorbitan monolaurate
UV	ultraviolet
v/v	volume/volume
V	volt
w/v	weight/volume
w/w	weight/weight
X-Gal	5-bromo-4-chloro-3-indolyl-β-D-galactopyranoside

I
Basic techniques: obtaining and handling embryos

1

Drosophila embryos

ZANDY FORBES and PHILIP INGHAM

1. Growing healthy flies

Adult fruitflies of the species *Drosophila melanogaster* are easily maintained in the laboratory. At 25 °C, the generation time of *Drosophila* is 9 days and embryogenesis is completed in 22 h. Traditionally, large numbers of flies are grown in half-pint milk bottles with a solid nutrient medium in the base; alternative containers include plastic fruit juice bottles or urine sample containers. Stocks of flies can be maintained at 18 °C in smaller tubes with the same medium. For 25 bottles or 120 tubes containing nutrient medium, boil 8 g agar in 1 litre tap water for 3 min, add 80 g corn (maize) meal, 18 g dried yeast, and 10 g soya flour (premixed with some of the water), plus 80 g malt extract and 40 g molasses. Boil for 20 min, cool to 50–80 °C, add 6.6 ml acid mix (a mix of 500 ml propionic acid and 32 ml phosphoric acid), and pour.

In order to produce healthy flies from which to collect large numbers of staged embryos, it is worth taking some care over the culture conditions. To avoid crowded cultures, which lead to small, growth-retarded flies with low fecundity, the parental stocks should be transferred to fresh bottles every 1–3 days, depending on the number of eggs that are laid; there should never be so many larvae in a bottle that the medium becomes sloppy. Once the flies from which eggs will be collected have eclosed from their pupae, transfer them to fresh bottles of medium supplemented with a sprinkling of live bakers' yeast and containing some folded tissue paper (this helps prevent them sticking to the yeast); keep them on this food for two or three days before starting to collect eggs.

Stocks of recessive lethal mutants are invariably maintained over balancer chromosomes. Although these stocks are usually reasonably healthy and reproduce themselves without much problem, it is worth outcrossing them to wild-type flies to produce really vigorous parents from which to collect eggs.

Egg-laying chambers can be made from any suitable container; we use plastic jars with a diameter of about 5 cm. For efficient ventilation, a small hole should be cut in the bottom and sealed with a fine mesh. The flies (between 100 and 200) are emptied into these chambers and a 5 cm Petri dish containing apple juice agar medium with a little live yeast paste in the centre

(*Protocol 1*), is taped to the open end of the container. The cages should be inverted or left on their side in a dark, quiet place at 25 °C for the appropriate time; the flies will lay most or all of their eggs on the apple juice agar. Disturbing the cages will cause the females to stop laying and withhold their eggs, thus reducing the number laid and making the staging of the embryos less accurate.

If large numbers of eggs are required (several thousand per hour), for instance for extraction of proteins or nucleic acids, the whole procedure can be scaled up: use Perspex (Plexiglass) population cages and large petri dishes of apple juice agar.

Protocol 1. Making apple juice plates[a]

1. Add 2 × 17.5 g agar (Oxoid No. 3 Technical) to 2 × 500 ml distilled water in 2 × 1 litre flasks. Autoclave for 20 min. Take care not to swirl after autoclaving as the agar may boil over.

2. Mix 25 g sucrose with 500 ml distilled water in a 1 litre flask. Warm in a microwave oven at full power for 4 min, then add 500 ml apple juice.

3. In a 4 litre beaker mix (in order): (a) the molten agar from step 1, (b) the sucrose/apple juice from step 2, (c) 4 g Nipagin (Sigma) dissolved in 20 ml 95% ethanol. Stir with a glass rod, avoiding bubbles; remove any froth with a tissue.

4. Using a 500 ml beaker, pour the mixture into 5 cm Petri dishes, or larger plates as appropriate, to half fill them.

5. When the mixture sets, put the plates back in their plastic sleeves, date them and store at 4 °C. They will keep for several weeks.

6. Before use, allow plates to warm to room temperature, remove condensation with a paper towel, and put a smear of yeast paste[b] in the middle of each plate.

[a] Sufficient for approximately 200 × 5 cm Petri dishes.
[b] Paste made from dried bakers' yeast dissolved in water. Keeps at 4 °C for 3–4 days.

---

2. Making larval cuticle preparations

Preparations of larval cuticle are used to identify morphological changes resultant from disrupted embryonic patterning. These preparations are usually made from larvae collected just before they hatch from the egg, otherwise yeast ingested by the feeding larvae can spoil the preparations.

Protocol 2. Cuticle preparations

1. Introduce flies to egg-laying chambers with 5 cm apple juice plates, and incubate at 25 °C in the dark. Change the plates after 5–6 h or when sufficient eggs have been laid.

2. For preparations of wild type larvae: use plates with *unhatched* eggs 20–22 h after collecting. For preparations of homozygous mutants from a stock carrying an embryonic lethal mutation: use the unhatched eggs remaining 24 h after laying (wild-type larvae will have hatched).

3. Arrange a drop of 14% sodium hypochlorite solution (BDH) and two drops of water in a Petri dish. Collect the unhatched eggs required using fine forceps, and place in the drop of sodium hypochlorite; leave for about 2 min or until the chorions have dissolved (watch under a dissecting microscope).

4. Rinse the larvae by transferring them through the two drops of water. Either mount directly (go to step **6**) or, to make perfect preparations, remove the vitelline membranes before mounting by including step **5**.

5. Using a piece of tissue, remove all the water so that the larvae are in direct contact with the plastic of the Petri dish. The vitelline membranes should adhere to the plastic. Gently poke each larva with fine forceps or a tungsten needle to free it from the membrane, then place the larvae in another drop of water to prevent them drying out prior to mounting.

6. Transfer the larvae into a drop of 1:1 lactic acid/Hoyer's medium[a] on a glass microscope slide. Arrange the larvae using fine forceps; do not use too much medium.

7. Place a coverslip gently onto the drop of medium and press firmly using two pairs of forceps to squash the larvae. This should rupture the vitelline membranes (if they were not removed in step **5**), but it is important to avoid all lateral movement otherwise the cuticle will be crushed.

8. Incubate the slides at 60 °C for at least 2 h; once the larvae have cleared, seal around the edge of the coverslip with nail varnish.

9. To analyse the preparations, use a compound microscope with phase contrast, Nomarski DIC, or dark field optics. To see the whole larva, a 16× objective is optimal. For more detailed analysis, 25× and 40× objectives are useful.

10. To photograph the specimens, we use Kodak Technical Pan film. Dark field preparations produce high contrast images; to ensure good resolution, underexpose the film and develop for low contrast (for Technical Pan, use the film at a rating of 400 ASA and develop with Kodak HC110 developer at a dilution of 1:19 for 8 min at 20 °C). For

Protocol 2. *Continued*

> phase contrast and DIC images it is desirable to increase the contrast in
> the photographic process: assume a film rating of 100 ASA and develop
> in Kodak D19 developer for 4 min at 20 °C.

---

[a] Hoyer's medium. Add 30 g gum arabica to 50 ml distilled water, stir overnight to dissolve,
then gradually add 200 g chloral hydrate (anhydrous) and 20 g glycerol. Centrifuge at 25 000 *g* for
30 min. The medium is stable for years at room temperature.

---

3. Collection and preparation of *Drosophila* embryos for immunocytochemistry, *in situ* hybridization, and electron microscopy

3.1 Collection and dechorionation of embryos

To obtain embryos for either immunocytochemistry, *in situ* hybridization, or
electron microscopy, eggs are collected on apple juice plates as described in
Section 1. If a broad range of embryonic stages is required, collect eggs
overnight and fix the following morning. Note that the frequency of laying
declines over time, so the distribution of stages will be skewed. If carefully
staged embryos are required, collect eggs for 2 h periods and age
appropriately. Before fixation, the chorion must be removed from the
embryos (*Protocol 3*). If the eggs or embryos are to be used for micromanipulation or microinjection, do not use *Protocol 3*; manual dechorionation
should be performed (see Chapter 10).

Protocol 3. Dechorionation using sodium hypochlorite

1. Squirt a few drops of water onto the apple juice plate and detach the eggs
   using a fine paint brush. Transfer them into an egg collecting basket[a].

2. Rinse the eggs thoroughly with water to remove excess yeast, blot dry on
   a paper tissue, and place in a small dish of 14% sodium hypochlorite
   solution to dissolve the chorions.

3. Using a pasteur pipette, run the sodium hypochlorite solution through the
   basket a few times, to ensure even dechorionation of the embryos.

4. Once the chorions have dissolved, remove the sodium hypochlorite and
   rinse the embryos well with distilled water. Go straight to fixation
   (*Protocol 4* or *Protocol 5*).

---

[a] These are easily made from plastic caps of scintillation vials. Cut the end off the cap and
replace it with fine stainless steel mesh, stuck on by quickly melting the rim of the cap onto the
mesh on a hot plate.

3.2 Fixation and methanol removal of vitelline membranes

For immunocytochemistry and *in situ* hybridization, embryos are dechorion-ated as described in *Protocol 3*, then the embryos are fixed in formaldehyde and the vitelline membranes removed using methanol (*Protocol 4*). Altern-atively vitelline membranes can be removed manually (*Protocol 5*); this is more difficult, slow, and generally only used if it is necessary to avoid the embryos coming into contact with methanol.

Protocol 4. Standard fixation method

1. Mix equal volumes of 4% formaldehyde[a] in PBS[b] and *n*-heptane (1 ml of mixture in a microcentrifuge tube for small collections of eggs, or 10 ml in a 15 ml centrifuge tube for several thousand eggs). This forms a biphasic mixture: the heptane acts to permeabilize the vitelline membrane that surrounds the embryo, the formaldehyde is the fixative.

2. Using tissue paper, remove excess water from the dechorionated embryos from *Protocol 3*, step **4** (but do not let the embryos get dry). Scoop the embryos up with a fine paint brush and transfer them to the fix/heptane prepared above.

3. Constantly agitate the embryos in the fix/heptane mix. For RNA detection, fixation for 20 min at room temperature works well. For immunocytochemistry, fixation times vary for different antigens and a range of times between 5 min and 1 hour should be tried.

4. Either[c]: (a) remove the aqueous (lower) layer and add an equal volume of methanol. Shake or vortex the tube for 30 sec and then leave for a further 30 sec. Or: (b) using a Pasteur pipette, transfer the fixed embryos to a 1:1 mix of methanol/heptane, precooled to −70 °C. Intermittently shake and chill at −70 °C for a few minutes. In both (a) and (b), the embryos will pop out of their vitelline membranes and fall to the bottom of the tube. Tipping the tube onto its side will help this process[d].

5. Collect the devitellinized embryos from the bottom of the tube with a pasteur pipette and transfer them to a fresh tube. Rinse two or three times with 100% methanol. The embryos can be stored indefinitely at −20 °C[e].

---

[a] Made by dilution from 40% formaldehyde stock (BDH). Formaldehyde is a good fixative for RNA and many antigens, but optimal fixatives will vary between antigens.
[b] PBS, phosphate-buffered saline. Alternative buffers which can be tried include BBS (10 mM Tris-HCl pH 7.5, 50 mM NaCl, 40 mM MgCl$_2$, 5 mM CaCl$_2$, 20 mM glucose, 50 mM sucrose); PEM (0.1 M Pipes, 2 mM EGTA, 1 mM MgSO$_4$ pH 8.0) and PEM/PLP (PEM plus 10 mM sodium periodate, 0.75 M lysine).
[c] Alternative (a) is the standard method; (b) gives better preservation of cellular morphology and is used if specimens are subsequently to be sectioned.

Protocol 3. *Continued*

^{d} De-vitellinization using methanol may fail if the embryos are poorly fixed or if all the chorions had not been completely dissolved prior to fixation; if there are too many embryos a greater proportion may fail to pop out.

^{e} Some antigens are lost or denatured if the embryos are left in methanol at room temperature; it is advisable to rinse devitellinized embryos rapidly and put them at $-20\ °C$ immediately if they are not to be used directly.

3.3 Fixation and manual removal of vitelline membranes

It is sometimes important that embryos do not come into contact with methanol, for example, for detection of particular antigens or if they are to be used for electron microscopy. In these cases, vitelline membranes must be removed by hand. After dechorionation (*Protocol 3*), follow *Protocol 5.*

Protocol 5. Fixation with manual removal of vitelline membranes

1. Fix embryos as described in steps **1** to **3** of *Protocol 4*, but in 5 ml glass bottles. This allows the embryos with their vitelline membranes to be removed more easily than from plastic tubes.

2. Remove the aqueous layer and, using a pasteur pipette, transfer the embryos (in heptane) into a glass embryo dish. Allow the heptane to evaporate (this can be speeded up by blowing gently on it).

3. Pick the embryos up with a paint brush and transfer them to a piece of double-sided sticky tape (Sellotape or Scotch tape) stuck to the inside of a small Petri dish lid. Do this rapidly since the embryos are very prone to desiccation once their membranes have been permeabilized. As soon as the embryos are on the tape cover them either with buffer or more fixative.

4. Rupture the vitelline membranes with fine tungsten needles and gently push the embryos out. Transfer the embryos to a microcentrifuge tube, rinse in PBS, and proceed directly with immunocytochemistry or other technique.

4. *In situ* hybridization and immunocytochemistry

Prior to performing *in situ* hybridization or immunocytochemistry, on either sections or whole mounts, *Drosophila* embryos should be dechorionated (*Protocol 3*), fixed and devitellinized (either *Protocol 4* or *Protocol 5*). Protocols for *in situ* hybridization for *Drosophila* embryos are given in Chapter 26; general protocols for immunocytochemistry are given in Chapter 21.

5. Detection of β-galactosidase activity in embryos

Reporter gene technology is widely used in *Drosophila* developmental biology. The most widely used reporter gene is *lacZ*, which produces the enzyme *β*-galactosidase. Detection of this enzyme in whole mounts of *Drosophila* embryos is described in *Protocol 6*.

Protocol 6. X-Gal staining to reveal *β*-galactosidase activity

1. Collect eggs and dechorionate as described in *Protocol 3*.

2. Rinse the eggs thoroughly with water and blot the basket dry. Place the basket in *n*-heptane that has been saturated with 12.5% glutaraldehyde in 1 × PBS[a]; leave for 15 min.

3. Blot the basket dry, pick up the embryos with a paint brush, and place them on a piece of double-sided sticky tape (Sellotape or Scotch tape) on a small Petri dish lid. Immediately cover the embryos with a drop of PTW (1 × PBS, 0.1% Tween-20).

4. Remove the PTW and replace with X-Gal staining solution[b].

5. Replace the solution with a drop of staining solution containing 0.2% X-Gal, and incubate at 37 °C until blue staining starts to appear (from a few minutes to several hours, depending on the level of expression).

6. When sufficiently stained, remove the embryos from their vitelline membranes (rupture the membranes with fine tungsten needles and gently push the embryos out).

7. Transfer the embryos to a microcentrifuge tube, rinse in PTW, then either mount for analysis or store at 4 °C in PTW.

[a] Saturate by making a 1:1 mixture of the two liquids and shaking vigorously for several minutes.
[b] X-Gal staining solution: 10 mM sodium phosphate buffer pH 7.2, 150 mM NaCl, 3.1 mM potassium ferricyanide, 3.2 mM potassium ferrocyanide.

---

6. Mounting stained embryos for microscopic analysis and photography

Embryos stained by whole mount *in situ* hybridization, immunocytochemistry, or X-Gal, can be mounted as either permanent (*Protocol 7*), semi-permanent (*Protocol 8*), or temporary (*Protocol 9*) preparations.

Protocol 7. Permanent preparations of *Drosophila* embryo whole mounts

1. Dehydrate stained embryos quickly through 70%, 95%, and 100% ethanol.
2. Remove the ethanol and add 1 ml of JB-4 (Polysciences) 'A + catalyst' solution. Allow to infiltrate until all the embryos have sunk to the bottom of the tube.
3. Mix 1 ml of 'A + catalyst' with 40 μl of JB-4 'solution B'. Keep this cool to slow down polymerization.
4. Remove 'A + catalyst' from the embryos and replace with the 'A + catalyst + B' mixture. Mix thoroughly and immediately pipette onto slides; the solution will not polymerize once on the slides due to inhibition by air.
5. Using forceps or a needle, arrange the embryos as required and cover with a coverslip. As air is now excluded, polymerization will occur. Leave overnight.

Protocol 8. Semi-permanent preparations of *Drosophila* embryo whole mounts

1. Dehydrate stained embryos quickly through 70%, 95%, and 100% ethanol or propanol.
2. Replace solution with GMM^a for 5 min and pipette embryos onto slides.
3. Place slides on a hot plate. The specimens should clear in 15–30 min and the medium should set as a semi-fluid within a few hours.

^{a} GMM: Add 1.5–1.75 g crushed, dried Canada balsam per 1 ml methyl salicylate. Stir until the Canada balsam dissolves completely (1–2 days), but do not heat. The medium may be thinned by the addition of small amounts of methyl salicylate.

Protocol 9. Temporary preparations of *Drosophila* embryo whole mounts

1. Transfer the embryos to a watch glass containing 80% glycerol in PBS.
2. Transfer an individual embryo in a drop of glycerol to microscope slide, between two strips of coverslip glass.
3. Lower on a coverslip. The embryo can be rolled into the desired orientation, or flattened by careful removal of supporting coverslip glass.

2

Nematode embryos

LOIS G. EDGAR and WILLIAM B. WOOD

1. Introduction

The free-living soil nematode *Caenorhabditis elegans* has many advantages for investigation of development and physiology in a simple animal. Its three-day life cycle, simple growth requirements, hermaphroditic mode of development, and indefinite survival at liquid nitrogen temperature are superbly suited to genetic analysis. Moreover, its transparency throughout the life cycle, small size, simple anatomy, and invariant cell lineage have made possible detailed descriptions of normal and mutant development at the level of individual cells (reviewed in ref. 1). This chapter describes methods for handling and cultivating this nematode and for collecting and studying its embryos.

Although the procedures given were developed for *C. elegans*, specifically the wild-type N2 strain var. Bristol (2), most are applicable to other free-living nematodes and those dealing with embryo manipulation may also apply to parasitic nematodes. For more information on *C. elegans* biology see ref. 1, especially the section by J. Sulston and J. Hodgkin for additional methods. Strains are available from the Caenorhabditis Genetics Center, Department of Genetics and Cell Biology, 250 Biological Sciences Center, University of Minnesota, St Paul, MN 55108-1095, USA, which also publishes the Worm Breeder's Gazette, a quarterly research newsletter.

2. Cultivation and handling of *C. elegans*

2.1 Preparing agar plates

C. elegans is conveniently grown on 6 cm NGM agar plates, seeded with a lawn of *Escherichia coli* bacteria. To make NGM agar, autoclave a mixture of 975 ml water, 3 g NaCl, 17 g agar, 2.5 g peptone, 1 ml cholesterol (5 mg/ml in ethanol), then, using sterile technique, add 1 ml 1 M $CaCl_2$, 1 ml 1 M $MgSO_4$, and 25 ml 1 M potassium phosphate pH 6, mixing after each addition.

2.1.1 Seeding plates with bacteria

A slow-growing uracil-requiring *E. coli* strain, OP50, is used to keep the lawn thin for optimal observation; auxotrophic strains can be used on larger plates for growing larger populations of worms. Bacteria are grown overnight from a single colony in any rich broth medium; the culture can be stored at 4 °C for up to several months. Keep NGM agar plates for a few days at room temperature, to allow evaporation of excess moisture, and spread with a few drops of bacterial suspension using sterile techniques, taking care not to damage the agar surface. Incubate at room temperature overnight before nematodes are added.

2.1.2 Transferring animals

Individual nematodes are conveniently transferred under a dissecting microscope using a 1 to 3 cm length of platinum wire (32 gauge) sealed into the end of a Pasteur pipette and flattened at the tip. The wire pick is flamed between transfers (to avoid cross-contamination) and touched to the lawn to scoop up bacteria, which are sticky and allow worms to be picked up more easily. If an animal is not immediately active after transfer to a room temperature plate, it may have been injured in transit. Again, care should be taken not to nick the surface of either of the agar plates during transfer. From an overgrown plate that is to be discarded, it is often convenient to transfer many animals by cutting out a chunk of the agar surface with a sterilized scalpel and placing it face down on a new plate.

2.1.3 Incubation of nematode cultures

The wild-type strain N2 will reproduce between 12 °C and 26 °C but produces maximum brood sizes (250 to 350 progeny) at about 20 °C. Generation times are 50 h at 25 °C, 70 h at 20 °C, and 100 h at 15 °C. Stock plates can be kept at 15 °C for at least 2 months between transfers. After about two generations on a small plate the nematodes deplete the bacteria and revert to 'dauerlarvae', which do not feed and can survive for several months if the plate does not become too dry.

2.1.4 Dealing with bacterial and fungal contaminants

Normal sterile technique on the open bench is generally adequate to prevent contamination. Mycelial fungi arising from airborne spores can be eliminated by one or more transfers of nematodes to new plates. Contamination with foreign bacteria or yeasts, which can decrease fertility and promote burrowing, can sometimes be removed by repeated transfer of a few animals every 30 min through fresh lawns of OP50. If this fails, eggs can be sterilized by following the alkaline hypochlorite procedure (*Protocol 1*) and plating the collected eggs. A simple procedure is to place 15–20 gravid hermaphrodites in a drop of freshly mixed 1:1 1 M NaOH/4–6% NaOCl at the edge of a plate with a central spot of bacteria. The animals largely dissolve, the more resistant eggs hatch, and the larvae crawl into the bacteria by the next day. The larvae are then transferred to a freshly seeded plate.

2.2 Growth in liquid medium

For preparation of large quantities of nematodes or their embryos, bulk growth on bacteria in liquid culture is more convenient than growth on plates. Media and techniques for large scale preparations are described in ref. 3.

3. Isolation of embryos

3.1 Bulk collection of mixed-stage embryos

Protocol 1. Collection of mixed-stage embryos by alkaline hypochlorite[a]

1. From a 6 cm plate containing gravid adults[b] and eggs, wash the worms into 1 ml M9 buffer[c]. A 'rubber policeman' is useful for dislodging eggs from the agar.

2. Add 0.5 vol. of alkaline hypochlorite (fresh 2:3 mix of 4 M NaOH/4–6% NaOCl).

3. After 5 min at room temperature, the animals will be largely dissolved (monitor with the dissecting microscope). Collect the released embryos by centrifugation (2 min, 350 g), wash (with sterile M9), and centrifuge twice more. Collect embryos in a minimal volume using a Pasteur pipette.

[a] This procedure selects against one- and two-cell embryos, which are more sensitive to the alkaline hypochlorite.
[b] Preparations enriched in pregastrulation stages can be obtained from young well-fed hermaphrodites, which generally do not retain embryos beyond this stage. Use populations which have been fertile for one day or less, and avoid collecting laid embryos.
[c] M9 buffer: 3 g KH_2PO_4, 6 g Na_2HPO_4, 1 ml 1 M $MgSO_4$ in 1 litre water.

3.2 Bulk collection of staged embryos

Obtaining a collection of embryos at precisely the same stage is impossible except by dissection (Section 3.3). Larger collections of approximately staged embryos can be prepared from a synchronized population of hermaphrodites, which will begin producing fertilized oocytes at about the same time (3). Obtain eggs from a well-fed population, either by hypochlorite treatment (*Protocol 1*), or by eluting and discarding the animals on the plate and recovering the embryos remaining on the agar surface using a 'rubber policeman'. Wash the collected eggs in M9 two or three times, distribute them onto a fresh plate containing no bacteria, and incubate overnight. Without a food source, the animals arrest as first-stage larvae; wash these off the plate, leaving behind any unhatched embryos, and replate onto seeded plates to obtain a synchronized population. Monitor the animals using a dissecting microscope and treat with alkaline hypochlorite when the first gravid adults appear. This procedure should yield primarily 4-cell to 16-cell embryos, which can be incubated to obtain approximately synchronised later embryos as desired. An alternative method is to use young hermaphrodites of a mutant strain (3), which when well-fed lays most embryos before the 8-cell stage.

3.3 Obtaining early embryos by dissection

Precisely staged embryos in small numbers can be obtained by dissection. Five to ten young well-fed hermaphrodites are placed in a small volume of egg salts (sterile 118 mM NaCl, 40 mM KCl) in a watch glass or depression slide and cut in half near the vulva with a fresh scalpel blade under the dissecting microscope. Among the released embryos, the 1-, 2-, and 4-cell stages are easily distinguishable. They can be transferred individually to slides for observation or to other containers for further treatments by mouth pipetting, using a capillary that has been heated and drawn out, connected to tubing with a mouthpiece. Early embryos can be cooled (4–10 °C) for up to at least 2 h to arrest development at a desired stage while many are collected, and will then resume normal embryogenesis synchronously when raised to growth temperature.

4. Observation of embryos

4.1 Microscopy of living embryos

Nomarski microscopy is an invaluable tool for observations and experimental procedures using *C. elegans* embryos, as the transparent eggshell allows *in vivo* observations of nuclei over time. Live embryos are best observed with Nomarski DIC optics at 40✕ (dry) or 63✕ to 100✕ (oil immersion) on agar pads (*Protocol 2*).

Protocol 2. Mounting embryos on agar pads

Requirements:

- blank slides and 'spacer' slides (the latter are microscope slides surfaced with sufficient tape to give 0.4 mm added thickness)
- autoclaved 5% Bacto agar in H_2O, molten
- egg salts (sterile 118 mM NaCl, 40 mM KCl), with and without 100 μg/ml tetramisole (Sigma T1512)
- mouth pipette (drawn-out capillaries, tubing, mouthpiece)
- siliconized depression slide

- gravid hermaphrodite worms
- platinum wire pick for worm transfer (see Section 2.1.2).

1. Make 0.4 mm agar pads by putting a drop of liquid agar on a blank slide lying between two spacer slides. Immediately cover it cross-wise with a second blank slide. After 1–2 min, slide the two non-spacer slides apart; store in a damp box until use (for up to several hours at room temperature, several days at 4 °C).

2. Transfer up to 80 gravid hermaphrodites to a 100 µl drop of egg salts on a siliconized depression slide. Add 100 µl of egg salts with tetramisole to paralyse the worms. Under a dissecting microscope, cut the worms in half with clean No. 15 scalpel blade.

3. With a mouth pipette, transfer eggs of the desired stage to an agar pad, and draw off liquid until the meniscus barely covers the eggs. Eggs can be grouped tightly using an eyelash mounted on toothpick with glue. Gently lay on a coverslip using forceps.

4. Cut the agar with a razor blade at the edges of the coverslip. Seal with silicone oil (Dow Corning) or 3S Voltalef oil (BDH). If using an oil objective, cut the agar smaller than the coverslip before sealing to avoid mixing with immersion oil.

After gastrulation, birefringent gut granules, identifying the gut cells, can be observed with polarized light by removing the Nomarski prism and changing the microscope condenser setting. Pregastrulation embryos almost invariably orient right or left side up (the first polar body, lying outside the vitelline membrane, is anterior), between gastrulation and comma stage the dorsal or ventral surface is up, and after elongation begins the lateral surface is again presented. Developmental sequences can be observed directly (4), or recorded by time-lapse video photography using an automatic focusing controller (5) or recorded on videodisc using the 4D microscope recently developed by J. White, Cambridge (6). The system records complete sets of serial optical sections at predetermined intervals and allows playback of consecutive images from any desired section to facilitate lineage mapping.

4.2 Microscopy of fixed embryos

4.2.1 General fixation methods

Microscopy of fixed embryos is used for nuclear staining, antibody staining, or detection of tagged reporter gene expression. Methanol (*Protocol 3*) or paraformaldehyde (*Protocol 4*) fixation is generally used, following permeabilization of the egg shell and vitelline membrane by freezing or pressure.

Protocol 3. Methanol fixation

1. Onto slides subbed with 2% gelatin[a], collect embryos obtained by alkaline hypochlorite treatment (*Protocol 1*) or dissection (Section 3.3) in a minimal volume of M9. Older embryos can be washed from Petri dishes and dropped onto slides with a Pasteur pipette.
2. Gently lay on a coverslip to just flatten the eggs as the liquid spreads. Wick off excess liquid with paper.
3. Freeze the slide by immersing in liquid nitrogen for a few seconds or by placing slide on dry ice for 10 min.
4. Pop off the coverslip with a scalpel and immediately place slide in cold 100% methanol for 10–20 min. Air dry slide.
5. Before use, rehydrate for 5 min in PBS.

[a] See Chapter 21, *Protocol 2*.

Protocol 4. Paraformaldehyde fixation

1. Collect eggs by alkaline hypochlorite treatment (*Protocol 1*) or dissection (Section 3.3), in a depression slide or microcentrifuge tube. Add an equal volume of dilute NaOCl solution[a] and mix. This softens the shell and gives a better final morphology.
2. After 1 min, add an equal volume of 1 mg/ml BSA in egg salts.
3. Transfer the embryos in minimum volume to a small drop of para-formaldehyde/glutaraldehyde fix[b] on a subbed slide (3 μl fix for 100 eggs).
4. After the eggs settle, gently lay on a cut-down coverslip (sized so that the drop spreads to the edges and flattens the embryos). Either freeze as in *Protocol 3*, or flood with 30 μl fix. If frozen, pop off coverslip and place in fix.
5. Leave in fix for 3–5 min.
6. Rinse for 5 min in PBS. Do not allow to dry out.

[a] 1:9 dilution of commercial 4–6% NaOCl in egg salts (118 mM NaCl, 40 mM KCl). Dilution stable for 3 h on ice.
[b] Fresh 2.5% paraformaldehyde (dissolved at 60 °C), 0.1% glutaraldehyde, in 125 mM phosphate buffer pH 7.0–7.4. Cool before use. Toxic.

4.2.2 DAPI staining

DAPI (diamidinophenylindole; Sigma D1388) is a fluorescent DNA-binding compound used for staining nuclei. The standard procedure is to stain slides for 5 min in 0.5–1.0 μg/ml DAPI in PBS, and mount in PBS or Gelvatol (Chapter 21). Observe with epifluorescence using UV-excitation filters.

4.2.3 Antibody staining

There is a large repertoire of *C. elegans* tissue-specific or gene product-specific antibodies available from various researchers (see *Table 1*). Procedures for antibody staining must be adapted to the antibody being used; one of the above fixations generally works for most antibodies. Following paraformaldehyde fixation, an additional permeabilization step (5 min in 0.5% Triton X-100 or Nonidet P-40 in PBS, or cold acetone) is needed. DAPI staining can be done during the final rinsing steps.

Table 1. Antibodies binding to *C. elegans* embryos

Antibodies etc.	Tissue specificity	References
Anti-tubulin	Polymerized tubulin (spindles)	9, 14
Rhodamine-phalloidin	Cytoplasmic F-actin	15
Anti-myosin (heavy chain):		
HCA	Body wall muscle myosin	16
HCB	Body wall muscle myosin	16
HCC	Pharyngeal myosin	17
Monoclonal 3NB12	Early pharynx	18
Monoclonal MH27	Belt desmosomes of hypo-dermal cells	19
Monoclonal SP37	Gut	S. Strome, personal communication
Monoclonal J130	Early hypodermal extra-cellular coat	20
FITC-soybean agglutinin	Early hypodermal extra-cellular coat	21
Monoclonal K76	Germ line-specific P-granules	22
Monoclonal 1CB4	Subset of neurons; gut	23

4.2.4 Other staining methods

Standard X-Gal staining methods for β-galactosidase (e.g. Chapter 20) work well following either fixation method; the usual transformation fusion construct carries a nuclear localization signal so that staining cells are clearly defined (7). A gut-specific esterase appearing at about 200 cells can also be stained for enzyme activity (8).

5. Preparation for experimental embryology

5.1 Embryo permeabilization

Permeabilization of the chitinous egg shell and the inner vitelline membrane presents the major problem in experimental embryology using living *C. elegans* embryos; several methods allow partial or complete permeabilization. One-cell embryos cut from hermaphrodites are still permeable as the shell is not fully hardened. Hypochlorite-softened embryos (*Protocol 4*, steps **1**, **2**) can be permeabilized by mounting on a gelatin-subbed slide under a coverslip with vaseline or tape as a spacer and applying brief, gentle pressure with a dissecting needle until the embryos flatten slightly and rebound. Eggs mounted on agar pads or a subbed coverslip can also be individually permeabilized by using a laser to punch holes through the shell (9).

5.2 Embryo devitellinization and culture

For small numbers of embryos, the egg shell and vitelline membrane can be removed by treatment with chitinase followed by drawing into a narrow pipette. Such 'naked' embryos, if cultured in EGM (embryonic growth medium), will continue cell division and cellular differentiation to the full number of approximately 550 cells (10), although they will not undergo normal morphogenesis.

Protocol 5. Permeabilization and culture of embryos

1. Collect embryos by dissection in a depression slide (preferably with two wells), following *Protocol 2*. If collecting delicate pronuclear stages, add 50 µl EGM[a] to the egg salts.

2. By mouth pipette, transfer embryos of desired stages (in minimum volume, 10–20 µl) into a 100 µl drop of dilute NaOCl (*Protocol 4*, footnote a) into a second well. Leave for 3 min. Meanwhile, wipe the first well dry and set up a 15–20 µl drop of chitinase solution[b] under silicone oil (3 µl Dow Corning or 3S Voltalef oil pipetted over drop). Between the wells, set up two 30 µl drops, the first of EGM solution[a], or 1 mg/ml BSA in egg salts, and the second of egg salts.

3. Collect embryos after the 3 min and rinse quickly through the two drops.

4. Pipette the embryos into the chitinase. Leave for up to 8 min. Early embryos will round up as the egg shell is digested, and pretzel stages may hatch. If digestion does not work well, try older worms, which seem to make thinner egg shells.

5. Rinse through a drop of EGM into a drop of EGM under oil.

6. Shear off the vitelline membrane by drawing the embryos individually in and out of a narrow capillary[c]; if the egg shell is well digested and the pipette is the right diameter it takes only 1–2 min to do 100 embryos. Blastomeres in embryos of less than eight cells can be successfully separated with an eyelash tool (*Protocol 2*) and a steady hand under the dissecting microscope.

7. Pipette desired embryos or blastomeres to EGM (30 µl) in incubation chamber[d] (if using immunology slides, cover with 3 µl oil). For overnight incubation, keep in damp box.

[a] For 10 ml EGM (embryonic growth medium): 50 mg PVP-40 (Sigma; dialysed against H_2O and lyophilized); 588 μl of 1 M NaCl; 252 μl 1 M KCl; 1 ml 0.25 M Hepes pH 7.4; 1 ml 5 mg/ml inulin (Sigma I3754); 600 μl Grace's amino acids (Sigma G0273); 1.97 ml H_2O; 4 ml fetal bovine serum, heat-treated; 100 μl penicillin-streptomycin (Sigma P3539); 100 μl 100 mg/ml galactose; 100 μl base mix (= 100 mg adenine, 10 mg ATP, 3 mg guanine, 3 mg hypoxanthine, 3 mg thymine, 3 mg xanthine, 3 mg uridine, 5 mg ribose, 5 mg deoxyribose, 100 ml H_2O); 100 μl fresh 14 mg/ml L-glutamine; 50 μl BME vitamins (Sigma B6891); 40 μl 0.5 M Na_2HPO_4; 20 μl 1 M $MgSO_4$; 20 μl fresh 14 mg/ml pyruvic acid; 10 μl lactic acid syrup (Sigma L4263); 50 μl chicken egg yolk (1:1 dilution in H_2O). Mix, incubate on ice several hours, centrifuge, and filter-sterilize (0.2 μm). Keep at 4 °C for up to 1 month. Use tissue-culture grade H_2O (e.g. Sigma W3500).

[b] 3 mg/ml chitinase (Sigma C1525), 10 mg/ml chymotrypsin (Sigma C4129) in egg salts.

[c] Permeabilization pipettes should be drawn-out to have a very narrow bore (Kwik-fil, World Precision Instruments, 1B100F-4), cut to size just smaller than the embryos using a scalpel on Parafilm under dissecting microscope. Attach the capillary to a mouth tube and load the tip with EGM using a syringe. Store between sessions with the tip in H_2O.

[d] Incubation chamber (allows use of inverted microscope): 25 × 75 × 6 mm aluminum or plastic slide with central 16 mm diameter hole with cover slip sealed to bottom with silicone grease, and coverslip sealed to top with oil. Teflon-coated immunology multi-well slides (e.g. Cel-Line 10-141, HTC Super-Cured) are also suitable.

6. Molecular methods adapted for *C. elegans* embryos: general considerations

Standard molecular methods must be modified to deal with small quantities, if accurately staged embryos are required. Run-on transcription from nuclear extracts (3); RNA preparation from less than 100 embryos (L. Edgar, unpublished results); RT-PCR amplification from a few embryos (M. Yandell, personal communication); and *in situ* hybridization for moderate to abundant messages (11) have been successfully accomplished. Transformation, by injection of plasmid or cosmid DNA into the adult syncytial germ line, has been used to identify sequences rescuing mutant phenotypes (12, 13), and for analysis of gene expression in *lacZ* fusion constructs. A number of tissue-specific promoter constructs have been made available as 'cassettes' by A. Fire (7).

References

1. Wood, W. B. (ed.) (1988). *The Nematode* Caenorhabditis elegans. Cold Spring Harbor Laboratory Press, Cold Spring Harbor, NY.
2. Brenner, S. (1974). *Genetics*, **77**, 71.
3. Schauer, I. and Wood, W. B. (1990). *Development*, **110**, 1303.
4. Sulston, J., Schierenberg, E., White, J., and Thomson, J. N. (1983). *Dev. Biol.* **100**, 64.
5. Deppe, U., Schierenberg, E., Cole, T., Krieg, C., Schmitt, D., Yoder, B., and von Ehrenstein, G. (1978). *Proc. Natl. Acad. Sci. USA*, **75**, 376.
6. Wood, W. B. (1991). *Nature*, **349**, 536.
7. Fire, A., White-Harrison, S., and Dixon, D. (1990). *Gene*, **93**, 189.

8. Edgar, L. G. and McGhee, J. D. (1986). *Dev. Biol.*, **114**, 109.
9. Priess, J. and Hirsh, D. (1985). *Dev. Biol.*, **107**, 337.
10. Edgar, L. G. and McGhee, J. D. (1988). *Cell*, **53**, 589.
11. Costa, M., Weir, M., Coulson, A., Sulston, J., and Kenyon, C. (1988). *Cell*, **55**, 747.
12. Fire, A. (1986). *EMBO J.*, **5**, 2673.
13. Mello, C., Kramer, J., Stinchcomb, D., and Ambros, V. (1991). *EMBO J.*, **10**, 3959.
14. Kemphues, K., Wolf, N., Wood, W. B. and Hirsh, D. (1986). *Dev. Biol.*, **117**, 156.
15. Strome, S. (1986). *J. Cell Biol.*, **103**, 2241.
16. Miller, D. M., Ortiz, I., Berliner, G. C., and Epstein, H. F. (1983). *Cell*, **34**, 477.
17. Epstein, H. F., Miller, D. M., Gossett, L. A., and Hecht, R. M. (1982). *Muscle development* (ed. M. Pearson and H. Epstein). Cold Spring Harbor Laboratory Press, Cold Spring Harbor, NY.
18. Priess, J. and Thomson, N. (1987). *Cell*, **48**, 241.
19. Schierenberg, E. (1984). *Dev. Biol.*, **101**, 240.
20. Cowan, A. E. and McIntosh, J. R. (1985). *Cell*, **41**, 923.
21. Mello, C. C., Draper, B. W., Krause, M., Weintraub, H., and Priess, J. R. (1992). *Cell*, **70**, 163.
22. Strome, S. and Wood, W. B. (1983). *Cell*, **35**, 15.
23. Okamoto, H. and Thomson, J. N. (1985). *J. Neurosci.*, **5**, 643.

3

Embryos and larvae of invertebrate deuterostomes

NICHOLAS D. HOLLAND and LINDA Z. HOLLAND

1. Introduction

The recent information explosion on pattern-determining genes has re-awakened interest in the nature of segmentation, the universality of developmental mechanisms, the phylogenetic relations between animal phyla, and the origin of the vertebrates in particular. Much insight can be expected to come from developmental studies of invertebrate deuterostomes: a group which includes the closest relatives of the vertebrates. All invertebrate deuterostomes are marine, and they comprise (discounting the enigmatic arrow worms) echinoderms, hemichordates, tunicate chordates, and acraniate chordates.

Here we summarize general methods for raising marine invertebrates, and discuss culturing of selected species of invertebrate deuterostomes, all with indirect development (small egg, distinctive larval type, and definite metamorphosis).

2. General principles

Several subjects summarized here are treated more extensively in ref. 1, which, in spite of its title, has worldwide relevance.

2.1 Sources of ripe adults

At coastal laboratories, ripe adult invertebrates are usually obtained from nearby habitats by the researchers or in-house collectors. At inland laboratories, breeding stock is usually supplied by commercial shippers (e.g. Pacific Biomarine Laboratories).

2.2 Labware

Use non-toxic containers (glass or many, but not all, plastics) for storing sea water, maintaining adult invertebrates, and culturing developmental stages.

Reserve repeatedly used containers exclusively for culture purposes and wash them carefully (2) or use disposable labware.

2.3 Sources of sea water

Most coastal laboratories have flow-through, natural sea water for maintaining adult invertebrates and raising their young. However, in some flow-through systems (e.g. Station Zoologique, Villefranche-sur-Mer) and in recirculating systems (e.g. Stazione Zoologica di Napoli), the water quality is only good enough for maintaining adults, and purer sea water must be collected offshore and brought to the laboratory for culturing embryos and larvae. As soon as possible after collection, such water should be filtered through coarse paper or glass-fibre filters and stored in a cold, dark place.

At most inland laboratories, marine invertebrates are maintained and their developmental stages are cultured in artificial sea water. Either use a commercial salt mixture (e.g. 'My Sea' from Jamarin Laboratories) or start with reagent grade salts and mix according to published formulae (1).

2.4 Temperature control

Most semi-tropical and tropical species will develop normally at room temperature (which remains adequately constant in many air-conditioned laboratories). In contrast, many temperate species must be kept cooler than room temperature. Also, development rate is temperature sensitive (the Q_{10} is a little more than 2 until the lethal maximum is approached), and a constant culture temperature is needed for a reproducible rate of development.

At many coastal laboratories, the water temperature in flow-through systems is suitably cool and constant; however, at inland laboratories, refrigeration is usually needed.

(a) Aquaria with built-in chillers are available (e.g. from Universal Marine Industries).

(b) For raising embryos and larvae below room temperature, a refrigerated display cabinet (e.g. from a soft-drink wholesaler) can be modified by the addition of an adjustable thermostat (D5-model 9001, B071 from United Electric Controls Company), allowing the temperature to be controlled between 0 °C and 20 °C (with ± 0.5 °C accuracy).

2.5 Salinity and light

All developmental stages should be kept at a salinity similar to that of the natural environment for the species. Light has less influence than salinity or temperature on development. Day length is not critical, but do not expose cultures to direct sunlight. If cultures are being fed on phytoplankton (Section

6.4), the quality and quantity of the light should be appropriate to keep the algae healthy. Light may influence late larval behaviour and metamorphosis.

2.6 Initiating development

Methods for obtaining gametes vary with species and will be given in Sections 3–6. The fertilizable female gametes of invertebrate deuterostomes may be primary oocytes, secondary oocytes, or ova; it is often useful to refer to all these collectively as eggs. During natural spawning, eggs and sperm are shed into the sea water, where fertilization takes place. This natural process is mimicked in the laboratory by obtaining eggs and sperm separately and then mixing them to initiate *in vitro* insemination.

After eggs and sperm have been mixed, let the eggs settle for about 10 min and decant or aspirate most of the water containing excess sperm; replace with clean sea water, and repeat if necessary. The culture vessel should contain at most a monolayer of settled eggs. It is often desirable to stir cultures gently (mechanized stirring devices are described in ref. 1). For unstirred cultures, reduce the density of embryos and change the greater part of the water several times each day. A method for changing the water on actively swimming larvae has been described (3).

2.7 Later larvae of many species must be fed

Many marine invertebrates with indirect development begin to feed in their larval stages, and development falters at this point unless they are provided with suitable food (diets reviewed in ref. 1). After feeding starts, the number of cultured larvae per millilitre *must* be decreased from concentrations of the order of hundreds or thousands to just a few (in some instances, less than one!). Thus, for later larval stages, it may be difficult to accumulate enough material for some purposes (e.g. the construction of cDNA libraries).

2.8 Manipulating embryos and larvae

It is frequently necessary to remove extraembryonic coats, because they interfere with microinjection, blastomere dissociation, determination of precise fate maps, and the entrance of large molecules such as antibodies. Removal methods vary between species and are given in Sections 3–6. Additional methods for handling embryos and larvae are listed with recent references:

- microsurgery and microinjection of living developmental stages (1, 2, 4, 5)
- conventional optical microscopy (6) and confocal microscopy (7)
- scanning (8, 9) and transmission electron microscopy (10, 11)
- blastomere lineage tracing (12–14)
- blastomere dissociation, isolation, and reassociation (15–17)

- *in situ* hybridization (18, 19)
- electrophysiology (10, 20, 21)

3. Phylum Echinodermata

3.1 Introduction

Only the class Echinoidea (sea urchins) is treated here; reviews of development for the other echinoderm classes are available (22). Sea urchins are common on most coasts and are extremely fecund. Their eggs are stored in the ovary as haploid ova and are spawned as such, ready for immediate fertilization. A commonly used echinoid in developmental studies (the example described here) is *Strongylocentrotus purpuratus*, the purple sea urchin, from the Pacific coast of North America. The adults may be collected in the intertidal or shallow subtidal zones and are usually ripe throughout the winter and spring.

In aquaria, adult sea urchins live well at densities of just under one animal per liter in 10–15 °C sea water that is vigorously aerated. At coastal laboratories,they can be maintained indefinitely on a diet of seaweeds if the accumulating faeces are removed every few days. Even in recirculating sea water aquaria, these omnivores readily accept boiled potatoes or frozen shrimp, but the danger of fouling the water is considerable. Another option is not to feed the animals, since they remain ripe and healthy during a fast of several weeks. Land-locked biologists might also consider investing in more frequent shipments of animals instead of expensive aquarium systems.

3.2 Obtaining gametes

S. purpuratus lacks obvious external sexual dimorphism, and the ratio between males and females is usually close to 1:1. For inducing sea urchins to spawn without harming them, administer mild electric shock (circuit diagram and instructions in ref. 2) or inject them with 0.5 M (isotonic) KCl or 10 mM acetylcholine (made up in sea water immediately before use). For an average-sized adult urchin, inject 0.5 ml of the stimulating solution through the soft membrane around the mouth. With practice, the dose can be adjusted to result in either a heavy spawning or a light one; when spawning is light, one can usually obtain repeated small spawnings from a single animal over a period of several days.

After stimulation of a ripe sea urchin, the gametes rapidly begin to emerge from the pole of the body opposite the mouth. The mass of sperm emerging from a male is white, and the eggs emerging from a female are yellowish-orange (individual eggs are about 80 μm in diameter). Do not put spawning individuals in containers with other ripe animals, because gametes in the sea water may trigger an epidemic spawning. Keep the surfaces of the animals moistened with sea water during gamete collection.

To collect sperm, remove a spawning male from the water and orient him mouth-down. For approximately the next 15 min, the sperm mass accumulating on top of the shell is collected 'dry' (i.e. undiluted with sea water) in a Pasteur pipette and transferred to a small container that is covered and stored refrigerated. The stored sperm should remain usable for at least a couple of days. To prevent premature insemination of eggs, keep spawning males well away from other urchins. After handling a spawning male, wash your hands in tap water to kill adhering sperm.

To collect eggs, invert a spawning female, mouth-up, and support her on the rim of a small beaker filled to the brim with sea water. The eggs are emitted directly into the sea water and fall to the bottom of the beaker for about 15 min. Wash the spawned eggs several times by decanting and replacing the overlying sea water. Use the washed eggs within a few hours of spawning. All sea water contacting the eggs should be filtered and maintained at 10–15 °C.

3.3 Fertilization and development

When inseminating sea urchin eggs, keep sperm concentrations low to avoid polyspermy, which results in abnormal embryos at first cleavage. Dilute one drop of the refrigerated, 'dry' sperm with 5 ml sea water (any of the diluted, motile sperm not used within about 10 min should be discarded). Add 0.1 ml diluted sperm for each millilitre of settled eggs in 100 ml sea water, and stir the culture gently for about 20 sec. After 1 min, remove a few eggs from the culture and examine them microscopically at low power in a drop of water. Fertilization envelopes should be rising on over 90% of the eggs. A schedule of development is given in *Table 1*.

Table 1. Development schedule (12 °C) for *S. purpuratus*, a sea urchin

Time	Stage
0 h	Insemination
2.5 h	2-cell stage
4.0 h	4-cell stage
5.5 h	8-cell stage
27 h	Blastula hatches
72 h	Early prism larva
96 h	Early pluteus larva (feeding begins)

Do not exceed an initial density of about 0.3 ml settled eggs (roughly 3×10^5 eggs) per 100 ml when raising a culture to the early pluteus stage. Raising the subsequent feeding stages through metamorphosis is labour intensive and results in low yields. Methods are given in ref. 23.

3.4 Removing extraembryonic coats

The extracellular coats around fertilized sea urchin eggs and embryos are, from outside to inside: the jelly layer, the fertilization envelope, and the hyaline layer. Normally, the jelly coat slowly dissolves, the fertilization envelope is shed when the blastula hatches, and the hyaline layer is converted into the cuticle. Some procedures require premature removal of one or more of these coats.

- Remove the jelly coat by brief treatment with acidified sea water (1) or by stripping away the fertilization envelope.

- Remove the fertilization envelope by first softening it chemically with Ca^{2+}/ Mg^{2+}-free sea water, with urea, with peroxidase inhibitor, or with protease (1) and then stripping it away by passing the fertilized eggs several times through a screen with an approximately egg-sized mesh. To prevent the envelope-free eggs clumping when returned to sea water, soften the hyaline layer in a sulfhydryl-reducing reagent and wait 15 min before stripping away the fertilization envelope (6).

- After removal of the hyaline layer plus fertilization envelope (15), the embryos can be dissociated into their constituent blastomeres.

4. Phylum Hemichordata

4.1 Introduction

The two classes of hemichordates are the Pterobranchia and the Enteropneusta (acorn worms). We will not review pterobranch development, because these minuscule animals produce few eggs and develop without passing through a distinctive larval stage. In contrast, acorn worms tend to be large, fecund animals, and about one-half of the species develop indirectly.

The best-studied acorn worm with indirect development is *Ptychodera flava*, which is ripe during the late fall in Hawaii and during the winter in India. These worms can be collected by shovel from the sandbar at the mouth of Kaneohe Bay, Hawaii. Place animals in dishes of sea water at 24–25 °C (the approximate field and laboratory temperature) with 3 cm of sand at the bottom so that they can bury themselves.

4.2 Obtaining gametes

Natural spawning is necessary. Worms that are about to spawn will project their anterior ends out of the sand between about 16:30 and 17:30 hours (M. G. Hadfield, personal communication). The pale yellow eggs adhere to mucus on the anterior part of the trunk of spawning females for up to 30 min.

Each egg is about 110 μm in diameter and surrounded by a conspicuous jelly layer. When a male spawns, the whitish sperm rapidly disperse in the surrounding sea water.

4.3 Fertilization and development

If males and females spawn together in the same bowl, the eggs are soon fertilized; if not, add sperm-filled water from the bowl of a spawning male to the bowl of a female spawning in isolation. The sperm concentration is not critical, since the eggs resist polyspermy. Transfer fertilized eggs to clean sea water in a sand-free container. After fertilization, an inconspicuous fertilization envelope gradually appears interior to the jelly layer; a schedule of development through the early tornaria larva is given in *Table 2*.

Table 2. Development schedule (29 °C) for *P. flava*, an acorn worm

Time	Stage
0 h	Insemination
4.5 h	2-cell stage
5.3 h	4-cell stage
6.0 h	8-cell stage
14.5 h	Blastula
22.5 h	Gastrula
42.5 h	Late gastrula hatches
2.5 d	Early tornaria larva (feeding begins)

The tornaria larvae will feed and develop slowly in the laboratory, but have not been raised through metamorphosis. Advanced larval stages (meta-tornariae) are available in the spring plankton in Hawaii; if placed in a dish of sea water with a layer of sand they will metamorphose into juvenile acorn worms during the next few hours. No methods have yet been developed for fertilization envelope removal or other manipulation of acorn worm embryos.

5. Phylum Chordata: subphylum Tunicata

5.1 Introduction

The subphylum Tunicata comprises the ascidians (covered here), thaliaceans, and appendicularians. Most ascidians develop indirectly and are divisible (with a few exceptions) into two groups: colonial species having internal fertilization and retaining their embryos internally; and solitary species spawning their gametes into the surrounding sea water. All ascidian species are hermaphrodites. Although some species are self-fertile, as a general rule development is most successful when sperm and eggs from two different individuals are mixed.

More than one dozen species of ascidian are in common use by embryologists; here we we will cover *Ciona intestinalis*, a shallow water ascidian found almost worldwide. Adults are available throughout the year, although in colder climates the adults lack gametes during winter. Adults maintained in the laboratory should be kept under *constant illumination* to prevent spawning. In a flow-through system, these filter feeders eat natural plankton in the sea water; in a recirculating system, feed with Liquifry Marine (Liquifry Ltd) or not at all.

5.2 Obtaining gametes and fertilization

After ripe animals have been exposed constantly to light for several days, transfer them individually to small containers of sea water and place in complete darkness for 30 min. When the light is turned on again, each animal should spawn both eggs and sperm within about 30 min. Eggs are about 150 μm in diameter and surrounded by conspicuous follicle cells. When spawning ceases, remove the adult and wash the eggs free of non-fertilizing sperm by several changes of clean sea water. A few individuals are self-fertile, and their egg batches should be discarded as follows: wait 2 h and dispose of any culture with cleavage stages. Self-sterile egg batches will lack cleavage stages and yield good cultures when fertilized by sperm dissected from another animal (see below).

Alternatively, dissect eggs from the oviduct, which runs parallel to the nearby (whiter) sperm duct. The latter duct is clearly visible through the transparent body wall of a ripe animal. Cut open the body wall, remove a long segment of oviduct, and gently squeeze out the eggs. Then dissect the sperm duct from another individual, and squeeze the sperm mass into approximately 10–15 volumes of sea water. Add 2 ml diluted sperm (which may be kept for several hours) to each batch of eggs in 30 ml seawater (relatively concentrated sperm suspensions are used because ascidian eggs are rarely prone to polyspermy). Gently stir the eggs for a few minutes, and then wash away the non-fertilizing sperm with two changes of clean sea water.

5.3 Development

In *C. intestinalis*, as in all ascidians, the spawned eggs are surrounded by a chorion (also called the vitelline layer) with its outer surface covered by a layer of follicle cells (mentioned above) and its inner surface associated with test cells. When a fertilized or an unfertilized egg contacts sea water, the space underlying the chorion widens. Polar body emission cannot be seen in living material; the first obvious sign of fertilization is the advent of the two-cell stage. A schedule of development is given in *Table 3*.

Table 3. Development schedule (16 °C) for *C. intestinalis*, an ascidian

Time	Stage
0 h	Insemination
1.0 h	2-cell stage
2.0 h	8-cell stage
3.5 h	Blastula
4.5 h	Early gastrula
12.0 h	Tail bud elongation
22.0 h	Tadpole larva hatches[a,b]

[a] At hatching, all follicle and test cells are shed with the chorion.
[b] Tadpoles, which do not feed, swim for 12–24 h before metamorphosis.

5.4 Removing extraembryonic coats

For most ascidians, including *C. intestinalis*, the chorion and its associated cells must be stripped away manually using sharp tungsten needles, limiting the numbers of naked eggs that can be produced. For *Phallusia mammillata*, however, a technique applicable to batches of eggs has been developed (24). The fragile, naked embryos should be reared in unstirred sea water in a container with its inner surface coated with 0.9% agar.

6. Phylum Chordata: subphylum Acrania

6.1 Introduction

The Acrania include the lancelets (also called amphioxus). They are relatively inconvenient subjects for developmental studies, because their breeding seasons each year are limited to a few weeks, during which the animals spawn sporadically every few days with an irregular pattern not predictable from environmental factors.

The three lancelet species that have been most studied by developmental biologists are: *Branchiostoma belcheri* in the Far East (especially on the China coast at Xiamen and Qingdao), *B. floridae* in Florida and the Caribbean (described here), and *B. lanceolatum* of the Atlantic and Mediterranean coasts of Europe (e.g. near Kristineberg, Sweden; near Helgoland, Germany; Banyuls-sur-Mer, France; and near Plymouth, England; although the once famous populations in the Bay of Naples and at Faro in Sicily have vanished).

6.2 Collection of *B. floridae*

Adults of *B. floridae* can be collected in large numbers by shovel and sieve (1 mm mesh window screen) from the sandy bottom in water about 1 m deep just off the southern shore of the Courtney Campbell Causeway in Old

Tampa Bay, Florida. During the spawning season, which lasts from early August to early September, collect lancelets during the late afternoon for spawning the same evening. Collect several hundred animals because not all are ripe, and frequently only a small percentage, if any, of the apparently ripe females can be induced to spawn. Each ripe animal contains several dozen gonads that are clearly visible through the body wall; the testes are greyish-white, and the ovaries are light to dark yellow. Unlike some species, *B. floridae* never spawns during collection or transport.

Fill a carboy with sea water from the collecting site, where salinities are always somewhat less than full strength sea water. In the laboratory, filter this water and use it for maintaining both adults and developmental stages. Although water temperature in the field is about 30 °C during August, it is convenient to raise the embryos and larvae at an air-conditioned room temperature of 25 °C. Wash the newly collected lancelets free of debris, and distribute them individually in small dishes of sea water. Leave the room lights on constantly, because darkness may stimulate inopportune spawning (in nature, the animals spawn shortly after sundown).

6.3 Obtaining gametes and fertilization

Starting at about 20:00 hours on the evening of the collecting day, induce spawning electrically. Via two platinum electrodes connected to a stimulator (Grass Instrument Co.) inserted in a dish with a female, administer a non-lethal shock of direct current (50 V in 10 msec pulses) for 2 sec. If time permits, quickly change the sea water to remove chlorine generated by the electric current. Females spawn within 5 min of shocking or not at all. The more anterior and posterior ovaries only rarely spawn out, but they may sometimes do so if stimulated 30 min later.

The eggs, which are actually secondary oocytes arrested at the stage of the second meiotic spindle, emerge from the atriopore of a spawning female. A typical spawning episode usually lasts less than 5 min, during which a couple of thousand pale yellow eggs (about 140 μm in diameter) accumulate on the bottom of the dish. Low power microscopy reveals the first polar body (at the animal pole) and a distinct vitelline layer, which is sometimes slightly elevated from the egg plasma membrane. As soon as a female is through spawning, transfer her eggs by pipette to a 5 cm Petri dish about one-half full of clean, filtered sea water.

Although ripe males can be induced to spawn electrically, motile sperm are more reliably obtained by the following method. Place a ripe male in a 5 cm Petri dish containing freshly prepared 10 mM NH_4Cl in sea water adjusted to pH 8. Stroke his ventral surface from anterior to posterior with a Pasteur pipette to express sperm out of his atriopore and use the same pipette to collect them as they begin to swim, but before they become too diluted. Immediately add one or two drops of this sperm suspension to several thousand eggs in a 5 cm Petri dish. The eggs are not prone to polyspermy.

Within a couple of minutes after insemination, each fertilized egg should be surrounded by an elevating fertilization envelope. If no fertilization envelopes rise and the surface of the eggs is covered with bound sperm, the male was not ripe enough. In such cases, the eggs can still be fertilized if sperm from a riper male are added. A few minutes after fertilization, remove most of the excess sperm by changing the water three or four times. Avoid pipetting the fragile embryos, and do not stir the cultures.

6.4 Development

The schedule for development in *Table 4* is based on our unpublished observations and those of T. H. J. Gilmour. For lancelets, there is some disagreement on where to divide the embryonic period from the larval period; we will define the larval stage as starting with the formation of the mouth.

Table 4. Development schedule (25 °C) for *B. floridae*, a lancelet

Time	Stage
0 h	Insemination
55 min	2-cell stage
75 min	4-cell stage
95 min	8-cell stage
3.5 h	Blastula
5.5 h	Gastrula
10.0 h	Neurula hatches
38.0 h	Larval mouth opens
18 d	Larval metamorphosis beginning
23 d	Metamorphosis complete

After the neurulae hatch, they swim slowly near the surface of the water. At this point, there will be some dead and abnormal embryos on the bottom of the 5 cm Petri dish; so transfer the normal, swimming neurulae by pipette to a 9 cm Petri dish containing clean sea water. Thereafter, change the sea water daily. After the first day of development, the embryos and early larvae spend most of their time crawling slowly on the bottom of the culture dish by means of their epidermal cilia, although they sometimes swim briefly.

When the larvae begin feeding (60 h after insemination), feed them unicellular phytoplankton such as *Dunaliella*, *Pyramimonas*, *Amphidinium*, and *Rhodomonas*. Centrifuge the algal cultures in a clinical centrifuge and resuspend the algae in sea water before feeding them to the larvae at a concentration of approximately 10^6 cells/ml. Keep dishes of feeding larvae near a cold fluorescent light, and change the sea water daily. Maintain the number of larvae in each 9 cm Petri dish at about two thousand during the first week, at a few hundred during the second week, and finally at a few dozen after the start of the third week.

6.5 Removing extraembryonic coats

Remove lancelet fertilization envelopes individually from fertilized eggs with sharp tungsten needles or fine forceps. Raise the naked embryos in dishes coated internally with 0.9% agar to prevent sticking, and add penicillin (about 50 units/ml).

References

1. Strathmann, M. (1987). *Reproduction and development of marine invertebrates of the northern Pacific coast*. University of Washington Press, Seattle.
2. Lutz, D. A. and Inoué, S. (1986). *Methods Cell Biol.*, **27**, 89.
3. Switzer-Dunlap, M. and Hadfield, M. G. (1977). *J. Exp. Marine Biol. Ecol.*, **29**, 245.
4. Rappaport, R. (1986). *Methods Cell Biol.*, **27**, 345.
5. Colin, A. M. (1986). *Methods Cell Biol.*, **27**, 395.
6. Harris, P. J. (1986). *Methods Cell Biol.*, **27**, 243.
7. Summers, R. G., Musial, C. E., Cheng, P. C., Lieth, A., and Marko, M. (1991). *J. Electron Microscop. Tech.*, **18**, 24.
8. Morrill, J. B. (1986). *Methods Cell Biol.*, **27**, 263.
9. Hirakow, R. and Kajita, N. (1991). *J. Morphol.*, **207**, 37.
10. Longo, F. J., McCulloh, D. H., Ivonnet, P. I., and Chambers E. L. (1992). *Microscop. Res. Tech.*, **20**, 298.
11. Holland, N. D. and Holland, L. Z. (1992). *Biol. Bull.*, **182**, 77.
12. Cameron, R. A., Fraser, S. E., Britten, R. J., and Davidson, E. H. (1990). *Dev. Biol.*, **137**, 77.
13. Nishida, H. (1987). *Dev. Biol.*, **121**, 526.
14. Tung, T. C., Wu, S. C., and Tung, Y. Y. (1962). *Scientia Sinica*, **11**, 629.
15. McClay, D. R. (1986). *Methods Cell Biol.*, **27**, 309.
16. Nishida, H. (1991). *Development*, **112**, 389.
17. Wu, S. C. (1986). In *Advances in science in China. Biology*. (ed. Zhao Ganguan) Vol. 1, p. 231. John Wiley, New York.
18. Saiga, H., Mizokami, A., Makabe, K. W., Satoh, N., and Mita, T. (1991). *Development*, **111**, 821.
19. Holland, P. W. H., Holland, L. Z., Williams, N. A., and Holland, N. D. (1992). *Development*, **116**, 653.
20. Okado, H. and Takahashi, K. (1990). *J. Physiol. (Lond.)*, **427**, 583.
21. Bone, Q. (1992). *J. Marine Biol. Assoc. UK*, **72**, 161.
22. Giese, A. C., Pearse, J. S., and Pearse, V. B. (ed.) (1991). *Reproduction of marine invertebrates*, Vol. 6. Boxwood, Pacific Grove, CA.
23. Leahy, P. S. (1986). *Methods Cell Biol.*, **27**, 1.
24. Zalokar, M. and Sardet, C. (1984). *Dev. Biol.*, **102**, 195.

4

Zebrafish embryos

RACHEL M. WARGA

1. Introduction

Zebrafish (*Brachydanio rerio*) embryos have gained considerable popularity in recent years because they offer several advantages for developmental studies. They are cheap and easy to breed, can be obtained all year round and the embryos are transparent. The latter are easy to manipulate and develop quite rapidly: from egg to hatching takes roughly 3 days.

Many genetic mutations are now becoming available. Zebrafish embryos lend themselves to saturation mutagenesis methods and the proportion of homozygote mutants is increased because it is possible to produce haploid embryos (1). Finally, because of the optical clarity of the embryos, it is possible to screen for mutants very effectively.

This Chapter summarizes methods for obtaining and handling embryos in preparation for observation or various manipulations. Other chapters contain methods specifically adapted for fish embryos: Chapter 9 describes techniques of lineage analysis, Chapter 15 methods for ablation of single cells, and Chapter 16 reviews techniques for continuous observation of cell behaviour in living embryos.

2. Fish maintenance and egg production

Zebrafish are easily obtained from most pet stores worldwide. They are fairly hardy and, with proper care, will produce eggs on a regular basis. They are most easily maintained in 40 litre aquaria with 25 fish each, kept at 28.5 °C. Either replace one-third of the water each day or use a filter and replace one-half of the water once a week. The water should be kept free of debris. Tap water aged for a day (to release chlorine) is adequate; more consistent conditions are obtained with 60 mg Instant Ocean (Aquarium Systems) per litre of distilled water. The fish should be fed twice a day, one food source preferably live (e.g. *Daphnia*, brine shrimp, *Drosophila* adults or larvae). Control the light/dark cycle with an automatic timer (14 h light, 10 h dark). For further details on zebrafish care, feeding, and breeding, consult (1).

However, a simple method for obtaining small numbers of eggs is given in *Protocol 1*.

Protocol 1. Obtaining zebrafish eggs

1. Set up a 40 litre tank with around 25 fish, an equal number of males to females[a].
2. The night before collecting eggs, cover the entire bottom of the tank with a layer of marbles. The marbles prevent the fish from eating their eggs after spawning.
3. Wait at least an hour after the lights come on before siphoning[b] the bottom of the tank for embryos.
4. Leave the marbles in the tank as long as embryos are needed but be sure to siphon off any uneaten food. The fish will usually lay a large number of eggs (several 100 per tank) on the first day, but the number can decrease down to 30–50 over the following days.
5. When embryos are no longer needed remove the marbles.
6. Immediately after collection, rinse embryos several times with fresh embryo medium (*Protocol 2*).

[a] Males are slimmer, with more yellow, especially on the belly; females are more silvery blue.
[b] A siphon can be made with 1 cm diameter stiff plastic or glass tube attached to flexible tubing.

Protocol 2. Preparation of fish embryo medium

1. Make the following stock solutions[a]:
   - solution A: 8 g NaCl, 0.4 g KCl, 100 ml H_2O
   - solution B: 0.358 g Na_2HPO_4, 0.6 g KH_2PO_4, 100 ml H_2O
   - solution C: 1.22 g $CaCl_2$, 100 ml H_2O
   - solution D: 2.46 g $MgSO_4(.7H_2O)$, 100 ml H_2O
   - solution E: 0.35 g $NaHCO_3$, 10 ml H_2O
2. Mix 1 ml solution A, 0.1 ml solution B, 1 ml solution C.
3. Add 95.9 ml H_2O, then 1 ml solution D, and 1 ml solution E.
4. Adjust pH to 7.0–7.5.

[a] Solutions A–D may be autoclaved for storage. Solution E cannot be autoclaved but may be filter-sterilized.

---

3. Preparation for embryo manipulation

Embryos should be reared to the desired stage in embryo medium (*Protocol 2*) at 28.5 °C. Keep them at a density below 50 embryos per 100 ml in a 250 ml beaker or in 10 ml in a 90 mm Petri dish. Rotting eggs should be removed daily. For most manipulations of zebrafish embryos it is necessary first to remove the chorion (*Protocol 3*). After dechorionation, embryos are mounted in either 3% methylcellulose (for pre-somite stage embryos; *Protocols 3* and *4*) or 1.2% agar (for older embryos; *Protocol 5*, also see Chapter 16) in order to prevent them from floating around and to control their orientation. Embryos older than about 18 somites should first be anaesthetized (*Protocol 5*).

Protocol 3. Dechorionation and preparation for manipulation of young embryos

1. Stage the embryos (see Appendix 1), clean them in embryo medium (*Protocol 2*) and put them into agar coated dishes[a] containing embryo medium.

2. Under low power on a stereo dissecting microscope, remove the chorions using fine sharpened watchmakers' forceps (Dumont No. 5). This is done by gently teasing open a hole in the chorion and carefully shaking out the embryo. Do not let the embryos touch the water surface.

3. Once the embryos are free of the chorion, transfer with a fire-polished pipette through several rinses in embryo medium to a clean agar dish. Embryo medium contains increased Ca^{2+}, which is necessary to keep cells in the blastoderm intact; agar provides a smooth non-sticky surface for the embryos to rest upon.

4. Mount pre-somite stage embryos in 3% methylcellulose (*Protocol 4*) on a glass depression slide. The embryo is gently inserted into a drop of methylcellulose and a loop made from fine nylon thread is used to orientate the embryo.

5. Cover the drop of methyl cellulose with several drops of embryo medium to prevent dessication.

6. Perform manipulation or observation as necessary.

7. To release the embryos from methylcellulose, immerse the entire slide into embryo medium. Leave 15–30 min for the methylcellulose to absorb medium and soften, and tease out the embryos with a fine nylon loop

[a] Agar dishes. Coat the bottom of Petri dishes thinly with 2% Bacto-agar in embryo medium (*Protocol 2*). Dissolve the agar by boiling, wait for solution to cool to just above gelling temperature (about 45 °C) and quickly pipette into the dishes.

Protocol 4. Preparation of 3% methylcellulose

1. Add methylcellulose (3% w/v) to embryo medium (*Protocol 2*) preheated to around 60 °C.
2. Shake the solution and heat for a further 10 min at 60 °C. After the methyl cellulose seems to be in solution, cool in a −20 °C freezer for 1 h, stirring every 20 min.
3. Remove and store overnight at 4 °C, to remove air bubbles.

Older embryos are best immobilized in 1.2% agar and they may also need to be anaesthetized. These procedures are described in *Protocol 5*. A variant on this method, specifically for recording cell movement in intact embryos, is described in Chapter 16.

Protocol 5. Preparation of older embryos for manipulation

1. Make a 1.2% solution of agar in embryo medium by boiling; cool to 37–40 °C and place an aliquot in a test-tube in a water bath at the same temperature.
2. Embryos older than 18 somites should be anaesthetized before mounting; use about 10 drops of Tricaine[a] in 5 ml embryo medium (*Protocol 2*).
3. Wash the embryos free of Tricaine, then carefully remove the chorions using fine sharpened watchmakers' forceps.
4. Transfer an embryo in a minimum amount of embryo medium into the test-tube; quickly suck up the embryo, transfer onto a glass slide and add several drops of cooled liquid agar.
5. Let the agar harden. Cut out a small piece of agar with a pointed scalpel. Pick up the block with forceps and orientate on a glass slide as required and cover with more liquid agar to hold it in position.
6. Perform experiment as required, through a small window cut in the agar over the region of interest.
7. To remove embryos from agar, immerse the slide with agar blocks in a dish of embryo medium and insert a closed pair of forceps next to the tail. By opening the forceps slowly the agar should break open following the long axis of the embryo.

[a] Add 0.4 g Tricaine methanesulfonate (MS222, Sigma) to 98.2 ml H_2O; add 1.8 ml 1 M Tris pH 9, and adjust pH to 7.2.

4. Maintaining zebrafish stocks

It is important to raise some embryos through to adults, to maintain the stocks and to ensure a constant supply of young, more fertile, fish for breeding. Newly hatched fry should be kept at a density of 20–30 per 500 ml beaker; the water must be changed daily. For the first two weeks, feed the fry on live *Paramecium*; this may be supplemented with Liquifry but live food gives the best results. After two weeks (or when the fry are large enough), feed with live brine shrimp twice per day; this can be supplemented with live nematodes or rotifers. Four weeks after hatching, transfer the fish to normal aquaria and feed them on live brine shrimp supplemented with finely ground dried flake food. Six weeks after hatching, feed the fish as adults (Section 2).

References

1. Westerfield, M. (1993). *The zebrafish book* (2nd edn). University of Oregon Press, Eugene, OR.

5

Xenopus

ARIEL RUIZ i ALTABA

1. Introduction

Amphibian embryos have a prominent place among the species used to study vertebrate embryonic development. There is a long and distinguished history of experimental manipulations that were carried out in salamanders and newts. Perhaps the best known example is the 1924 organizer transplant experiment of Hilde Mangold (1), then a student in Hans Spemann's lab. This experiment was not an isolated event, rather it was the culmination of a long search for the basis of the formation of embryonic tissues and their patterning in several laboratories, including that of Spemann. To understand the roots of present day molecular embryology, the reader is invited to read the fascinating book by Viktor Hamburger (2) written 50 years after the appearance of Spemann's magistral work (3).

While most of the problems in embryonic induction, cell fate, and patterning that we try to understand today were formulated a long time ago, the last decade has seen an explosion of information regarding the mechanisms controlling cellular behaviour. This is due to the advances of molecular biology and the choice of an amphibian species amenable for molecular analyses. The South African clawed frog *Xenopus laevis* (Daudin) is routinely used today as the amphibian experimental species of choice. The description and tabulation of the embryonic stages of *Xenopus* development by Nieuwkoop and Faber (4; see Appendix 1b) provides a reference point for all workers in the field. In this and other chapters (Chapters 9, 11, 14, 17) information is provided regarding basic techniques for the handling and manipulation of *Xenopus* embryos that should enable the novice to perform his or her first experiment and challenge the advanced student.

2. Frogs

Xenopus laevis can be kept in large tanks with approximately one frog per 2–4 litres. Carbon-filtered tap water in most places will be adequate although high salt or calcium content, or dirty water are not recommended. In these cases, a balanced salt solution such as Holtfreter's version of Ringer's solution should do. Frogs should be kept in clean water changed 1–2 h after feeding. Closed systems where the water is recycled and purified are highly recommended. Even though these frogs are aquatic and do not normally come out of the water for long walks, care should be taken to keep the containers covered at all times since they like to jump out and wander about. Careless researchers have unwillingly admitted to losing a few frogs in this way. *Xenopus* will eat anything that moves in front of their mouths, including their own tadpoles. However, a mixure of fish chow given twice or three times a week is a healthy diet and they will eat it willingly. It should be pointed out that some researchers extol the benefits of chopped liver. Female frogs can be kept for several years and recycled about every 4–6 months. Male frogs are killed to obtain homogenates of their testes. It is best to keep male and female frogs in separate tanks. Albino frogs are very useful for some techniques, such as whole mount *in situ* hybridization (Chapter 26), and can be obtained from a variety of suppliers.

A routine check for diseased frogs is recommended. Some of the symptoms include very thin, slow-moving frogs, rough skin (it is normally quite slimy), red patches in the belly or legs, and foamy skin when squeezed. Rough, red skin may be a symptom of 'red leg', a contagious frog disease that does not affect humans. Affected frogs should be isolated and discarded.

3. Oocytes

Mature female *Xenopus* frogs contain a large number of oocytes in their ovaries. A description and tabulation of *Xenopus* oocyte development is provided by Dumont (5). An average frog will have a few thousand oocytes at all stages of development, from young previtellogenic (stage I) to mature (stage VI) oocytes. *Xenopus* oocytes are used for a variety of experiments, including injection of nucleic acids and proteins for translation and/or processing (6) and destruction of specific maternal messages, by injection of antisense oligonucleotides, to test for their function later in development after reintroduction into a foster mother (7). Due to their large size ($\approx$1 mm) stage V–VI oocytes provide a good source of materials for biochemical experimentation. Developing oocytes can be cultured *in vitro* in the presence of vitellogenin for several days.

Protocol 1. Isolation of oocytes

1. Anaesthetize a female frog by placing it in a bucket with a small amount of water containing 0.1% MS222 (Sigma).
2. As soon as the frog is anaesthetized, place it on its back over moist paper towels. Make sure the frog is not responding to touch.

3. Make a small incision parallel to the main body axis in the lateral part of the belly with a sharp pair of scissors. Cut through both the skin and the muscles.

4. Carefully separate the muscles of the body wall and grab the ovary. If the frog is pigmented the mature oocytes will have a black or brown (animal) half and a white or grey (vegetal) half.

5. Pull out a portion of the ovary, its size depending on how many oocytes are required, and cut it free with scissors.

6. Place ovary portion in a Petri dish containing 1 × MBSH[a]. Tear and cut the ovary into small pieces to ensure that all parts are exposed to a well aereated salt solution. Some oocytes will come loose.

7. Carefully place the remaining ovary inside the frog and stitch the muscles and skin with fine surgical suture. Make two or three knots in different stitches. Non-dissolving suture is fine; it will become loose later.

8. Place the frog in a bucket with a small amount of water. This will keep it wet and prevent it from drowning before waking up.

9. Oocytes can be freed from the theca and surrounding follicle cell layers manually, using very fine watchmakers' forceps, or by treatment with 0.5–1% crude collagenase (Sigma type I) in 1 × MBSH.

10. For collagenase treatment, place ovary parts in a Petri dish in collagenase for 2–4 h on a moving platform. The oocytes should be removed periodically as they are slowly freed from the ovary. Transfer free oocytes to fresh 1 × MBSH and rinse several times. Isolated oocytes are kept at 18 °C (20 °C maximum) for up to a week.

[a] 10 × MBSH medium. Mix 51.3 g NaCl and 23.8 g Hepes. Add water to 700 ml and adjust pH to 7.5. Add 0.75 g KCl, 2 g $NaHCO_3$, 0.8 g $Ca(NO_3).4H_2O$ and 0.6 g $CaCl_2$. Bring final volume to 1 litre.

Oocyte viability depends, in part, on how carefully the dissection has been carried out and on whether or not follicle cells still surround the oocyte. For most purposes (e.g. injection of plasmids or RNA), follicle cells will not affect the experiment and it may be a good idea to leave the oocyte with a follicle cell layer. Reduced incubations in collagenase will retain more follicle cells. Injection of oocytes through follicle cell layers requires well-bevelled needles (see Chapter 9).

For translation assays, oocytes can be incubated for 6–20 h (depending on the rate of translation and equilibration of endogenous amino acid pools) in small volumes (e.g. 30–100 µl/oocyte) containing ^{35}S-methionine or other labelled amino acids. Concentrated radiolabelled amino acids can also be injected into oocytes. A plasmid vector to direct high levels of expression of cloned sequences in oocytes is pOEV (8). This simplifies *in vitro* RNA preparations, particularly of long transcripts.

The germinal vesicle, the oocyte nucleus, can be isolated manually with ease. Simply pinch the animal (pigmented) pole of the oocyte and gently squeeze the equatorial region. A translucent ball of 50–100 µm in diameter will appear with some whitish yolky cytoplasm. For nuclear localization assays of translated products, after injection of RNA or plasmids, remove as much as possible of the layers adhering to the germinal vesicle, as a large portion of the endoplasmic reticulum is found here.

4. Egg laying and fertilization

X. laevis matings display a typical posture. The smaller male grabs the larger female by holding his arms around her abdomen with the help of the sexual pads. These are rubbery regions in the inner side of forearms and hands of male frogs and are sometimes used to identify males unequivocally. Fertilization is external. The eggs (matured oocytes) are fertilized asynchronously as they leave the cloaca and encounter clouds of swimming sperm. In the laboratory, egg laying is induced with hormones and fertilizations are synchronous.

Protocol 2. Egg laying

1. Inject female *Xenopus* frogs with mare serum (≈200 units; Sigma) subcutaneoulsy in the upper leg region about 1–7 days before the day on which eggs are required. This stimulates oocyte maturation.

2. Inject human chorionic gonadotropin (≈800 units; Sigma) to induce egg laying. Females will start laying eggs after about 6–8 h at room temperature or after 10–15 h if frogs are placed in 16 °C water at the time of injection (water will slowly warm up). Injection at 10–11 p.m. will 'induce frogs to lay eggs the following morning.

3. Before collecting eggs, kill a male frog by anaesthetizing it first. Make an incision perpendicular to the main body axis at the level of the lower abdomen and locate the yellowish fat tissue on either side of the midline. Pull fat tissue gently and you will find the white testes next to it. Dissect the two testes and place them in a Petri dish with about 15 ml of 1 × MBSH (see *Protocol 1*) containing antibiotics (e.g. penicillin and streptomycin). They can be kept at 4 °C for up to 4–5 days. Dispose of the dead frog appropriately.

4. Collect eggs from the tank with a 25 ml pipette if the frogs are in 1 × ELS[a] or very gently 'squeeze' the females by rubbing the flanks and abdomen with your fingers if frogs are kept in tap water. To squeeze a frog successfully one needs some practice in holding it with one hand. Properly done, the frog will remain calm. Squeeze the eggs out into a dish with no liquid. Do **not** try to squeeze frogs forcefully. Collect eggs in a deep Petri dish. Eggs left in water (but not ELS) become activated and will not fertilize.

[a] 10 × ELS medium. Mix together 70 g NaCl and 18 g Tris base. Add water to 700 ml and adjust pH to 7.6. Add: 0.75 g KCl, 2 g NaHCO$_3$, 2 g MgSO$_4$.7H$_2$O, 0.8 g Ca(NO$_3$).4H$_2$O, and 0.6 g CaCl$_2$.2H$_2$O. Bring final volume to 1 litre.

Protocol 3. Fertilization

1. To fertilize eggs, drain all liquid from the Petri dish. Cut a small piece of testis with forceps and gently homogenize it (for example, use an Eppendorf tube and a loose-fitting plastic pestle).

2. Pour the testis homogenate onto the eggs and mix with forceps to ensure that sperm reaches all the eggs. Wait 2–5 min. Fill the dish with 0.1 × MMR[a]. The eggs will adhere to the bottom of the dish through their swollen jelly coats (which they acquire before they are laid).

3. Fertilized eggs will rotate so that the animal pole faces upwards in about 20 min and will undergo the first cleavage after about 90 min at room temperature. The site of germinal vesicle breakdown, the white area on the animal pole visible in pigmented mature oocytes and unfertilized eggs, will disappear as the pigmented area expands.

4. To slow down development, place the fertilized eggs or embryos at the desired stage at lower temperatures (e.g. 16 °C). The viable temperature range is roughly 15–24 °C. Embryos can also be kept at 4 °C for a limited period. For staging embryos, consult the normal table of Nieuwkoop and Faber (ref. 4; see Appendix 1a).

5. Before manipulation, 'de-jelly' embryos by incubation in 3% cysteine pH 7.6 for 3–10 min. Always use fresh cysteine solution. Change cysteine solution once or twice, depending on how rapidly the eggs lose their jelly coats. The rate of breakdown of the jelly coat is different for different embryo batches, so be careful and always watch the embryos while they are being de-jellied. If left loo long in cysteine, the embryos will die and turn white. De-jellied eggs will still be surrounded by the vitelline membrane.

6. When eggs are completely free from their jelly coats, wash them thoroughly several times with 0.1 × MMR. Development will proceed normally in this solution. Keep embryos in fresh, well aereated solutions. For best results do not put more than 100 embryos in a 150 mm Petri dish.

Protocol 3. *Continued*

Discard dead or dying embryos (whitish, soft) immediately, as they can affect the development of healthy neighbours. Embryos can be moved around with the help of a wide-bore Pasteur pipette.

---

[a] 10 × MMR medium: mix together 58.4 g NaCl and 12 g Hepes; add water to 700 ml and adjust pH to 7.6; add 1.5 g KCl, 2.4 g $MgSO_4.7H_2O$, and 0.4 g EDTA. Make up to 1 litre.

References

1. Spemann, H. and Mangold, H. (1924). *Wilhelm Roux Arch. EntwMech. Org.*, **100**, 599.
2. Hamburger, V. (1988). *The heritage of experimental embryology. Hans Spemann and the organizer.* Oxford University Press, Oxford.
3. Spemann, H. (1938). *Embryonic development and induction.* Yale University Press, New Haven.
4. Nieuwkoop, P. D. and Faber, J. (1967). *Normal table of* Xenopus laevis *(Daudin).* Elsevier/North Holland, Amsterdam.
5. Dumont, J. N. (1972). *J. Morphol.*, **136** 153.
6. Gurdon, J. B. and Melton, D. A. (1981). *Annu. Rev. Genet.*, **15**, 189.
7. Holwill, S., Heasman, J., Crawley, C.-R. and Wylie, C. C. (1987). *Development*, **100**, 735.
8. Pfaff, S. L., Tamkun, M. M., and Taylor, W. L. (1990). *Anal. Biochem.*, **188**,192.

6

Avian embryos

CLAUDIO D. STERN

1. Introduction

The avian embryo offers a very accessible system in which molecular studies can be combined with 'classical' embryology. Transplantation, cell labelling, immunocytochemistry, and chemical treatments can be combined with *in situ* hybridization and Northern analysis to study changes in gene expression after experimental manipulations. Moreover, with new techniques for producing transgenic birds now becoming available, the developmental effects of targeted mutations can be studied in cellular as well as molecular detail.

Avian embryos lend themselves very well to experimental embryology. The blastoderm is large, flat, and translucent, it grows well after isolation and the eggs are cheap and easy to obtain, can be used throughout the year, and the embryos easy to handle. Most importantly, unlike most of the other species discussed in this book, development of the embryo may be started or stopped as required simply by changing the incubation temperature from ambient or 10–14 °C (storage) to 38 °C (incubation), at least over the first two days of development. The embryo develops relatively fast, reaching the 10 somite stage at about 2 days' incubation. Chick and quail embryos hatch 19–21 days after laying if incubated at 38 °C; quail embryos develop slightly faster (19–20 days) than chicks (20–21 days). An interspecific chimaera system (quail/chick) is available and widely used, enabling the origin of transplanted cells to be followed without dilution of tracers, and several methods for lineage analysis and fate mapping are available. Two excellent staging systems exist and are widely accepted: Eyal-Giladi and Kochav (1), for pre-primitive streak stages of development, in Roman numerals (I–XIV), and Hamburger and Hamilton (2), after the appearance of the primitive streak, in Arabic numbers (2–46). Both are reproduced in this book (see Appendix 1c and d).

1.1 The egg

The avian egg consists of the yolk (mainly phospholipids), enveloped by a translucent, non-cellular *vitelline membrane*. This is surrounded by *albumen* (egg white) and the whole contained within a *shell*, from which it is separated by two *egg membranes*. At the blunt end of the egg is an *air sac*. All of these components are secreted by the mother. The embryo itself lies initially on the surface of the yolk, just under the vitelline membrane. The vitelline membrane is attached to two glycoprotein threads (*chalazae*), which allow the yolk to rotate so that the embryo always faces the top of the egg. During early stages of development, the edges of the single-cell thick embryo attach to the inner (yolk) face of the vitelline membrane, on which it expands. In addition to nutritive components, the albumen contains *lysozyme*, a bacteriostatic agent. For this reason, egg albumen is often used in embryo cultures to prevent the growth of microorganisms as well as for nutrition.

Eggs bought in shops do not usually contain viable embryos as they have not been fertilized. Many areas have local breeders/poultry farms who can supply suitable numbers of fertile hens' eggs, and in addition there are larger suppliers who can deliver country-wide. In the UK, for example, Ross Poultry will deliver quantities of fertile hens' eggs by courier or by parcel post, but this will increase the cost of the eggs. Fertility and embryo viability should be at least 90% in winter months but may drop to about 70% in summer. You should insist on this from your supplier. Once obtained, fertile eggs may be stored for up to 5–7 days before incubation; they keep best at 10–14 °C and should in any case not be allowed to fall below 6 °C. If storing eggs for a number of days before use, it may be useful to put them in their trays with the pointed end upwards (unlike the way they are usually delivered) and to stack the trays up tightly, in a box, so as to minimize the loss of CO_2. This is reported to improve the keeping properties of the eggs.

To start the incubation, simply place the eggs horizontally (long axis down) in a 38 °C incubator kept humid by inclusion of a suitably large volume of distilled water. If the incubation period desired is more than 48 h, then the eggs should be turned to prevent adhesion of the yolk membranes to the shell membranes, which impairs development. Incubators that do this automatically several times a day are available. If no such incubator is available, then turning may be done by hand, an odd number (e.g. three) of times per day so that they spend alternate nights on a different side. To turn them by hand, flip them over by 180° about the long axis of the egg so as to maximize the movement.

After incubation, the eggs can be allowed to cool to room temperature before use for up to a few hours, although younger embryos are more tolerant of this than older ones. After the third day of incubation, embryos should be used for most purposes within 2–4 h of removing them from the incubator. As eggs cool, the heart will slow down or even stop; this does not imply that the embryo is dead. Check by warming up the egg or embryo to 38 °C gently.

2. The embryo

By the time of laying, hens' eggs have spent some 20 h developing *in utero* (see Appendix 1). For this reason, freshly laid eggs are at the 'blastula' stage (usually stages IX–X in winter). In summer, eggs are retained inside the mother for longer and may be at later stages (typically stages XII–XIII). Younger embryos can be obtained by extraction from the mother, which may not require killing the bird. It is possible to stroke the belly of the bird at a known time in oviposition to coax it to drop its egg prematurely, but this requires some skill and, of course, access to suitable birds.

For the techniques described below, I will concentrate on young stages of development, where the embryo is small and some guidance is needed to handle the embryos successfully.

2.1 Explanting embryos for fixation or protein/RNA extraction

To explant embryos, you will need the following materials:

- dissecting microscope, preferrably one with transmitted light
- saline (either Tyrode's, CMF-Tyrode's, PBS or Pannett–Compton; see Section 4)
- two pairs of watchmakers' forceps, number 4 or 5
- one pair of coarse forceps, about 15 cm (6 in) long
- one pair of small, fine scissors, with straight blades about 2 cm (0.75 in) long.
- for young embryos (less than 2 days), a spoon/spatula
- Petri dish to collect embryos
- sylgard-coated Petri dish in which to dissect embryos[a] and fine steel insect pins, size A1 or D1.
- container for egg waste
- Pasteur pipette with the end cut off at the shoulder, stump flamed to remove sharp edges, and rubber teat
- as required, materials and solutions for fixation[b], protein extraction[c], or RNA extraction[d].

---

[a] Sylgard 184 (Dow Corning) is clear silicone rubber polymerized by mixing two components (9 parts rubber solution : 1 part of accelerator/catalyst). Mix the two well and pour to the desired depth (2–5 mm) into suitable containers, e.g. plastic Petri dishes. Allow the dishes to stand for about 1 h at room temperature for air bubbles to leave, then cure at about 55 °C until polymerized (3 h to overnight). The dishes can be stored indefinitely. Black Sylgard is also available.

[b] Fixatives. The most commonly used fixatives are: absolute ethanol (fix 48 h), absolute methanol (fix overnight at −20 °C) and buffered formol saline (mix 9 parts of phosphate buffered saline at pH 7–7.4 with 1 part of formaldehyde solution, to give 4%, final, of formaldehyde) (fix 30 min to overnight). Further details can be found in appropriate chapters for each method. It is a good idea first to pin the embryo in the desired position through the extraembryonic membranes whilst submerged in saline in a Sylgard dish, then remove the saline and replace with fixative. After a few minutes the pins can be removed and the embryo transferred to a vial containing either fixative or saline for storage.

c Protein extraction. Tissue is collected in an Eppendorf tube standing in solid CO_2. Dissect tissue with A1 or D1 insect pins mounted into the end of a Pasteur pipette while the embryo is pinned to the Sylgard dish and immersed in saline containing a mixture of protease inhibitors (see Chapter 21). As soon as each piece is dissected, pick it up with a Gilson micropipette set to 3–5 μl and transfer it to the dry Eppendorf tube sitting in solid CO_2. Keep frozen at −70 °C until needed.

d RNA extraction. Proceed as described above (footnote *c*) for protein extraction, except that no protease inhibitors are required. Collect the embryos or pieces in a minimal volume of saline, by transferring them with a Gilson micropipette to a dry Eppendorf tube standing in solid CO_2. In our experience, the best methods for RNA extraction are that of Krieg and Melton, described in Chapter 22 and that of Chomczynski and Saachi, described in Chapter 24. The second method will produce a higher yield but the first produces RNA of a better quality.

Embryos between 36 hours and 3 days' incubation can be explanted from the eggs quickly as detailed in *Protocol 1*.

Protocol 1. Explantation of 2–4 day embryos for fixation or protein/ RNA extraction

1. Remove eggs from incubator. With the coarse forceps, tap the blunt end of the egg so as to penetrate the shell. Use the tip of the coarse forceps to remove pieces of shell carefully so as to avoid damaging the yolk.

2. Allow egg white to pour into waste bucket, assisted by the forceps, taking care to avoid damage to the yolk.

3. Once most of the albumen has been poured off, make sure the embryo is uppermost; if not, turn the yolk by stroking it very gently with the sides of the closed forceps.

4. Make four cuts into the vitelline membrane around the embryo with scissors. If the embryo does not lie exactly in the centre of the egg, make the first cut on the side of the embryo nearest to the shell, and proceed in this way until all four cuts have been made. Make sure all the cuts meet each other.

5. With a pair of watchmakers' forceps, pick up a corner of the cut square and lift up the embryo and membranes. Transfer it to a dish with saline. Remove the vitelline (transparent) membrane with forceps and discard.

6. This procedure should produce reasonably clean embryos very quickly (less than 1 min per embryo with practice) and is therefore suitable for the collection of large numbers of embryos, e.g. for RNA extraction.

For embryos younger than about 2 days, a similar method can be followed except that the blastoderm will not survive if picked up with a pair of forceps. *Protocol 2* contains modifications as required for these younger embryos.

Protocol 2. Explanation of 0–2 day embryos for fixation or protein/
RNA extraction

1. Follow steps 1–4 in *Protocol 1*.

2. Pick up the square of embryos/membranes with the spoon/spatula, trying to collect only a minimal amount of yolk.

3. Transfer the yolk/embryo/membranes with the spoon into a Petri dish with saline under a dissecting microscope. With fine forceps, turn the square of yolk/membranes/embryo so that the embryo is uppermost.

4. After enough embryos have been placed into the Petri dish (about 8–15 depending on size of dish), use two pairs of forceps to separate the embryo from adhering yolk. Working at low magnification, pick up a corner of the square of vitelline membrane with one pair of forceps and slowly but steadily fold it back, steadying the yolk with the other pair of forceps. During the whole procedure the membrane and embryo should remain totally submerged in saline. The embryo should be attached to the membrane. If not, peel the membrane completely and then use forceps gently to remove the embryo from the underlying yolk.

5. Pick up the embryo, with or without adhering membrane, with the wide-mouth Pasteur pipette, and transfer it to a dish with clean saline for final cleaning or dissection. The edges of the extraembryonic membranes should be perfectly circular, provided that the embryo has not been damaged during the explantation procedure.

6. This method is a little slower than that in *Protocol 1*, but it should still yield embryos at a rate of about 3 min/embryo with some practice.

For embryos older than 4–5 days, it is possible simply to open the egg as described in *Protocols 1* and *2* and gently to lift the embryo with coarse forceps, without the need to cut the membranes first. To do this, place the tip of the forceps under the neck of the embryo and lift it out without clasping.

2.2 Dissection of tissues

Once embryos have been obtained as described in Section 2.1, the desired tissues may be dissected using mounted needles. For this, the best tools are:

(a) Tungsten needles, sharpened from very fine tungsten wire. The easiest way to sharpen them is to insert them repeatedly along their length into the side of a very hot flame after they have been mounted into a Pasteur pipette or needle holder. This will produce very sharp needles that can be re-sharpened repeatedly.

(b) Fine steel insect pins, mounted into the end of a Pasteur pipette or needle holder.

(c) Microsurgical knives. Some workers use platinum sewing needles, very carefully sharpened with oil on a very fine sharpening/honing stone and checking repeatedly under a microscope. These can be mounted into metal needle holders. We prefer to purchase very fine iridectomy knives (Weck, 45° angle, with long blades). If looked after, they should last a long time, since they are expensive.

Embryonic tissues can be dissected with these instruments, with the embryo submerged in saline, under a dissecting microscope. For embryos older than 36 h, it may be necessary to aid the dissection with trypsin (freshly made 0.1–0.2% trypsin 1:250, Difco, in saline). Pin one embryo at a time in a Sylgard dish, keeping the membranes taut, submerged in trypsin solution at room temperature; dissection can begin immediately. If the dissection takes longer than a few minutes the embryo may start to dissociate completely. In this case, wash off the trypsin solution before continuing. In some cases (e.g. for analysis of proteins), it is not desirable to use proteases during dissection. In this case we have found that double strength, calcium- and magnesium-free saline at 30 °C helps somewhat.

Pick up the dissected pieces with a Gilson micropipette set to 3–5 µl or appropriate volume, and transfer to a suitable container depending on the procedure to be followed.

3. Culture of young embryos

Embryos older than about 2 days' incubation can be accessed in the egg, by windowing. The techniques that follow concern embryos at earlier stages. The most successful methods for culture are all modifications of the one first described by Dennis New in 1955 (3). Two modifications will be described, both of which work well for embryos of less than about 36 h incubation at the time of setting up. The first allows embryos to be cultured for up to 30 h after setting up (e.g. stage 4 to stage 13), the second allows culture for up to some 60 h (e.g. stage 4 to stage 20).

3.1 New (1955) technique and modifications

The 'standard' method contains minor deviations from New's original description, with the main advantages that the embryo can be cultured in a flat plastic dish and that glass rings made as described here allow the embryo and ring to be transferred from one culture container to another. New originally described the use of rings made by bending and fusing glass tubing, and are therefore circular in cross-section.

3.1.1 Standard culture method

The following materials are required:

- finger bowl or Pyrex baking dish about 2″ (2 cm) deep
- Pannett–Compton saline stocks A and B (see Section 4). Before use, mix 120 ml A, 2700 ml H_2O and 180 ml B, in this order.
- 2 pairs small forceps (watchmakers' No. 4 or 5, but not too sharp)
- 1 pair coarse forceps for shell
- 1 pair small scissors
- 1 Pasteur pipette, end lightly flamed to remove sharp edges; rubber teat
- 35 mm plastic dishes with lid
- watch glasses
- rings cut from glass tubing, approx. 27 mm outer diameter, 24 mm inner diameter, 3–4 mm deep
- 1 small beaker
- 1 beaker or plastic bag for egg waste
- plastic box with lid for incubating culture dishes
- 38 °C incubator
- hens' eggs incubated 8–24 h (depending on stage needed)

Protocol 3. New culture (slightly modified)

1. Fill the large dish about three-quarters full with saline.
2. Open an incubated egg as described in steps **1–3**, *Protocol 1*, except that the thin, runny albumen is kept for culture. Collect the thin albumen in the small beaker.
3. When yolk is clean and free from adhering albumen, carefully tip it into the saline container, taking care not to damage the vitelline membrane on the edges of the broken shell. The blastoderm should face upwards. If not, carefully turn the yolk with coarse forceps.
4. Make a cut into the vitelline membrane enveloping the yolk just *below* the equator. Continue to cut all the way around the circumference of the yolk.
5. With two pairs of fine forceps, slowly but *steadily* 'peel' the North Pole of the vitelline membrane, all the way off the yolk. Do not stop during this process. The embryo should come off with the membrane. Let the membrane rest on the bottom of the dish, inner face (containing the embryo) pointing upwards.
6. Lower a watch glass and a glass ring into the container. Slide the vitelline membrane, preserving its orientation, onto the watch glass, and

Protocol 3. *Continued*

arrange the ring over it so that membrane protrudes around the ring. Pull out the assembly from the saline.

7. With fine forceps, work carefully to fold the cut edges of the vitelline membrane over the edge of the ring, all the way around its circumference.

8. Place the watch glass over a black surface. Suck off as much fluid as possible from the outside of the ring with the Pasteur pipette. If there is much yolk remaining over and/or around the embryo, wash it carefully with clean saline.

9. Remove any remaining saline, both inside and outside the ring. It is important that the embryo remains dry during incubation. Therefore discard any embryos in which the vitelline membrane has been damaged.

10. Put some thin albumen (about 1–2 mm thick layer) on the bottom of a 35 mm plastic dish. Slide the ring with vitelline membrane off the watch glass, and transfer it to the dish, over the pool of egg albumen. Press lightly on the ring with two forceps to allow it to adhere to the dish.

11. If the level of albumen comes close to the edge of the ring, remove the excess. Also aspirate any remaining fluid from inside the ring. It is best if the vitelline membrane bulges upwards, above a good pool of albumen. This will also help to drain off further fluid accumulated during culture to the edges of the ring.

12. Wet the lid of the plastic dish with albumen. Discard the excess, and seal.

13. Place the dish in a plastic box containing a piece of tissue paper or cotton wool wetted in distilled water, seal the box, and place it in an incubator at 38 °C.

3.1.2 A modification allowing longer culture periods

This method is based on a technique described recently (4). Here the vitelline membrane is draped *over* the glass ring, which allows the extraembryonic tissues to spread more freely. It also helps to use a larger (50 mm) plastic dish for culture. In this way, embryos may be cultured for up to 60 h after explantation, or stage 20, whichever occurs sooner.

Protocol 4. Modification for extended culture

1. Follow steps **1–5**, *Protocol 3*.

2. Submerge a 50 mm plastic dish directly into the large saline container, and place the glass ring into it.

3. Carefully slide the vitelline membrane, still with the embryo uppermost, onto the glass ring inside the plastic dish.

4. Remove this assembly from the saline container.

5. Pipette off the excess saline, allowing the vitelline membrane gradually to come to rest over the glass ring.

6. Working in stages, replace the saline under the ring with a pool of albumen, making sure that the ring contains this pool. The vitelline membrane should bulge up slightly at its centre.

7. Seal dish by smearing albumen on the lid and proceed as in steps **12–13**, *Protocol 3*.

This method has not yet been fully perfected and may be modified further for better results. It takes a little longer than the standard method to set up, mainly because of the difficulty in replacing the saline with albumen. It may help if the ring is first secured onto the bottom of the plastic dish using a thin film of silicon grease.

4. Salines for avian embryos

Two salines are most widely used: Pannett–Compton saline[a], used mainly for procedures in which the yolk is submerged completely in a large volume, and Tyrode's saline[b], used for dissection of parts of embryo. The reason for using either of these is probably mainly historical. American laboratories tend to use Howard–Ringer's saline, again for historical reasons. Alternatively, standard phosphate-buffered saline, containing calcium and magnesium, may be used. In my experience, embryos do grow better if either Pannett–Compton or Tyrode's are used rather than Howard–Ringer's.

[a] Pannett–Compton saline stocks (can be autoclaved for storage):
- solution A: 121 g NaCl, 15.5 g KCl, 10.42 g $CaCl_2.2H_2O$, 12.7 g $MgCl_2.6H_2O$, H_2O to 1 litre;
- solution B: 2.365 g $Na_2HPO_4.2H_2O$, 0.188 g $NaH_2PO_4.2H_2O$, H_2O to 1 litre.

Before use, mix (in order): 40 ml A, 900 ml H_2O, and 60 ml B.

[b] Tyrode's saline stock (10×; can be autoclaved for storage):
- dissolve: 80 g NaCl, 2 g KCl, 2.71 g $CaCl_2.2H_2O$, 0.5 g $NaH_2PO_4.2H_2O$, 2 g $MgCl_2.6H_2O$ and 10 g glucose in 1 litre H_2O.

Before use, dilute 1:10 with distilled water. The working solution may be buffered with bicarbonate, but we usually omit this.

For Ca^{2+} and Mg^{2+}-free Tyrode's (CMF), omit the Ca^{2+} and Mg^{2+} salts.

5. Mounting embryos for photography

For photography, it helps to keep embryos flat and surrounded by glass rather than plastic. We mount them in chambers constructed from a glass slide and one or more pieces of plastic tape (e.g. PVC insulating tape). Stick a length of tape onto the slide (more pieces required for older, and therefore thicker, embryos). With a scalpel blade, cut a square into the tape. Transfer the embryo to this, in suitable medium (e.g. glycerol or saline; the tape will not stand organic solvents). Seal with a glass coverslip, avoiding air bubbles; remove any excess mounting fluid. If the embryos are already in an organic mounting medium, use pieces of coverslip instead of plastic tape to define the chamber.

References and further reading

1. Eyal-Giladi, H. and Kochav, S. (1976). *Dev. Biol.*, **49**, 321.
2. Hamburger, V. and Hamilton, H. L. (1951). *J. Morphol.*, **88**, 49.
3. New, D. A. T. (1955). *J. Embryol. Exp. Morphol.*, **3**, 326.
4. Kučera, P. and Burnand, M. -B. (1987). *Teratogenesis, Carcinogenesis Mutagenesis*, **7**, 427.

7

Postimplantation mammalian embryos

GILLIAN M. MORRISS-KAY

1. Introduction: development of the culture technique

Techniques for culturing whole postimplantation mammalian embryos were pioneered by D.A.T. New in Cambridge (1, 2). Originally, rat embryos were cultured on plasma clots in watch glasses. Better development was found to result from the use of blood serum as a culture medium and by the use of 'circulators', in which continuously gassed serum passes as a constant stream through a chamber containing the embryos. The serum was obtained after allowing blood to stand and clot; embryonic development was quite good in that somites formed and the trunk elongated, but the two heart tubes often failed to fuse, and the trunk was excessively rigid, often protruding out from the curve of the yolk sac. These problems were overcome by centrifugation of the blood immediately after removal from the animal, so that the clot did not form in contact with blood cells (3). The serum is also heat-inactivated (30 min at 50 °C) to remove complement.

The gas phase of the cultures was also found to be important. The gas mixture used for routine cell and tissue culture is 5% CO_2 in air (i.e. 20% O_2), and this had also been used for rat embryo culture. Use of this gas mixture for embryo culture during the period of cranial neurulation resulted in failure of closure of the cranial neural tube, so that studies aimed at elucidating the mechanism of this process were confined to its early stages (4). Increasing the oxygen concentration exacerbated the problem, but decreasing it to 5% or 10% resulted in normal neurulation (5). Transmission electron microscopy of rat embryos during this period of development (day 9.5 to day 10.5) revealed that mitochondria at this stage are of the anaerobic type, with few cristae (6). This is true of embryos freshly explanted from the uterus, and of embryos cultured in equilibrium with a gas phase of 5% O_2. In contrast, embryos cultured with a gas phase of 20% or 40% O_2 have mitochondria of the aerobic type. The apparent use of glycolysis as an energy source during neurulation coincides with a period of rapid embryonic growth, during which the foregut and hindgut form and the expanding yolk sac comes to surround the embryo (see *Figure 11*). Diffusional distances for oxygen supplied from the maternal blood become maximal, until, towards the end of cranial neurulation on day 10, the yolk sac circulation develops and the embryonic heart starts to beat.

The apparently perfect development that can be obtained *in vitro* from day 9 to day 11 or 12 of pregnancy makes this an ideal system for experimental manipulations aimed at elucidating the mechanisms of various aspects of morphogenesis, including neurulation, neural crest cell migration, early heart development, and somite development. From this stage onwards, cultured embryos require a gas phase with a higher oxygen content: 20% gives good development up to day 11; later stages require higher oxygen levels to compensate for the fact that *in vivo*, the haemochorial placenta is progressively taking over the functions of the yolk-sac placenta: 40% O_2 from late day 11 to late day 12, and 95% from late day 12 onwards. From the afternoon of day 12 (day 12.5), access of oxygen (and nutrients) to the embryos is facilitated by opening the yolk sac in addition to using a 95% O_2 gas phase (7).

2. The maternal–embryonic relationship prior to culture

Early postimplantation-stage rodent embryos are dependent on the yolk sac for their nutrition. Unlike the haemochorial placenta, whose intimacy and complexity cannot at present be mimicked in culture, the yolk sac is able to function as efficiently *in vitro* as *in vivo*. *Figure 1a* shows a section of a day 8.5 pregnant uterus (rat). Small capillaries within the decidual tissue provide blood to the ectoplacental cone trophoblast and to a blood sinus surrounding the embryo. *Figure 1b* shows that the decidual tissue surrounding the blood sinus is lined by trophoblast cells. The parietal yolk sac (Reichert's membrane), on the embryonic side of the sinus, consists of a thick basement membrane with sparsely distributed cells on its embryonic aspect. This acts as a coarse filter, excluding blood cells but allowing the transfer of macro-molecules, including antibodies. The transferred molecules are taken up by the visceral yolk-sac endoderm. Phagocytosed proteins are converted to amino acids by lysosomal enzyme action, before being made available to the embryo.

In culture, the filtration function of Reichert's membrane is mimicked by the method of serum preparation, during which red blood cells and fibrin are removed, and complement inactivated. Reichert's membrane does not grow if left on the cultured embryos, so must be opened.

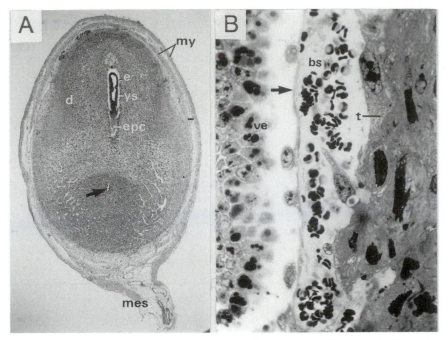

Figure 1. The maternal–embryonic relationship in the rat on day 8.5. (A) Transverse section of the pregnant uterus. The embryo (e), yolk sac (ys), and ectoplacental cone (epc) are only loosely attached to the surrounding decidual tissue (d), which contains a capillary network connected to branches of the uterine artery and vein in the mesometrium (mes). The uterine wall is highly muscular (my = myometrium); the uterine lumen is marked by an arrow. (B) The maternal blood sinus (bs) is separated from the phagocytic cells of the visceral yolk sac endoderm (ve) only by the thickened basement membrane of the peripheral yolk sac (arrow). A layer of trophoblast (t) lines the sinus, anchoring the embryo loosely to the surrounding decidual tissue. A: ×35; B: ×300.

3. Materials required

(a) Culture medium: 50% serum (*Protocol 1*), 50% Tyrode's saline[a]

(b) Gas phase: 5% CO_2, 5% O_2, 90% N_2 (supplied to the bottles through a sterilized Pasteur pipette plugged with sterilized cotton wool at the wide end)

(c) Culture bottles: 60 ml Pyrex bottles (sterile), with silicon bungs

(d) Silicon grease and cling-film for sealing

(e) Incubator, the bottles must be rotated at approximately 30 r.p.m. and the temperature maintained at 37–38 °C

(f) Dissecting instruments: these should be clean but not necessarily sterile, for opening the abdomen and removing the uterus. The kit should contain a pair of strong scissors, a pair of medium forceps, a pair of fine scissors and a pair of fine forceps

(g) In a sterile pack (with non-sterile pipette bulbs), assemble two pairs of sharpened watchmakers' forceps (check that the points are fine and that they appose exactly), one cataract knife and two Pasteur pipettes (shortened and fired to give a wide opening for transferring dissected embryos to the culture bottle)

(h) Pregnant rats. They should be at day 9.5 of pregnancy (day of positive vaginal smear = day 0.5). In the early afternoon the embryos are usually at the late presomite stage (*Figure 11a*), but there is some variation both within and between strains

^a Tyrode's saline. Add the following to a small volume of distilled water in the order given: 8 g NaCl, 0.2 g KCl, 1 g Glucose, 0.05 g $NaH_2PO_4.2H_2O$. When these have dissolved, add water up to 975 ml, then 0.2 g $CaCl_2$ and 0.1 g $MgCl.6H_2O$. Dissolve, transfer to 100 ml bottles (97.5 ml per bottle), and autoclave. Millipore filter 2.5 ml of 4% $NaHCO_3$ into each bottle of the cooled solution. Stock tubes of sterile 4% $NaHCO_3$ can be kept at 4 °C.

Protocol 1. Preparation of rat serum

1. Anaesthetize rat with diethyl ether vapour.

2. Place the anaesthetized rat on its back and dampen the ventral surface with 70% alcohol.

3. Make an incision into the lower abdominal wall and expose the abdominal aorta, taking care to strip away excess fat from around the aorta.

4. Using a 15 or 20 ml syringe and suitable needle (21G), withdraw as much blood as possible, using gentle suction.

5. Withdraw the needle, remove it from the syringe, and carefully empty the syringe contents into a chilled sterile centrifuge tube (not containing anticoagulant).

6. Immediately centrifuge the blood at 2500 *g* for 10 min.

7. After centrifugation, break the clot with a sterile needle, squeezing against the side of the tube, and remove it.

8. Centrifuge the tube at 2500 *g* for 10 min.

9. Pipette the serum into a sterile universal container.

10. Heat-inactivate the serum (to destroy complement) by standing it in a 56 °C water bath for 30 min.

11. Store at −20 °C until required.

Serum prepared in this way is referred to as 'immediately-centrifuged, heat-inactivated serum'.

---

4. Preparation of embryos for culture

The procedure given in *Protocol 2* describes the preparation of embryos from rats on the afternoon of day 9 of pregnancy (day 9.5). All illustrations are for a right-handed person.

Protocol 2. Obtaining embryos for culture

1. Kill the pregnant dam by means of chloroform vapour or CO_2. Swab the abdomen with 70% alcohol and open it with strong sharp scissors to give a broad U-shaped incision, cutting through both the skin and the muscular abdominal wall.

2. Displace the gut in order to uncover the pregnant uterus. This has the appearance of two strings of beads joined to the vagina in the midline, and with an ovary at each end. Each 'bead' contains an embryo surrounded by decidual tissue. A fine mesentery, the mesometrium, runs along the whole length of each uterine horn, attaching it to the pelvic cavity (*Figure 2*).

3. Using fine forceps and small scissors, hold the ovarian end of one uterine horn and cut carefully through the mesometrium along the whole length of the uterine horn to detach it from its anchorage (*Figure 2*). Take care to cut close to the uterus, leaving behind any fat.

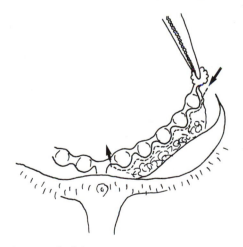

Figure 2. Location and removal of the uterus.

4. Cut the uterus transversely to separate it from the ovary and from the vagina. Place it in a dish of Tyrode's saline.

5. Repeat this procedure to remove the second uterine horn.

Protocol 2. *Continued*

6. Place the dish containing the gravid uterus under a dissecting micro-scope with dark-field illumination or epi-illumination.

7. Using two pairs of watchmakers' forceps, carefully open the uterus by tearing apart along the anti-mesometrial side, leaving the decidual swellings *in situ* (*Figure 3*).

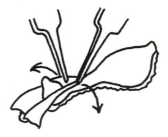

Figure 3. Opening the uterus.

8. Grasp the opened uterus with one pair of forceps close to a decidual swelling (using your non-dominant hand) and gently separate the base of the pear-shaped swelling from the uterus by pushing the other pair of forceps between the two tissues (*Figure 4*). Take care not to tear the decidual tissue.

Figure 4. Separating a decidual swelling from the uterine wall.

9. Transfer the decidual swelling(s) to a fresh dish containing Tyrode's saline. Use a spoon or lift by grasping the *blunt* end with forceps.

10. Examine the decidual swelling under the dissecting microscope. Identify a small depression (the uterine lumen–arrow in *Figure 5*). Bear in mind the position of the embryo (outlined with a broken line in *Figure 5*) during the next stage of the procedure.

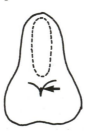

Figure 5. Locating the uterine lumen in the decidual swelling.

11. Using one pair of forceps to steady the whole structure, insert the closed points of the other pair of forceps into the depression (*Figure 6a*), so that when they are opened a small slit is made (*Figure 6b*). The points should make gentle contact with the dish before being opened. The bright red ectoplacental cone should now be seen.

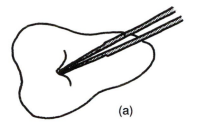
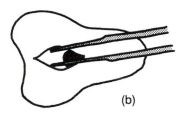

(a)   (b)

Figure 6. Making a slit in the uterine lumen.

12. Keeping the slit open, complete it towards the blunt end by cutting with the cataract knife (*Figure 7*).

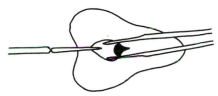

Figure 7. Extending the slit with cataract knife and forceps.

13. Using two pairs of forceps, gently and slowly extend the opening up to the pointed end of the decidual swelling. Take care not to touch the embryo, which should remain attached to one half (*Figure 8*). Discard the empty half.

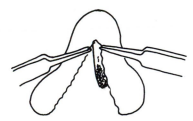

Figure 8. Exposing the embryo.

14. Holding the decidual tissue close to the embryo with forceps (in your non-dominant hand), gently separate the embryo from it, using the cataract knife to break the fine strand-like connections made by the trophoblast (*Figure 9*). These loose connections between embryonic and maternal tissue traverse a blood sinus containing maternal blood (*Figure 1b*). The embryo is surrounded by Reichert's membrane, through which

Protocol 2. *Continued*

nutrients pass before being taken up by the endodermal (outer) cells of the yolk sac.

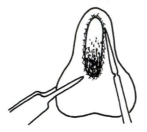

Figure 9. Isolating the embryo.

15. Transfer the embryo to a clean dish of Tyrode's saline and change to bright field illumination. Examine the embryo, referring to *Figure 11a*. Is Reichert's membrane still present? Has the visceral yolk sac been punctured or squashed? The red structure at one end is the ectoplacental cone (trophoblast plus maternal blood). This must remain attached to the embryo for the culture to succeed.

16. Reichert's membrane must now be removed (unless it has already come off). Look for a nipple-like region at the non-ectoplacental cone end (*Figure 10a*). Grasp this with both pairs of forceps and gently tear open the membrane up to the ectoplacental cone (*Figure 10b*).

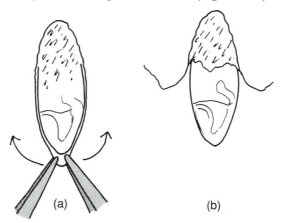

Figure 10. Opening Reichert's membrane.

17. Check that you have not damaged the embryo. The visceral yolk sac should be smooth and taut. The embryo, if perfect, is now ready for culture. It may be slightly more or less advanced than that shown in *Figure 11a*. Discard any damaged embryos. Minor injuries often heal during culture.

18. Transfer the dish of embryos to be cultured to a laminar flow cabinet. Using a sterile blunt-ended Pasteur pipette, transfer the embryos, in a minimum volume of saline, to a culture bottle containing 5 ml culture medium. Up to 10 embryos can be put into one bottle for a 24 h culture.

19. Introduce the Pasteur pipette attached to the duct from the gas cylinder into the culture bottle and adjust the flow of gas so that the jet makes a small depression in the surface of the medium. Gas for 3 min.

20. Close the bottle with a greased silicon bung as soon as the pipette is withdrawn. Turn off the gas, and wrap a piece of cling-film around the bung and bottle neck to secure the bung. Place the bottle on the rollers in the incubator at 38 °C.

21. If the embryos are to be cultured for longer than 24 h, gas again with 5% CO_2 in air after 24 h (or earlier if the cranial neural tube has closed). The embryos can be observed within their glass bottles using a dissecting microscope with bright-field transmitted light. Ideally, this should be within an enclosed 'hot box' at 37–38 °C.

---

5. Development during the period of culture (day 9.5 to day 10.5)

During the 24 h from day 9.5, the heart tubes form, fuse and descend, with concomitant formation of the foregut. As the foregut and hindgut form, the expanding visceral yolk sac is drawn down over the embryo (*Figure 11b*). The cranial neural folds form as a pair of enlarging convex structures, with the growing forebrain region projecting over the descending heart (*Figure 11b*). These become V-shaped and then concave in the transverse phase. Closure of the neural tube begins at the seven-somite stage, at the level of somites 5–6 (upper cervical level). Cranial neural tube closure begins as a rostral extension of the cervical neural tube, but at the 10-somite stage a second point of apposition is formed at the forebrain/midbrain junction, dividing the cranial neuropore into two. Both parts complete closure at the 14-somite stage. The first clear transverse landmark in the cranial neural folds (neural epithelium) is the preotic sulcus (*Figure 11b*). This forms at the two-somite stage, so is present in some embryos prior to culture. By the 14-somite stage a series of seven sulci and gyri (rhombomeres) have formed in the hindbrain, rostral to the level of the first somite. These can be seen in living embryos (try both bright field and dark-field illumination). The preotic sulcus disappears as the rhombomeres form; its position corresponds to the gyrus between rhombomeres 2 and 3. The first four somites are occipital somites, and are part of the head. When counting somites, take care to include the first, which is rather ill-defined at this stage. By the 12-somite stage the otic pit can be

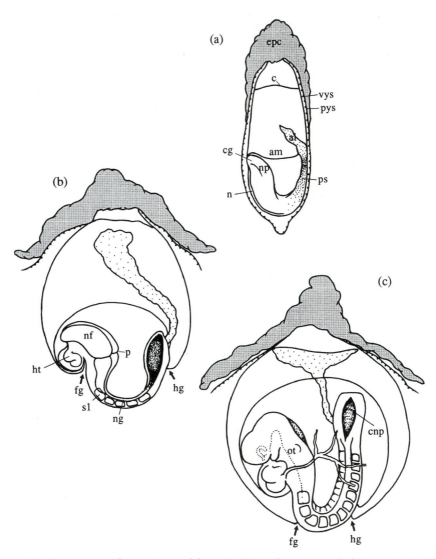

Figure 11. Appearance of rat embryos (a) on day 9.5, before removal of the parietal yolk sac (Reichert's membrane); (b) at the five-somite stage (late evening); (c) at the 12-somite stage (day 10.5). al, allantois; am, amnion; c chorion; cg, cardiogenic region; cnp, caudal neuropore; epc, ectoplacental cone; fg, foregut; hg, hindgut; ht, heart in pericardial cavity; n, notochord; nf, cranial neural folds; ng, neural groove; np, cranial neural plate; ot, otic pit; p, preotic sulcus; ps, primitive streak; pys, parietal yolk sac; s1, first somite; vys, visceral yolk sac.

clearly seen in the surface ectoderm, level with the position of the future second pharyngeal arch (*Figure 11c*). The first pharyngeal arch begins to form at the eight-somite stage, as neural crest cells from the rostral hindbrain region migrate down into it (their precise level of origin is from the midbrain/hindbrain junction to the preotic sulcus, i.e. the future rhombomeres 1 and 2) (8).

The yolk sac is the first haemopoietic tissue of the embryo. Blood islands form in it and coalesce to form vessels, which join up with independently-formed intraembryonic blood vessels. The heart starts to beat at approximately the seven-somite stage, and blood cells can be clearly seen circulating within the yolk sac vessels after this.

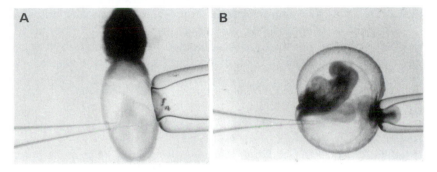

Figure 12. Microinjection into the amniotic cavity of embryos prior to or during culture: (a) on day 9.5; (b) on day 10.5 (see reference 9 for details). (a) ×20; (b) ×40.

6. Culture of younger postimplantation rat embryos

Younger embryos at stages prior to and during gastrulation (days 7 and 8) can also be cultured for up to 48 h with good results. The technique for dissection and culture is as described above, except that the gas phase is 5% CO_2 in air. These 'egg cylinder' stage embryos are very delicate, and require even more care during dissection than do day 9 embryos. Removal of Reichert's membrane presents a particular problem, since it is closely adherent to the embryo, and so is difficult to grip with forceps without damaging the embryo itself.

7. Culture of postimplantation mouse embryos

Mouse embryos of equivalent stages can be cultured using similar techniques to those described above (including the use of rat serum), but being smaller are slightly more difficult to manipulate than rat embryos.

8. Use of the whole embryo culture technique for experimental purposes

The technique described above enables embryos to be investigated and manipulated during the period of neurulation, neural crest cell migration, somitogenesis, and early cardiogenesis. Substances such as enzymes that degrade specific substrates of morphogenetic significance (e.g. extracellular matrix components) can be added to the culture medium or (to minimize any effect on yolk sac function) into the amniotic cavity (*Figure 12*) (9). Labelled cells, carbocyanine dyes to label small groups of cells *in situ*, or lineage tracers to follow the descendants of single cells can be microinjected (see Chapter 9). The embryo is then allowed to undergo further development *in vitro* prior to analysis of the effects of the injected substance, or of the position of the labelled cells. Cell cycle time in specific embryonic tissues can be analysed (10). For each of these approaches, the culture method enables embryos to be observed directly, so that they can be obtained for manipulation at specific stages of development, and the cultures terminated at specific stages.

References

1. New, D. A. T. (1978). *Biol. Rev.*, **53**, 81
2. New, D. A. T. (1990). In *Postimplantation mammalian embryos: a practical approach* (ed. A. J. Copp and D. L. Cockroft), pp. 1–15. IRL Press at Oxford University Press, Oxford.
3. Steele, C. E. and New, D. A. T. (1974). *J. Embryol. Exp. Morphol.*, **31**, 707.
4. Morriss, G. M. and Steele, C. E. (1974). *J. Embryol. Exp. Morphol.*, **32**, 5.
5. New, D. A. T., Coppola, P. T., and Cockroft, D. L. (1976). *J. Reprod. Fertil.*, **48**, 219.
6. Morriss, G. M. and New, D. A. T. (1979). *J. Embryol exp. Morphol.*, **54**, 17.
7. Cockcroft, D. L. (1990). In *Postimplantation mammalian embryos: a practical approach* (ed. A. J. Copp and D. L. Cockcroft), pp. 15–40. IRL Press at Oxford University Press, Oxford.
8. Tan, S. S. and Morriss-Kay, G. M. (1985). *Cell Tissue Res.*, **240**, 403.
9. Tuckett, F. and Morriss-Kay, G. M. (1989). *Anat. Embryol.*, **180**, 393.
10. Morriss-Kay, G. M., Tuckett, F., and Solursh, M. (1986). *J. Embryol. Exp. Morphol.*, **98**, 59.

8

Simple tips for photomicrography of embryos

CLAUDIO D. STERN

1. Introduction

Illustrating the results of experiments both appropriately and attractively is an important aspect of embryological (as of other) research. Here we give some guidelines that should help to decide how to photograph embryos and improve the resulting photomicrographs, concentrating on some common errors. Good microscopy is as important as the camera techniques themselves and we will start by quickly reviewing some of the basic techniques, giving their advantages and disadvantages for some common applications. At the end of this chapter we will give some advice on how to visualize and photograph embryos after specific techniques.

2. Tips for good microscopy

There are six basic techniques in light microscopy, each with advantages and disadvantages for specific applications: bright-field transmitted light (Köhler illumination), reflected incident light, dark field, phase contrast, Nomarski differential interference contrast (DIC), and epifluorescence. All photomicroscopes are capable of the first two, and most research departments will have each of the other four on at least one microscope. Specimens stained with histological dyes are best studied by conventional bright-field microscopy; more advanced techniques usually lead to loss of detail. For this (as well as for DIC), it is important to set up the microscope correctly for Köhler illumination.

2.1 Setting up Köhler illumination on your microscope

All photomicroscopes are capable of correct Köhler illumination. The object of this procedure is to optimize the illumination of the microscope to the numerical aperture of the objective. To begin with, you need to identify five controls on your microscope:

(a) the microscope focus knobs (fine and coarse)

(b) the condenser focusing knob

(c) the appropriate condenser lens(es) for the objective(s) required

(d) the field aperture

(e) the iris diaphragm

Of the latter two, the field aperture will be closer to the light source and when closed almost completely can be focused using the condenser.

When you have identified all of these, follow the procedure in *Protocol 1*.

Protocol 1. Setting up Köhler illumination

1. Select an objective (see Section 3.2) and appropriate condenser lens (c).

2. Focus the objective (a) roughly on the specimen.

3. Close down the field aperture (d) completely.

4. Focus the condenser (b) until you can see the smallest and sharpest possible image of its edges. Centre the field aperture, if necessary.

5. Open the field aperture (d) until it just illuminates the whole field of view.

6. Remove an eyepiece. Look into the tube from a few inches away from the end of the tube; you should see a bright circle. When you open the iris diaphragm (e) there should be a point at which the bright circle does not open further; this is the maximum iris aperture.

7. Now close the iris (e) so that about three-quarters[a] of this maximum aperture is illuminated.

8. Replace the eyepiece and adjust the light intensity.

Since each objective has its own characteristic numerical aperture, the adjustment will also differ between objectives. It is therefore important to repeat at least steps **6–8** *each time you change objectives*. Once used to it, it should only take about 20 sec to go through the whole procedure, and this time will be very well spent.

[a] With most modern objectives, this can be increased to about seven-eighths.

2.2 Incident reflected light and dark-field illumination

Some stained specimens, such as embryos stained for alkaline phosphatase (for example, from whole-mount *in situ* hybridization procedures; see Chapter 26) which have not been cleared are notoriously difficult to photograph by conventional bright-field optics. This is because transmitted light sheds dark shadows which can obscure the staining patterns and because it can render the staining too dark. One solution is to use an incident light source. To do this, simply point light beam(s) from a cold-light source (such as a fibre-optic light source) onto the specimen at a shallow angle to the stage of the microscope, to avoid reflections from the glass slide. The specimen should be covered with a glass coverslip, otherwise the meniscus of the fluid above it will cause unwanted reflections; it sometimes helps to place a piece of black paper under the specimen to avoid reflections from the condenser lens, and in any case the condenser should be lowered as far as possible from the specimen. Exposure should be about one to two stops longer than that indicated by the camera meter (see Section 3.2). If using colour film, then it may be of advantage to combine a low level of transmitted light with the incident illumination, preferably using a coloured filter (green is usually best) to help distinguish the structure of the specimen from the staining pattern. In this case, exposure should be only about one-half of a stop above that indicated by the meter.

Reflected light from an incident fibre optic or other source can also be used successfully through a simple ('dissecting') microscope. Again, a matt black background is advisable unless combined with transmitted light. As with specimens photographed through the compound microscope, it is important to avoid uncovered liquid surfaces. They will tend to transmit vibrations and cause reflections. Avoid photographing specimens floating simply in a deep dish full of liquid, as they move almost imperceptibly, just as a result of the vibrations caused by the camera shutter.

Whether using a compound or a simple microscope, it is always a good idea to place the fibre-optic light source on a surface other than that supporting the microscope, so that vibrations generated by the cooling fan of the light source are not transmitted to the microscope. Such vibrations are often the cause of photographs that do not appear sharp.

Radioactive *in situ* hybridization and other autoradiographs, as well as samples labelled with colloidal gold, also benefit from reflected light methods. Here it is of advantage to use coloured filters both for the incident light source (typically red) and for the transmitted light (typically green or blue). The same effect (red grains on blue or green background), but with more control over the contrast, can also be obtained with a confocal laser scanning microscope: the silver grains are imaged by reflection of the laser and the resulting image 'merged' with a transmitted light image of the specimen. Unfortunately, very low power objectives do not produce very good results by this method.

Autoradiographs and gold-labelled specimens can also be visualized using dark-field illumination. To do this properly, the correct dark-field condenser is required; the instruction manual with the microscope and this attachment should give instructions for its correct use, which varies from one microscope to another. If such a condenser is not available, then it is still possible to produce a dark-field effect using a simple trick. Using an ordinary (not phase contrast) objective of about 16× or less, use a small phase-contrast ring in the condenser to generate a dark-field-like effect in the centre of the field. The edges of the field may have bright rings or other undesirable effects and they are best avoided during photography, but the results can be very good. Although it is not usually possible to use the trick with higher power objectives, this is rarely needed.

2.3 Phase contrast and Nomarski differential interference contrast (DIC)

These are specialized techniques designed to enhance the contrast of specimens which, as a rule, have not been stained with histological procedures. Both have specific limitations. In brief, phase contrast cannot be used on specimens that have been cleared; they should be in aqueous medium. DIC cannot be used if the optical path crosses a plastic surface, and at least with normal condensers, cannot be used to image very thick specimens. In addition, phase contrast generates thick, bright rings around the edges of objects which obscure their true edges. In general, phase contrast is used mainly for live specimens (especially cells in tissue culture). DIC can be used both for live specimens and for histological sections; it is also very good for photography of whole mount *in situ* hybridization specimens processed with digoxigenin-labelled probes and alkaline-phosphatase detection systems (see Chapter 26). However, DIC is a more specialized technique and fewer departments possess microscopes with this facility. Both techniques are more successful with higher power objectives than with low magnifications (10× objectives and higher).

Very briefly, phase contrast requires a special condenser and special objectives. Each has a ring in it, one bright, one dark. Successful phase contrast requires precise alignment of the two rings, which must be matched in size. Alignment is done either through a special 'phase telescope' or simply by removing the eyepiece. Although DIC does not require special objectives, it does require several other components (polarizer, ¼ wavelength plate, Wollaston prism, and analyser), and special positions in the microscope capable of accommodating them. The exact layout and procedures for adjusting them vary between different microscopes (as does the degree to which the technique is generally successful for all objectives) and the reader should refer to the manual for detailed instructions.

One advantage of DIC over the other techniques is that, with correct adjustment, it can generate striking colours (the background becomes blue and the specimen acquires red shades where its curvature changes), which can allow advantageous colour photography in specimens that are colourless. This helps to compose plates combining stained and unstained specimens.

There is a simple optical subterfuge, using oblique illumination of the specimen, which can be used to produce acceptable DIC-like images (but without the colours), even on microscopes not equipped with Nomarski optics. To do this, first set up proper Köhler illumination as described above, but open the iris diaphragm just a little more than normal. The condenser assemblies on most microscopes have a disk with different apertures and/or lenses that can be rotated to suit objectives. While looking at the image, slowly rotate this disc slightly away from its normal position (which may be labelled '0' or with the magnification of the objective for which it is intended), until you see the edges of cells or other objects sufficiently enhanced.

Both with low power DIC and with the simple alternative above, one problem is that the illumination in the field of view is not even. It is important to be aware of this when composing photographs. One solution is to increase the magnification, but if this is not possible, then a smaller portion of the resulting negative may have to be printed for publication.

2.4 Epifluorescence

This technique has become widespread and therefore needs little introduction. Light from a strong source passes through a first filter which selects the *excitation wavelength* (usually not ultraviolet, as some state), then is made to go through the objective onto the specimen. This incident light is reflected back from the specimen through the objective. It will include both the excitation wavelength and light at any other wavelength (*emission wavelength*) due to special properties of the fluorochrome used. After the objective, the original excitation wavelength light is cut out by a *narrow bandpass* filter which only allows the emitted light to pass through.

With most common specimens, epifluorescence will produce pictures consisting mainly of a black 'background' and a variable proportion of bright 'objects'. Cameras with automatic exposure meters will tend to overestimate the exposure times required, and exposure adjustments often have to be made (see Section 3.3 below).

3. Photomicrography

3.1 Choosing a film: black-and-white or colour?

Journal publishers do not seem to like printing plates in colour, even if the author pays for their production, which they will often insist on. In some journals it may slow down publication. The question that the author has to address is: 'Would using colour add information to the illustration, or would it just make it look prettier?'. If the latter is the case, then use black and white. A simple rule is that if the picture is really monochrome (such as immunofluorescence with a single antibody, etc.) then colour adds nothing, and may even reduce contrast significantly. However, one often wants to have both a colour slide (for talks, etc.) and a black-and-white negative. In this case, it helps to have a microscope fitted with two cameras (e.g. Zeiss Axiophot or Olympus Vanox). If not, there are three possibilities:

(a) interchange between two camera backs, although this is cumbersome

(b) shoot all the pictures in colour, then change to a black-and-white film and reshoot

(c) take all the pictures on colour slide film, then copy the required slides onto black-and-white film using a camera with a macro lens

The major disadvantage of the latter procedure is that some resolution is always lost. Whichever solution has been chosen, then the next task is to choose a suitable film for the purpose in hand.

3.1.1 Colour film

It is usually better to use colour slide film than print film; the tolerance is greater and it is easier to choose what to print from slides (positive) than from colour negatives. After many tests, we have settled for Fuji 1600P for fluorescence and Fuji 64T for most other purposes. For photographing histological sections and specimens with little contrast, we rate the Fuji 64T film two stops underexposed (as if it were 250 ISO/ASA film) and push-develop it by two stops, which increases grain size and therefore contrast. It is easy to process your own colour slides (process E6). Development is done at 38 °C and requires three chemical baths. A kit can be bought commercially, and the cost of developing a film is about three-times cheaper than commercial processing. The results are often better, and you have complete control of the sensitivity of the film. It takes about 40 min to process a colour film with these kits.

3.1.2 Black-and-white print film

For most purposes, we use only Kodak TMAX100 (normal photomicrography) or TMAX400 (fluorescence, push-developed to 800 or 1600 ISO/ASA if necessary), which are very flexible films which can be pushed or pulled two or three stops without appreciable change in resolution or grain. For exposures requiring particularly high contrast and good resolution, we use Kodak Technical Pan, rated at 80 ASA and develop it either in Ilfotec LC29 or Acutol (for high definition and fine grain) or in HC110 (for higher contrast still).

3.2 Good framing

Assuming that the correct film has been chosen, that the microscope is set up to give optimal images (failure to do this is probably *the* most common cause of poor photographs), and that the specimen is optimally suited for photography (a rare event), then the key to good photographs lies in two main parameters: good framing and correct exposure. Neither is as trivial as it sounds, and both require a little experience.

The first thing to remember is that eyepiece magnification should be kept to a minimum, and objective magnification (and therefore numerical aperture of objective) to a maximum. This will give the best resolution for the magnification wanted. The object of interest should fill the frame with the minimal background (to avoid the biological equivalent of holiday snaps of your friend standing 100 yards away from the camera, in front of the Eiffel Tower). The long axis of the specimen or portion of interest should be aligned as well as possible with the long axis of the frame.

A good rule is always to orientate specimens the same way (for example, anterior to the left, as those working with *Drosophila* do), which helps in composing plates for publication afterwards and certainly helps unenlightened listeners at seminars.

3.3 Good exposure

3.3.1 How to expose film correctly

It is important to realize that you have to compensate for the brightness of the region of interest with respect to the *average* brightness of the whole frame. Exposure meters in cameras and microscopes usually measure the average brightness over most of the frame, and provide a guide to the exposure required to make this an even grey colour. This rarely agrees with what is needed: fluorescent specimens will have a dark background, which you want to appear *black*, and sections viewed by bright field may have large expanses of empty backgrounds which you want to see as *white*. The light meter will want to make these grey, so you have to compensate. As a rule of thumb, it usually works well to decrease the exposure by two whole f stops (i.e. ¼ of the meter exposure) for fluorescence or dark field, and to increase the exposure by two whole f stops (i.e. 4× meter exposure) for bright field photography. But the exact compensation will depend on the proportion of the frame taken up by the region of interest, and you have to experiment a little. In general, DIC and phase-contrast optics will expose correctly for the meter reading because the background is really grey in these cases.

The next most important thing to realize about exposure is that it is not enough simply to expose a film correctly for the amount of light available. The absolute length of exposure considerably affects the quality of the image obtained. Ideally, for non-moving/non-changing specimens, exposures should be in the region of *1 sec*. In addition to effects on the film, much longer exposures may be subject to drift of the focus or specimen, to vibration and to complicated compensation for *reciprocity failure* being required (look this up in a photography book). Much shorter exposures tend to affect both contrast and definition, as well as to have an effect on the colour temperature (see below) because of the higher voltage supplied to the lamp. However, it is not always possible to keep to 1 sec exposures, especially in the case of fluorescence; compensation for reciprocity failure (which is more serious for colour than for black-and-white film) must therefore be made. This is too complicated to explain here; refer to the manual for your microscope, and always take a test film to investigate these parameters for the film/specimen type you use.

Some modern photomicroscopes and cameras are equipped with 'spot metering' systems. This allows a measurement of the exposure necessary to produce mid-grey to be made in a small, central portion of the specimen. By measuring the light intensity within the area of interest, one can circumvent some of the guesswork when adjusting exposures for very bright or very dark backgrounds.

3.3.2 Correct exposure for control specimens

It is important, when using automatic camera meters, to be aware that experimental and control specimens may give different readings just because of the different nature of the specimens themselves. If taking alternate pictures of experimental and control samples at the same magnification and settings, and the camera has an 'exposure lock' facility, then use it to lock onto the exposure used for the experimental samples and take the control pictures with the same settings. If the camera does not have this, then make a note of the actual exposure times (after necessary adjustments up or down due to background, etc.) and then expose the controls manually using the same settings. Remember that settings will vary with lighting conditions, the objective used, and the opening of the iris diaphragm; it is always best to take alternate sets of experimental and control pictures.

3.3.3 Exposure for photo-montages

The same caveats as for controls apply to producing a photo-montage where several exposures of different regions of the same specimen are assembled to give a panoramic view. Use the camera meter to measure different regions of the specimen before you start to take shots. After that, use *manual* exposure control for all the shots, so that the background intensity does not vary. When taking the individual pictures, make sure that adjacent shots overlap by at least 1/5 of the frame so that they are easier to assemble. To assemble the montage, overlap the individual prints on a flat surface as perfectly as possible, two at a time. Tape them to the bench. Using a scalpel blade and a metal ruler, cut through *both* prints simultaneously in the region of overlap. This will give a perfect match at the cut surface. For final assembly, use photo-spray mount to attach the individual trimmed prints to a stiff card backing.

3.3.4 Double exposures

Double exposures (taking two photographs onto the same frame) are often worth a try with fluorescence and autoradiographs, to overlap the fluorescence image with the bright field. The idea is to expose a frame to the fluorescence image and then to the bright field/Nomarski, or whatever, image on the same frame, changing optical set-up between exposures. It is important to check that the two images are aligned perfectly. You can only do this successfully with colour film, and you have to remember that, because you expose the frame twice, you have to shorten the exposure in each shot even further than the normal amount required. Usually the exposure of the bright-field shot has to be decreased by one f stop (one-half of the time needed for a single exposure) but the fluorescence will probably not need to be adjusted from the settings for a single exposure. This will give bright fluorescence visible over the embryo. For autoradiographs, take one shot with either dark field or incident illumination, and a second with Nomarski or bright field (under-exposing about 0.5–1 f stop).

3.3.5 Filters

Filters can improve the quality of photomicrographs, if used correctly. In all cases, it helps to have a filter to remove the infrared component of the light source because this affects the meter more than most films; most photo-microscopes will have this already. For most colour photography, keep a 'light blue daylight' filter in place, even using tungsten-balanced film like Fuji 64T (see below). Most microscopes have such a filter. For fluorescence, do not use additional filters, other than those required for the wavelength of the fluorochrome, and do not worry about colour temperature.

For black-and-white photography, the choice of filter is more difficult and more important. In general, however, you will need a green filter. This is useful particularly for increasing the contrast of immunoperoxidase-stained specimens and sometimes for phase contrast (according to most people, although I have never found that green filters improve phase-contrast images). Do not add coloured filters to Nomarski images. They usually mess things up.

3.3.6 Colour temperature

A very common fault with colour photomicrography is failure to match the colour temperature of the film to that of the lamp illuminating the specimen. Thus, snapshots taken using ordinary ('daylight') film indoors will tend to have a yellow-orange tint to them, or may be greenish if the subject is illuminated mainly by a fluorescent source. Such films will have the correct-looking colour, however, if shot either in daylight or using flash. The same is true for photomicrography. Most microscope lamps are tungsten, which have a lower 'colour temperature' than required by daylight films, so bright-field images taken on ordinary outdoor film will tend to appear orange. To compensate for this you will need a fairly dark blue filter. Alternatively, use a tungsten balanced film. This is the easiest solution to avoid losing further light. This is why I recommend 64T ('T' for tungsten) film above. However,

Table 1. Photomicrography applications

Application	Light source	Microscope/technique	Film type	Comments
Sections processed with histological dyes/stains	Transmitted bright light (Köhler illumination)	Bright-field (Köhler illumination)	B/W[a]: Kodak Technical Pan, Ilford PanF Colour: tungsten (e.g. Fuji 64T)	Usually need to over-expose by 1–2 stops
Unstained sections	Transmitted light	Nomarski DIC is best	Kodak Technical Pan	If Nomarski, expose as shown by camera meter
Live cell cultures	Transmitted light	Phase contrast or Nomarski DIC	B/W: Ilford PanF, Kodak TMAX100 Colour (for DIC): Fuji 64T	Expose as indicated by camera meter
Stained or unstained whole mounts of embryos	Incident or transmitted light	Dissection or compound microscope, bright-field or reflected light	B/W: Kodak TMAX or Technical Pan Colour: Fuji 64T	Bright-field: normal exposure Incident: may need to overexpose 1–2 stops

Immunogold or auto-radiographs (including radioactive *in situ* hybridization)	Incident or dark-field (or confocal microscopy)	Compound microscope: dark-field or reflected Dissecting microscope: reflected	B/W: Kodak TMAX100 or Ilford Pan-F Colour: Fuji 64T	Can use double exposure mixed with transmitted light (with different colour filters)
Whole-mount *in situ* hybridization (DIG-labelled probes)	Incident or transmitted light or mixture	Compound or dissecting microscope; incident and/or bright-field or DIC	B/W: TMAX100 or Technical Pan Colour: Fuji 64T	May need to push contrast by under-exposing and over-developing
Fluorescence: single label	Epifluorescence or mix with transmitted light	Epifluorescence	B/W: TMAX400 Colour: Fuji 1600P	Usually need to under-expose fluorescence 2 stops
Fluorescence: double label	As above	As above; double or triple exposures	As above	Usually need to under-expose 2.5 stops each
Immunoperoxidase or alkaline phosphatase	As for whole mount *in situ*	As for whole mount *in situ*	As for whole mount *in situ*	As for whole mount *in situ*

[a] B/W, black and white.

the voltage supplied to the tungsten lamp also affects its colour temperature: the higher the voltage the higher the colour temperature and the more it will approach daylight (blue), the lower the voltage the more orange the effect on daylight film. Bluish backgrounds are usually slightly more tolerable than orange ones, so try to keep up the colour temperature.

4. Some examples of specific applications

Table 1 gives a summary of the techniques used in this laboratory for some specific photomicrography applications, which we have found to give consistent and successful results. However, it should only be taken as a rough guide; the best choice of film and technique will depend on the type of material being studied, the microscope and cameras available, the method of film processing used, and, of course, the skill of the operator(s).

Acknowledgement

I am grateful to Dr Savile Bradbury for his helpful comments on the manuscript and for teaching me a thing or two about microscopes.

Further reading

Bradbury, S. (1989). *An introduction to the optical microscope.* Oxford University Press, Oxford.

II
Experimental embryology

9

Fate maps and cell lineage analysis

ARIEL RUIZ i ALTABA, RACHEL M. WARGA, and
CLAUDIO D. STERN

1. Introduction

Knowledge about the fates of cells during development is of fundamental importance to our understanding of developmental mechanisms (1). In organisms like the nematode *Caenorhabditis elegans*, cell number is sufficiently reduced and the embryo is small and transparent enough to make lineage analysis by direct observation of cell divisions relatively straightforward. However, in systems where cell number is greater, special techniques have to be introduced to follow cell fates.

Lineage analysis requires the introduction of markers that allow the descendants of specific cells to be identified. The marker itself must not affect developmental fate, it must be easily identified, it should be cell autonomous (not transferred between lineally unrelated cells) and, ideally, it should not become diluted by cell division. Lysinated fluorescent dextrans (2, 3) or horseradish peroxidase can be injected into single cells and the labelled progeny followed for up to about 10 cell divisions. Groups of cells can also be labelled using lipophilic carbocyanine dyes (e.g. *DiI, DiO*) (4); these are intensely fluorescent and therefore can be used to follow cell fate over very long periods. Markers that do not become diluted include the construction of inter-specific (e.g. quail/chick, *Xenopus laevis/Xenopus borealis*) or inter-strain (e.g. between different strains of mice) chimaeras, the induction of mosaicism by genetic recombination (e.g. the *yellow* or *minute* loci in *Drosophila*), and the introduction of marker genes into single cells (e.g. *lacZ*; see also Chapter 20).

Here, we review some of the most widely used of these methods. The advantages and disadvantages of each are dependent on the system being studied, and their suitability for the question being addressed should be decided by the investigator, but the account given here should help in choosing an appropriate technique.

2. Preparation of embryos of different species for fate mapping or cell lineage analysis

In principle, embryos of any animal species can be used for cell lineage analysis with the methods described here, provided that the cells of interest are accessible at the relevant developmental stage and that embryos with labelled cells can develop to the stage of interest either in whole embryo culture or *in vivo*. Methods for achieving these aims are described for most groups in Chapters 1–7.

3. Fate mapping with lipophilic carbocyanine dyes

A simple way to construct fate maps makes use of the recently introduced carbocyanine dyes (4), DiI (red), DiO (green), and related compounds (marketed by Molecular Probes, Inc). These are applied extracellularly and become incorporated into cell membranes because of their lipophilic nature, and they are not generally transferred between neighbouring cells. They are also intensely fluorescent. These properties make them very suitable as general markers for fate mapping. Usually the dye is made up in some organic solvent and applied through a microelectrode by air pressure so that a small group of cells (between two and 100 cells, depending on the species, stage, location, and injection conditions) becomes labelled. In our experience, descendants of the labelled cells can still be found in the chick embryo up to one week after labelling. The availability of dyes of different colours (contact Molecular Probes for the latest news on available dyes) makes it possible not only to label two or more groups of cells in the same embryo but also to control in each case for the possible transfer of dye between groups: two adjacent groups of cells are labelled, one with each of two dyes, and embryos are scored after the desired incubation period for doubly-labelled cells. We will describe the simplest method for constructing fate maps which is suitable for most species of animals.

Protocol 1. Constructing a fate map with DiI and DiO

1. Make up a stock of dye. Three methods are particularly successful:

   (a) Make up DiI or DiO at 0.3% in dimethylformamide[a].
   (b) Make up DiI at 0.5% and DiO at 0.25% in absolute ethanol at room temperature. Store at −20 °C[b].
   (c) Make up DiI at 0.2–0.3% in vegetable cooking oil (Wesson vegetable oil in the USA, Mazola in the UK)[c].

2. Make electrodes. A simple way is to use thick-walled 50 μl capillary glass tubing (e.g. Sigma P1049), pulled with any standard electrode puller so that they have a very long, gradually tapering tip (very low heat and, in a vertical puller, using gravity only).

3. Break the extreme tip of the electrode (done most easily by touching the tip with a hard surface such as the back of a pair of forceps). It should still look very fine under a dissecting microscope.

4. Prepare embryos of the species and stage of choice as described in the chapter for the relevant species (see Chapters 1–7). Most species may be either in whole embryo culture or *in vivo*, but any system that allows easy access to the embryo and region of interest can be used. Avian embryos may be labelled either in New culture or *in ovo*. Mammalian embryos may be labelled in culture or *in utero* (the latter only for relatively late postimplantation stages when the embryo can be seen through the wall of the uterus or when the uterine wall may be opened without leading to abortion).

5. For most purposes, injection does not require special equipment. Although a pressure injection apparatus may be used (e.g. PicoSpritzer, General Valve Corp.), and the amount of dye expelled controlled very accurately in this way, this is not necessary and injection can be done through a mouth tube (e.g. Sigma A5177). If using the latter, aspirate some of the dye of choice to fill the tip of the electrode. Use separate electrodes for each dye used, if more than one, to avoid contamination.

6. Carefully expel a very small amount of dye to the region of choice in the embryo, looking under the dissecting microscope. There should be a faintly coloured, very small pool of dye at the site of injection. A very small spot will usually be sufficient. Always check under fluorescence optics to assess how many cells were labelled and their position.

7. Incubate the embryo to the desired stage and fix it in 4% formaldehyde in PBS containing 0.25% glutaraldehyde. Store them in the fixative at 4 °C in the dark but analyse labelled progeny as soon as possible. DiO, in particular seems to leach out over time, even in fixative.

8. Examine labelled progeny with conventional epifluorescence optics or using intensifying electronics (see below).

[a] Although this has been used successfully in some systems, we have found that this solution is toxic to cells in early avian embryos.
[b] It is also possible to dilute ethanol-based stocks of dye in aqueous solution, just before use. To do this, first heat up 45 μl of 0.3 M sucrose made up in crude distilled water (not deionized water) and the ethanol stock of the dye, in separate Eppendorf tubes, to 45 °C. Take up 5 μl of the warm solution of dye and quickly pipette it directly into the warm sucrose solution (not the sides of the tube). Immediately vortex for a few seconds. It should look clear and a pellet will not be seen if briefly centrifuged. This solution must be used the same day as it is made as it will come out of solution within several hours.

Protocol 1. *Continued*

  [c] This requires repeated heating of the dye/oil to about 45 °C for 30 min each time, alternating with vortexing for one whole minute each time. Continue this until dye has dissolved completely, which may take up to 4 h. The stock can be kept in the dark at room temperature for up to about 3 weeks without visible deterioration.

Because these dyes are lipophilic and water insoluble, embryos containing labelled cells cannot be sectioned easily without the dye spreading. Some techniques to do this have been reported in the literature, in which frozen sections of gelatin-embedded labelled embryos are cut and mounted in aqueous medium as soon as possible after cutting. However, there is a better technique which stabilizes the dye by photoconverting the fluorescence to an insoluble precipitate of diaminobenzidine, allowing conventional wax sections to be cut (*Protocol 2*).

Protocol 2. Photooxidation of carbocyanine-dye-labelled cells

1. Wash embryos containing labelled cells (previously fixed in 4% form-aldehyde in PBS containing 0.25% glutaraldehyde) three times (10 min each or longer if large embryos) in 0.1 M Tris pH 7.4. Subsequent processing has to be done one embryo at a time.

2. Incubate the embryos in a large volume (e.g. 10 ml) of a freshly made, 500 µg/ml solution of diaminobenzidine (DAB)[a] in 0.1 M Tris pH 7.4, in the dark at room temperature for 1–3 h, depending on their thickness, so as to infiltrate the tissue very well.

3. Place one embryo in a fairly deep ($\approx$3–5 mm) glass cavity slide immersed in fresh DAB solution and place a coverslip over it. Carefully remove any excess with paper tissues.[a]

4. Place slide with embryo on the stage of an epifluorescence microscope. Using a 10× or 20× objective, illuminate the regions containing labelled cells, focussing every 10 min or as required, until all fluorescence disappears completely in the field being illuminated and for a few minutes thereafter. This may take up to 1 h for each field, higher power objectives (with higher numerical aperture) being more efficient and producing less background.

5. Move the slide to illuminate all regions containing labelled cells as described in step 4 until no fluorescence remains visible in the embryo.

6. If steps 4 and 5 take more than about 2 h, replace the solution in the cavity slide with fresh DAB solution and repeat this every 2 h or so, as the solution becomes exhausted.

7. Remove embryo carefully[a] from the cavity slide and transfer to a small bottle containing tap water (which helps to intensify the DAB precipitate

and to blacken it). Wash embryo three times (30 min each) in tap water, then once in distilled water (10 min).

8. Dehydrate through a series of alcohols (70%, 95%, absolute twice; 20 min each). Clear in xylene (Histoclear gives less good results) and either view/photograph with transmitted light or proceed for conventional wax embedding and sectioning.

[a] DAB is a suspected carcinogen. Exercise appropriate care. Anything that has come into contact with DAB solution or powder must be oxidized in a solution of household bleach before discarding it.

---

4. Single cell lineage analysis by injection of fluorescent dextrans, horseradish peroxidase, or RNA

4.1 Intracellular pressure-injection of large cells (e.g. amphibian embryos)

4.1.1 Introduction

Amphibians have very large cells during the early stages of their development, which allows intracellular injection using air or water pressure. Fate maps and lineage analysis have been performed in the frog by injection of stable enzymes, such as horseradish peroxidase (5, 6), or mRNAs encoding these, such as *lacZ* trancripts encoding β-galactosidase (7), or inert tracers such as LRD or LFD (2, 3, 8). It is possible, however, that cells may take up small amounts of inert, stable, intracellular lineage tracers from dying neighbours and this possibility needs to be controlled for in any new type of experiment.

Protocol 3. General method for intracellular injection into *Xenopus* embryos

1. Place oocytes, eggs, or embryos in 3% Ficoll, 1× MMR[a] (IS injection solution) for at least 5 min before injection. IS helps the embryo to cleave in a stereotyped manner and the hypertonicity of IS will prevent the outward flow of cytoplasmic material from the injection site. Embryos should be treated with cysteine, to remove the jelly coat, prior to injection (see Chapter 5).

2. Place embryos in IS drops on siliconized and autoclaved microscope glass slides, 1–15 embryos per drop.

3. Make injection micropipette. Simple injection pipettes can be made with a basic puller and borosilicate glass micropipettes[b]. For injections into relatively young embryos, when cells are large, the tip must not be too fine. Bevelling the needles[c] will facilitate penetration and will diminish injection injury.

Protocol 3. *Continued*

4. Fill the micropipette. This can be done manually by introducing the material to inject through the back end of the pipette with a Pasteur pipette pulled to a fine tip. It is easier to use a pressure-driven injector (e.g. Narishige) and apply reverse pressure to fill the needle. If you have excess material to inject, fill the needle with a maximum of 3 μl. It is always better to keep materials to inject in closed tubes at low temperature rather than inside the needle at room temperature for a long period.

5. Calibrate the injection pipette. The method of calibration will depend on the injection machine used but generally division of 1 μl in the needle by injection time or needle length will do. Use sterile or DEPC-treated water[d] to calibrate needles.

6. Place slides with embryos in IS drops on a dissecting microscope stage. Focus on the embryos and carefully bring the needle close to the embryos to inject. Depending on the pressure, some material may leak out, so equalize the pressure and keep the needle out of the IS buffer when not injecting.

7. Inject embryos by puncturing the vitelline and cellular membranes with the needle at a an angle of about 90° to the vertical. Different angles of penetration and exit of the needle can cause serious damage so keep the same angle of injection on the way in and on the way out. The embryos will display a small resistance to the needle. You can provide some support by holding a pair of tweezers in the back of the embryo being injected (*Figure 1*). Apply pressure to release materials to inject. Do not move the needle while materials are being delivered. Dead embryos will appear very soft. If the needle becomes clogged, try shaking it gently in the IS buffer. If this does not clear it, break back the tip and recalibrate, or replace the needle.

8. After injection, collect the embryos in a Petri dish containing 10–15ml of IS. Embryos can be transferred with a wide-bored pipette.

9. Progressively exchange the IS buffer for 0.1 × MMR over a period of about 2–3 h. The first buffer change must never be made before 3–4 cleavages occur. The hypertonicity of IS prevents the outward flow of cytoplasmic material from the injected embryo or blastomere while it heals. However, IS can cause gastrulation defects if it is left for too long. A small protrusion of tissue may be observed at the site of injection. This will not affect the rest of the embryo and can sometimes be cut out. If injecting oocytes these can be brought back to 1 × MBSH in a similar manner.

10. Normally, 10–20% of injected embryos die. Keep track of the injected embryos and discard dead ones as soon as these are recognized (large,

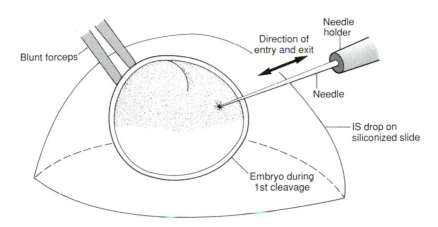

Figure 1. Embyro injection.

whitish, soft). Dead and decomposing embryos can affect the development of their neighbours. Make sure that embryos are kept in clean,
sterile, and well aereated buffers.

[a] 10 × MMR medium. Mix together 58.4 g NaCl and 12 g Hepes. Add water to 700 ml and
adjust pH to 7.6. Add 1.5 g KCl, 2.4 g MgSO$_4$.7H$_2$O and 0.4 g EDTA. Make up to 1 litre.

[b] For RNA injections, treat micropipettes with DEPC and autoclave prior to pulling them.

[c] Bevellers are commercially available but can also be made from the motor of an old record
player with a 4–5 cm diameter Perspex (Plexiglass) disc on the turntable, covered with very fine
abrasive paper.

[d] DEPC-treated water. Add diethylpyrocarbonate to water to 0.1%. Shake. Allow to stand
overnight at room temperature. Autoclave 20 min. *Note*: DEPC is highly toxic.

4.1.2 Notes on injection in frogs

Injection of large numbers of developing *Xenopus* embryos requires a fast
injection technique. A simple microinjector, such as a Singer macro-
manipulator, allows, with some practice, for the injection of several hundred
embyos of the same batch (*Figure 2*). The development of embryos to be
injected can also be slowed down by decreasing the temperature of the
buffers, thus allowing for a longer period over which to inject. A pedal to
activate the pressure machinery for injection is recommended since this leaves
the hands free to hold a pair of tweezers and the needle holder.

Injection volumes should be of 20–30 nl maximum per embryo. If injecting
single blastomeres, the volume will have to be reduced accordingly (but you
can increase the concentration of whatever you are injecting!). The time of
injection is not critical but very short bursts are not recommended. Injection
into one-cell embryos is best if done just after the first cleavage furrow is
visible. One can inject equatorially, animally, or vegetally. Alternatively, one

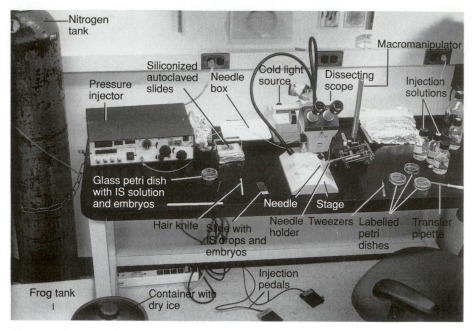

Figure 2. Injection/dissection set-up.

can inject into one or two blastomeres at the two-cell stage, a time when the embryo appears to take injection trauma better.

Practice and the appropriate concentration of injected materials will increase the percentage of injected embryos that develop normally. Toxicity of different materials should be carefully considered and monitored. Dextran tracers can be injected routinely at 10–50 mg/ml, plasmid DNAs and synthetic mRNAs at 50–200 µg/ml, and purified antibodies (e.g. IgG; crude serum is very toxic) at 100–500 µg/ml. However, a variety of dilutions should be tried to find the optimal concentration. Synthetic mRNAs should be purified by running a small Sephadex G-50 column followed by ethanol precipitations. High concentrations of unincorporated nucleotides are toxic. The stability of injected RNAs (9) can be generally increased by cloning the desired cDNA into pSP64T (Promega), which provides 5′ and 3′ untranslated sequences and a polyA.polyC tail (10). The localization of injected and endogenous RNAs can be determined by whole mount *in situ* hybridization (see Chapter 26). In general, plasmids neither integrate into the host genome nor replicate to high levels. These are maintained extrachromosomally and display mosaic distribution patterns in the embryo. Antibodies are very stable in the cytoplasm of injected embryos and display quite homogeneous distribution similar to that observed for dextran tracers (11, 12).

4.2 Intracellular microinjection into single small cells (e.g. avian, mammalian embryos)

Successful methods for injection of tracers or nucleic acids into single cells in avian and mammalian embryos are essentially similar to those described above for other species, but the cells of these higher vertebrate embryos tend to be smaller than equivalent cells in frogs. It is therefore impossible to use air pressure to drive the dye or nucleic acid into individual cells, as the resulting increase in volume would kill the injected cell. For this reason, the material to be injected is usually expelled by iontophoresis: pulses of electrical current are used to drive the charged electrolyte into the cell, with no appreciable change in cell volume. For mammalian embryos, methods for cell lineage analysis by intracellular injection have been reviewed recently (13) and the following sections are intended to complement it.

4.2.1 Iontophoretic injection of tracers into single avian/mammalian cells *in vivo*

The equipment required for injection into single avian and mammalian cells, in addition to those for handling the embryos themselves, is also similar to that described above. The best type of glass for making microelectrodes for injection by iontophoresis is aluminosilicate capillary (1.2 mm outside diameter, 0.9 mm inside diameter, 10 cm long, with inside filament) manufactured by A-M Systems Inc. (catalogue No. 5820). Aluminosilicate glass requires more heat than borosilicate for pulling, and although any conventional electrode puller can be used, some (e.g. Narishige) do not seem to produce enough heat to pull this kind of glass. Any basic electrophysiological system can be used. Requirements are: a high input impedance headstage and preamplifier with current injection and resistance monitoring facilities (e.g. Digitimer Neurolog system with NL102G), modules for controlling frequency and duration of pulses to trigger the current injection module (e.g. Digitimer Period Generator NL304 and Digital Width NL401), and a suitable output device. The latter may be any digital storage oscilloscope but a less expensive alternative is to purchase a conventional (non-storage) oscilloscope and to use a digital storage adaptor (e.g. Thurlby instruments DSA511).

In addition to a manipulator to move the electrode, it is almost essential for both penetration and withdrawal of the electrode from cells to use a stepper motor controlled by a microcomputer. A cheap and convenient setup for this (SignifiCat; catalogue No. SCAT-01e) is marketed by Digitimer Ltd and includes the computer (Epson HX20), interface, software, stepper motor, and even a three-axis manipulator. The manipulator and/or stepper motor drive should be mounted on a conventional (upright) microscope equipped with epifluorescence optics even for injecting non-fluorescent compounds (e.g. RNA, DNA, peroxidase), because it is always useful to include some rhodamine-dextran to visualize the cell being injected. Most important, the microscope should be fitted with both a low power objective (e.g. 2.5× or 4×) to locate the embryo and with a suitable higher power objective with ultra-long working distance. For most applications, the best type of high power objective is a 20× with a working distance of 1–3 cm (e.g. Olympus 20× ULWD). It is also important that the microscope used should focus by movement of the nosepiece of the microscope rather than the stage. If the stage moves, then the micromanipulator must be mounted on it rather than on the bench. Finally, it is important for injection into small cells to control vibration, which can cause problems. Any vibration isolator can be used. The cheapest way of doing this, however, is to use an inflated inner tube from a car tyre on which is placed a smooth slab of stone or metal (e.g. a heavy slab of marble or granite from your local undertaker or churchyard). If metal, this must be connected to earth (e.g. a water pipe) independently of the electrical set-up to avoid an earth loop, which is dangerous.

Protocol 4. Iontophoretic injection into single avian or mammalian cells

1. For avian embryos, prepare embryos in New culture (*Protocols 3* or *4*, Chapter 6) or as if for *in ovo* operations (*Protocol 2*, Chapter 12). For rodent embryos, set up the embryo as if for culture (*Protocol 2*, Chapter 7) and place it in culture medium on the lid of a 3 cm or 5 cm Petri dish.

2. Pull electrodes from aluminosilicate glass as indicated in the notes above. Back-fill the tip with the dye or solution to be injected[a] by inserting a Pasteur pipette that has been pulled to a very fine tip into the back of the pulled electrode and expel enough liquid to fill the end of the electrode up to the start of the shoulder (a small fraction of a microlitre). Using a small piece of plasticine or Bluetak secured to the wall or other vertical object, stand the electrode upright, tip down, for a few minutes to allow any bubbles to clear. If more than 5–10 min, place electrodes upright in an air-tight container (e.g. jam jar) humidified with a pool of distilled water, but do not allow the electrode tips to come into contact with the water.

3. Use 1.2 M LiCl to fill the electrode holder (e.g. holder for 1.2 mm glass, with Ag/AgCl electrode, Clarke Electromedical) with a hypodermic syringe and the rest of the shank of the electrode itself with a pulled Pasteur pipette. Place the electrode in the holder, making sure to avoid any bubbles.

4. Place the embryo under the microscope. Illuminate it with a fibre optic source. For mammalian embryos, use a second micromanipulator, controlling a holding pipette with gentle suction applied to hold the embryo steady and to move it around the dish.

5. Connect the bath of saline containing the electrode to the ground connection on the headstage, using a thin, flexible wire or agar/silver wire bridge[b] immersed in the bath.

6. Close down the illuminating aperture of the epifluorescence source to the minimum and use the small light spot to locate the embryo in the dish and to focus the microscope roughly (no need to look into the microscope). Now position the electrode tip roughly using the manipulator to be close to the illuminated spot.

7. Both the desired region of the embryo (which will only be visible if there is enough light from the fibre optic source) and the electrode tip (visible mainly from the fluorescence once the aperture is opened) should be in the field of view of the 20× objective, but the electrode tip will be at a different focal plane from the embryo. If either electrode or embryo are not in the field, adjust as necessary.

8. Check electrode resistance (e.g. by passing 1 nA pulses of current and reading the voltage on the oscilloscope; instructions for this should be found in the manual accompanying the current injection module of the preamplifier; on the Digitimer NL102G, simply turn on the switch labelled: RESIST.CHK). Electrode resistance is calculated by Ohm's law: R (Ω) = V (V)/I (A). It should be between 50 and 110 MΩ.

9. Impale a cell in the region of interest using the manipulator and/or stepper motor. The oscilloscope may reveal a noisy signal when the outside of a cell is touched. If this is seen, use negative capacitance or a large current pulse to penetrate the cell (on the Neurolog NL102G, press the button labelled NEG.CAP for about 1 sec). The trace will now be much less noisy and there should be a shift in the baseline. The difference in level will reflect the resting potential of the cell, provided the embryo has been kept warm.

10. Apply a single current pulse of about 2 nA (on the Neurolog NL304, push the button labelled SINGLE). With the fibre optic illumination turned off and the fluorescence on, you should see a cell faintly filling up. If not, increase the current and apply another pulse. If maximum current is applied and the cell does not fill up, you are probably not in a cell, or the electrode may be blocked. If the latter, the trace on the oscilloscope will be noisy. Clear with pulses of current and bursts of negative capacitance.

11. Repeat steps 9 and 10 until a cell is impaled successfully.

12. Now apply 2–6 nA pulses of current, each 500 msec duration every 500 msec (1 pulse per second) for about 1–2 min or until the cell looks brightly fluorescent. Too much dye, however, can lead to compartmentation in the cell resulting in a spotty appearance of the fluorescence when the clone is examined later.

Protocol 4. *Continued*

13. Turn off pulses, wait 2 sec, and then withdraw the electrode as fast and as smoothly as possible. If using the stepper motor with the SCAT01e, press '0' on the Epson keyboard to do this. Look at the screen of the oscilloscope to monitor the resting potential again, which will probably be a little less than on first impalement.

14. Quickly check that the cell is still fluorescent and that only one cell has been injected. Discard any embryos in which more than one cell was labelled.

15. Incubate the embryo for the desired time (for fluorescent dyes or peroxidase, the maximum is generally 1.5–2 days, depending on the cell division rate of the cells in question; clones of up to 1000 cells can be seen with suitable image intensification) as suitable for the species being used.

16. Prior to analysis, fix the embryo as suitable for the marker being used. For peroxidase use 4% formaldehyde in PBS for 30 min at room temperature or absolute methanol overnight at −20 °C. For fluorescent dextran-amines (LRD, LFD) use 4% formaldehyde containing 0.25% glutaraldehyde in PBS for 1 h.

[a] Use high quality deionized water rather than a saline to dissolve the compound to be injected. If using fluorescent dextrans (e.g. rhodamine-dextran, neutral, molecular weight 10 000, lysine fixable, Molecular Probes Inc.), these are made up in deionized water at 10 mg/ml and 10 μl aliquots stored frozen until the day of use. Do not refreeze once thawed. Enzymes may be made up and stored in the same way, at a suitable concentration for each enzyme (usually around 1 mg/ml). RNA and DNA should also be made in water, which in the case of RNA should have been treated with DEPC. If injecting RNA, pretreat the electrode glass with DEPC for 1 h and autoclave it before pulling it. When injecting non-fluorescent compounds, include a small amount of any fluorescent tracer (e.g. rhodamine-dextran) in the stock solution so that injection can be monitored.

[b] Agar/silver bridges. Melt 3% agar in 0.9% NaCl by boiling. Fill a 15 cm length of flexible plastic tubing (≈0.35 cm diameter) with the molten agar/saline. Before it sets, insert a 5–7 cm piece of chlorinated silver wire half-way into it so that about 2 cm protrude. Allow to set at room temperature. Seal the joint with Epoxy resin and cover with heat shrink tubing. Connect the protruding end of the silver wire to the reference plug of the headstage using thin electrical cable fitted with a suitable plug for the equipment used. Silver wire can be chlorinated electrolytically in a saturated solution of KCl or dipped in molten AgCl (dipped wires will last longer than plated wires).

4.2.2 Analysis of clones in avian or mammalian embryos

The methods used will depend on the tracer selected. For horseradish peroxidase or alkaline phosphatase, the methods described for peroxidase- or alkaline-posphatase-coupled antibodies in whole mounts or sections (Chapter 21) can be followed. Embryos can be sectioned either before or after development of the colour reaction for the enzymes.

For fluorescent tracers, mount the embryo or sections in Gelvatol (see

Chapter 21) to reduce quenching. In most cases some form of image intensification will be required. This can be a silicon intensifier target (SIT or ISIT) camera, a multi-channel plate (video intensifier) or a charge couple device (CCD) camera with Peltier cooling (which reduces the dark current) and the ability to integrate. For clones of up to about 500 cells, however, it is possible to visualize them and even to photograph them on conventional high-sensitivity colour slide film (e.g. Fuji 1600P). It is also possible to use a confocal laser scanning microscope for imaging.

4.3 Injection of fluorescent dextrans into single cells in zebrafish embryos

Because zebrafish embryos are optically transparent, a single cell can be labelled with fluorescent dye and its clonal progeny mapped throughout development (14, 15). When screened in the one-day embryo, when many tissues have begun the process of differentiation, labelled cells are identified on the basis of their morphology within a tissue (14, 15). If a cell has not been identified at this time, it can be observed over the following several days until its tissue identity can be finally determined. Cells can be followed with this technique in the live embryo for up to a week.

4.3.1 General method for injection into single fish cells

The basic techniques used for dye injection into single zebrafish cells are similar to those given above for amphibians and amniotes. Electrophysiology equipment is required to penetrate the cells and to monitor the impalement, as in amniotes. Up to the blastula stage, air pressure is used to expel the dye from the electrode once impalement has been achieved, as for amphibian cells.

Protocol 5. Labelling single fish early blastula cells

Electrodes are usually made on the day before the experiment. The basic set-up for labelling requires basic electrophysiology equipment (pulse generator, preamplifier, and oscilloscope; see Section 4.2.1) and a source of air pressure[a].

1. Pull thin wall borosilicate glass tubing with internal filament using any suitable electrode puller, so that they have a resistance of 150–200 MΩ when filled with 0.2 M KCl.

2. Fill the tips of electrodes like these with a fresh, filtered solution of neutral tetramethylrhodamine dextran (RD), molecular weight 10 000 (Molecular Probes), at a concentration of 5 mg/ml in 0.2 M KCl[b].

3. Store RD-filled electrodes overnight in a moist sealed container at 4 °C. This removes air bubbles from the electrode tips.

Protocol 5. *Continued*

4. Mount embryos on glass depression slides in a 'bath' of embryo medium (see Chapter 4), and view them with Nomarski DIC optics.

5. Fill the shanks of RD-filled electrodes and the electrode holder with 0.2 M KCl.

6. Connect the bath to ground with a silver chloride ground wire embedded in agar (see footnote b, *Protocol 4*).

7. Break the tip of the electrode to a resistance of 30 to 100 MΩ (see *Protocol 4*) and apply a brief pulse of depolarizing potential (4–14 V) or of negative capacitance (see *Protocol 4*) to penetrate the cell membrane. Monitoring voltage with an oscilloscope helps to determine when the electrode is within a cell and whether the cell remains undamaged: a healthy resting potential is around −30 mV in fish cells.

8. Air pressure will gently expel dye into the cell. The cell should look faintly pink when viewed under Nomarski DIC optics, but at high concentrations the dye is toxic to cells.

9. Place the embryo containing the labelled cell in an agar-coated dish full of embryo medium (Chapter 4).

[a] A 'simple' set-up consists of: stimulator (model SD9, Grass Instruments), intracellular amplifier (model 5A, Getting Instruments Inc. or Intra 767, World Precision Instruments Inc.), oscilloscope (model 2245A, Tektronix), air pressure (PV820 Pneumatic Picopump, World Precision Instruments Inc.). The equipment described in Section 4.2.1 can also be used.

[b] In fish, lysinated dyes are toxic. If a fixable tracer is required, make a 3 mg/ml solution of biotin-dextran (10 000 molecular weight, Molecular Probes Inc.) in 0.2 M KCl and add to the RD (5 mg/ml) solution. After injection with this mixture and subsequent development, embryos can be fixed in 4% paraformaldehyde in PBS. Biotin is then detected using the ABC kit/Vectastain from Vector Laboratories and the DAB reaction (see Chapter 21). After staining, embryos can be dehydrated in alcohol, cleared in methyl salicylate, and mounted in Permount.

Cells in embryos older than the early blastula stage are small and do not tolerate pressure injection. Here, iontophoretic injection should be used, similar to the protocols given above for amniotes. We will point out some differences between the two.

Electrodes pulled from thick-walled borosilicate glass with internal fibre should have a resistance of about 100 MΩ when filled with 0.2 M KCl. To penetrate the cell membrane, negative capacitance or a depolarizing potential may be used. Apply pulses of current as described for amniotes or use the capacitance compensator (CAP COMP) button on the preamplifier, gently turning it up or down. The dye should be RD or RD mixed with biotin-dextran as described in *Protocol 5*. The cell should be visible but not brightly fluorescent.

After injection, place the embryo in an agar coated dish full of embryo medium to develop.

4.3.2 Screening for labelled clones in fish embryos

Orientate gastrula stage embryos in 3% methylcellulose in embryo medium (see Chapter 4, *Protocol 2*) on glass depression slides. For older embryos use either bridged coverslip chambers[a] or embed in 1.2% agar in embryo medium on microslides. Labelled cells can be viewed with a low light level video camera mounted on a compound microscope (see Section 4.2.2). For fish experiments, we prefer the red sensitive intensifier from Videoscope used with their Newvicon high resolution camera. Output from the video display monitor may also be relayed to a storage device, such as a high resolution video tape recorder (e.g. Panasonic AG6720), a computer (e.g. Macintosh II or Quadra, with a Quick Capture Board from Digital Design), or an optical disc recorder (Panasonic or Sony). Once an embryo has been screened, it should be removed to a dish full of embryo medium to continue development as required.

[a] Microcoverglass chambers: two 22 mm × 22 mm No. 1 glass coverslips are glued to each end of a 24 mm × 60 mm No. 1 coverslip. After placing an embryo in the centre in an excess of medium, a second 24 mm × 60 mm No. 1 coverslip is placed on top. The medium will hold the coverslips together by capillary action. Using this chamber, embryos can be observed from either side.

References

1. Slack, J. M. W. (1991). *From egg to embryo*. Cambridge University Press, Cambridge.
2. Gimlich, R. C. and Braun, J. (1985). *Dev. Biol.*, **109**, 509.
3. Gimlich, R. C. and Cooke, J. (1983). *Nature*, **306**, 471.
4. Honig, M. G. and Hume, R. I. (1989). *Trends Neurosci.*, **12**, 333.
5. Jacobson, M. and Hirose G. (1981). *J. Neurosci.*, **1**, 271.
6. Smith, J. C. and Slack, J. M. W. (1983). *J. Embryol. Exp. Morphol.*, **78**, 299.
7. Kintner, C. R. (1992). *Cell,* **69**, 225.
8. Dale, L. and Slack, J. M. W. (1987). *Development*, **100**, 279.
9. Harland, R. and Misher, L. (1988). *Development*, **102**, 837.
10. Krieg, P. A. and Melton, D. A. (1984). *Nucleic Acids Res.*, **12**, 7057.
11. Wright, C. V. E., Cho, K. W. Y., Hardwicke, J., Collins, R. H., and De Robertis, E. M. (1989). *Cell*, **59**, 81.
12. Ruiz i Altaba, A., Choi, T., and Melton, D. A. (1991). *Dev. Growth Differ.*, **33**, 651.
13. Beddington, R. S. P. and Lawson, K. A. (1990). In *Postimplantation mammalian embryos: a practical approach* (ed. A. J. Copp and D. L. Cockroft), pp. 267–292. IRL Press at Oxford University Press, Oxford.
14. Kimmel, C. B. and Warga, R. M. (1987). *Dev. Biol.*, **124**, 269.
15. Kimmel, C. B., Warga, R. M., and Shilling, T. F. (1990). *Development*, **108**, 581.

10

Microinjection into *Drosophila* embryos

PHILIP INGHAM and ZANDY FORBES

1. Introduction

Microinjection of *Drosophila* embryos is relatively easy to perform and has several important applications. The most common of these is injection of DNA for the production of transgenic flies; the same general approach can be used for injection of RNA, oligonucleotides, metabolites, lineage tracers, cytoplasm, and antibodies, and for transplantion of nuclei or cells between embryos.

2. General procedures for microinjection

2.1 Equipment requirements

The following equipment is required for microinjection of *Drosophila* embryos:

- inverted microscope (or compound microscope with long working distance objectives)
- a micromanipulator allowing fine control of movement in three directions (e.g. Leitz or Narishige)
- a length of narrow gauge plastic tubing attached to a needle holder and a 1 ml disposable syringe (an adequate alternative to an expensive gas-driven microinjector)
- needle puller (e.g. Narishige) and glass micropipettes
- needle beveller

The method of needle preparation depends on the puller available. We use a Narishige puller and pull from 1 mm outer (0.75 mm inner) diameter glass micropipettes (e.g. 25 μl Drummond Microcap). The final diameter of the pulled needle is varied: for injection of solutions it should be as narrow as possible (less than 5 μm), for cell transplantation, the internal diameter should be 15–20 μm.

When needles are pulled they tend to seal at the end. For the narrowest needles, immediately prior to use break the tip off by touching it against tape or the edge of a coverslip (use the micromanipulator). Then fill the needle with the solution to be injected by lowering the tip into a drop of the injection solution on a siliconized microscope slide and sucking it up using the attached syringe. Move the needle around during filling to avoid particulate matter sticking.

Needles for transplantation of nuclei and cells have a wider diameter; if the tip is simply broken off, it will be jagged and will damage the cells and/or embryo. Instead, the tips should be ground down to produce a smooth bevelled opening (see Chapter 9, *Protocol 3* for details). Suitably ground needles can be stored and used repeatedly.

2.2 Collection and preparation of *Drosophila* embryos for microinjection

To obtain sufficient eggs or embryos of a particular stage required for microinjection, eggs are collected from well fed, rapidly laying, flies over 30 or 60 min periods at 25 °C, and aged appropriately (Chapter 1). The first two egg lays of the day should be discarded as they will include eggs that have been withheld by the females and developed beyond the desired stage. Eggs for micromanipulation can be dechorionated using sodium hyprochlorite (Chapter 1), but is more satisfactory to use manual dechorionation (*Protocol 1*).

Protocol 1. Preparation of eggs for microinjection

1. Squirt a few drops of water onto the egg-collection plate and detach the eggs using a fine paint brush.

2. Transfer the eggs to a 15 mm square of damp filter paper (Whatman 3 MM). Blot dry.

3. Prepare a microscope slide with a 20 mm strip of double-sided Scotch tape stuck in the middle (Sellotape is not recommended; it is too sticky for this step). Transfer the eggs to the tape by carefully lowering the slide, tape side down, onto the 3MM paper.

4. Prepare another microscope slide with a 20 mm strip of double-sided Scotch tape stuck in the middle; carefully lower this onto the eggs so that they are now sandwiched between the two pieces of tape. Press the top slide down very gently, watching through the dissecting microscope to see that most of the embryos have come into contact with the tape.

5. Lift off the top slide, thus tearing open the chorions and leaving the eggs exposed in their vitelline membranes. At this point start a timer.

6. Prepare a 5 mm wide strip of double-sided Sellotape (not Scotch tape; it is not sticky enough for this purpose) stuck to a 10×22 mm piece of glass coverslip. An alternative is to use glue dissolved from Sellotape[c].

7. Carefully but quickly, using a pair of fine forceps, pick up the embryos individually and arrange them in a line along the strip of Sellotape or glue, in an appropriate orientation[d]. Allow to dry for a total of 5–12 min[a] (timed from when the chorions were split open). Cover with 15S Voltalef oil[b] (a 1:1 mixture of 10S and 20S Voltalef oil from Produit Chimiques Egine-Kuhlmann).

[a] If the embryos become too dry, their survival rate declines; if not sufficiently dessicated, they will leak when the needle is inserted. The optimal drying will vary from day to day with changing atmospheric conditions and should be determined empirically.

[b] This protects the embryos from further desiccation but because it is gas permeable does not compromise their subsequent development.

[c] Dissolve the glue off Sellotape with chloroform, and use a Pasteur pipette to draw out a stripe of glue on a coverslip. This method is preferrable if the embryos are subsequently to be processed for immunocytochemistry or *in situ* hybridization since they can easily be removed from the coverslip by dissolving the glue with heptane. The adhesive on some types of tape is toxic to embryos, so test before beginning an experiment.

[d] Orientation depends upon the procedure to be performed. The anterior end of the embryo is less blunt than the posterior end and can be recognised by the micropyle at the end.

2.3 General considerations for microinjection

Embryos are injected at 18 °C as embryonic development is slower at this temperature and so more embryos can be injected at the desired developmental stage. Also, cytoplasm is more viscous at this temperature which helps to reduce leakage.

After microinjection, the pieces of coverslip holding the injected embryos are placed on small apple juice plates with yeast paste, in a sealed box containing moist filter paper, and incubated at 25 °C until the surviving embryos have hatched. The larvae will crawl out onto the apple juice agar and into the yeast. After 24 h, remove the apple juice agar from the Petri dish and place it in a bottle containing solid nutrient medium (Chapter 1); check to make sure that no larvae have been left behind in the Petri dish.

3. Injecting DNA and other solutes

In order to produce lines of transformed flies, the injected DNA must be incorporated into the genome of the pole cells; these form within 2 h of fertilization. It is therefore essential to inject DNA into embryos before this time. The injection procedure is given in *Protocol 2*; for additional details see ref. 1.

Protocol 2. Injection of DNA into *Drosophila* embryos

1. Collect eggs over 20–30 min periods and immediately prepare for injection as described in *Protocol 1*. Arrange the dechorionated embryos, under oil, in a row with their posterior ends protruding over the edge of the Sellotape.

2. Place the coverslip with the embryos on a slide on the inverted microscope, set up with a micromanipulator holding a fine injection needle (Section 2.1).

3. Lower the needle (filled with the DNA solution) into the oil and bring it in line with the embryo to be injected. Focus on the posterior end of the embryo and adjust the height of the needle so that it is in precisely the same plane as the embryo. The needle must be almost parallel to the slide, to avoid tearing the vitelline membrane.

4. Advance the needle using the fine control on the micromanipulator so that it inserts into the posterior part of the embryo.

5. Withdraw the needle partially so that the tip is as close as possible to the posterior pole of the embryo. Inject the DNA by applying gentle pressure with the syringe. Inject until a small clear blob is visible; if the contents start to leak out rapidly, you have injected too much![a]

[a] Conventionally, the amount of DNA injected is 1/20–1/10 the volume of the embryo, but smaller amounts produce better survival without lowering transformation rates.

This protocol can be adapted for the injection of any other solution; the main variable is the orientation of the embryos, which depends upon the desired site of injection.

4. Cell and nuclear transplantation

4.1 Pole cell transplantation

Somatic and germ line cells can be transplanted between embryos using similar procedures, but differing in the ages of the donor and host embryos used. For somatic cell transplantation, the donors must have completed cellularization; the precise ages will vary between experiments. For germ line transplantation, the operation is most succesfully performed when both donors and hosts are at particular stages of development. We describe below only the protocol for pole cell transplantation (*Protocol 3*); this can be adapted for transplantation of any cell at other developmental stages. For additional details see ref. 1.

Protocol 3. Pole cell transplantation

1. Ideally, the donors should be at the late cellular blastoderm stage, whilst the hosts[a] should be commencing gastrulation. Accordingly, collect the donors for 60 min and age for a further 2 h before use; collect the hosts for 60 min and age for a further 2.5 h before use.

2. Arrange donors and hosts, under oil, on a strip of double-sided Sellotape on a coverslip; the donors perpendicular to the edge of the tape with their anterior ends towards the needle and the hosts in the reverse orientation, at a slight angle to the donors (*Protocol 1*). It is convienient to lay out the embryos in alternating groups of five or six hosts followed by two or three donors. Because the needles used for pole cell transplantation are thicker than those for DNA, the host embryos have to be more dessicated or the contents will leak out after transplantation; 10–12 min desiccation at step **6** in *Protocol 1* is normally optimal.

3. Place on the microscope stage set up with a micromanipulator holding a thick, bevelled injection needle (Section 2.1). Partially fill the needles with oil before you are ready to inject; they can be re-used many times if all the oil and cytoplasm is expelled after each series of injections.

4. Viewing the embryos at high magnification (200×) align the needle with a single donor embryo. Insert the needle through the anterior pole of the embryo and push it right through to the posterior so that the tip is in contact with the pole cells.

5. Using very gentle suction, aspirate the pole cells into the needle; you should be able to see each cell entering the needle, at a rate of about one every two seconds: if you suck too quickly, the cells will be damaged and the transplantation will not be successful.

6. Draw the needle containing the pole cells out of the donor embryo and line it up with the posterior pole of a host embryo. Focusing on the end of the embryo, bring the needle into the same focal plane. Prod the end of the embryo with the needle to be sure that it is centered. When it is well aligned, carefully insert the needle through the egg cell membrane. Do not push the needle too far into the embryo; the aim is to deposit several cells into the polar plate which should be starting to migrate dorso-anteriorly, so the opening of the needle should be only just inside the membrane. Carefully expel about five or six pole cells into the polar plate, alongside the host pole cells, and then withdraw the needle.

7. When all the host embryos have been injected, destroy the remaining donor embryos by puncturing with the needle. Leave the hosts to develop

Protocol 3. *Continued*

on apple juice plates as in *Protocol 2*, and then transfer the agar and hatched larvae into bottles of food. Leave to grow at 25 °C, collect the virgin female flies as they hatch and cross them to the appropriate males[a].

[a] To aid identification of successful transplants, the hosts should carry a dominant female sterile mutation, so that only chimaeric females will be fertile (2).

4.2 Nuclear transplantation

In principle, the nuclear transplantation technique (*Protocol 4*) is similar to that described for cell transplantation. The main difference is in the age of the embryos: donors should be at the syncytial blastoderm stage, but hosts should be younger, prior to the migration of nuclei to the periphery of the embryo.

Protocol 4. Nuclear transplantation

1. Arrange groups of donors and hosts as described in *Protocol 3*.
2. Insert the needle into a donor and move it to the edge of the embryo so that the tip is lined up with the nuclei.
3. Aspirate the nuclei into the needle by gentle suction.
4. Withdraw the needle from the donor and insert into a host. Expel several nuclei into the centre of the embryo. Follow step 7 of *Protocol 3*.

References

1. Roberts, D. B. (ed.) (1986). Drosophila: *a practical approach*. IRL Press at Oxford University Press, Oxford.
2. Ingham, P. W. (1984). *Cell*, **37**, 815.

11

Cell transplantation in *Xenopus*

ARIEL RUIZ i ALTABA

1. Introduction

Transplantation of cells in *Xenopus* embryos can be easily achieved. Early embryonic cells, blastomeres, can be separated, dissociated, and cultured with ease due to their large size and the fact that all cells contain a storage of yolk. This chapter describes simple procedures for transplanting single labelled blastomeres and for grafting small groups of cells into host embryos.

2. Single cell transplants

The fate of the transplanted blastomeres can be followed with the help of a lineage tracer injected before first cleavage into the donor embryo (see Chapter 9). The use of pigmented and albino embryos also helps in the identification of transplanted cells. After dissociation, single cells can be introduced into the blastocoele of blastula- and gastrula-stage host embryos. When host embryos reach the tadpole stage, these can be fixed and serially sectioned to localize the position and determine the fate of the transplanted cells. This procedure can be used, for example, to determine the influence of the transplanted cells on their new neighbours (1), the timing of commitment of single blastomeres to different fates, and the ability of committed single cells to differentiate fully in a different environment (2).

Protocol 1. Transplantation of labelled single cells

1. Inject a lineage tracer, such as rhodamine-lysine-dextran (10–20 nl at 25 mg/ml; Molecular Probes), into prospective donor embryos before first cleavage (as described in Chapter 9, *Protocol 3*). This will ensure that the cells used for transplantation will be labelled and hence recognizable from those of the host embryo.

2. Place the embryos in a plastic Petri dish laid with 2% agarose (made in water) containing 0.5 × MMR^a with antibiotics. Agarose dishes help prevent the embryos from sticking to the dish.

Protocol 1. *Continued*

3. Remove the vitelline membrane manually by pinching it with very fine watchmakers' forceps (Dumont No. 55). Use one pair in each hand and pull the membrane in opposite directions (*Figure 1*). The embryos can readily recover from small injuries. If a lot of cytoplasm oozes out, start again. Always try to pinch the vitelline membrane from a region that will eventually be discarded. Smaller cells tend to recover better from minor injuries than do larger cells. In some cases it may be easier to pinch the vitelline membrane at the animal pole.

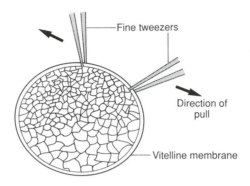

Figure 1. Vitelline membrane removal.

4. Isolate the appropriate region of the embryo from which the cells are to be taken (e.g. the equatorial region of early gastrula stage embryos; stage 10 embryos) by cutting the embryo with tungsten needles or hair knives[b]. Free the dissected part of any broken cell by moving it around in the dish or by gently pipetting it up and down with a wide-bored Pasteur pipette. Be careful to prevent the embryos or dissected parts to touch the surface; they will rapidly disintegrate due to the surface tension of the liquid. Transfer the isolated part to a new agarose dish containing fresh 0.5 × MMR.

5. After dissecting a few embryos, transfer the isolated parts to a new dish containing 0.5 × MMR—*without calcium and magnesium*. Place the dish on a rotating platform. The cells of young embryos will disaggregate in the absence of calcium and magnesium. Older embryos (past the gastrula stage) may require the presence of a protease such as pronase (Boehringer; use at 0.5% for 1–2 min). Monitor the embryo parts. After 10–60 min, most blastomeres will become loose and the dish will contain a heap of single cells in the centre. The outer cell layer is more resistant to dissociation and may remain more or less intact.

6. Place host embryos in an agarose dish containing 1 × MMR with

antibiotics. Transfer a small group of single cells from the dissociation dish to the host embryo dish. Keep single cells apart as the medium now contains calcium and magnesium and these will tend to reaggregate rapidly.

7. Remove the vitelline membrane from late blastula (stage 8–9) host embryos by pinching a hole in the animal pole and then tearing off the vitelline membrane. The hole in the animal pole should be as small as possible and no cells should be removed from the animal hemisphere. Although a few cells are usually lost, if too many cells are removed or the incision is too big, the embryo will not heal. The hole will reveal the blastocoele.

8. Push a single labelled cell through the liquid and bring it close to the hole made in the animal pole of a host embryo. Push the cell inside the blastocoele.

9. Brush off any cytoplasm extruded from around the silt and bring together the edges of the hole. Let the embryo sit for about 20 min before moving it. When the hole appears closed, carefully transfer the embryo to an agarose dish containing 0.1 × MMR. The higher salt concentration in the dissection dish helps the embryos to heal but it needs to be exchanged to a lower concentration before stage **6** or the embryos will partially exogastrulate.

10. Fix embryos at the appropriate developmental stage with 4% form-aldehyde or paraformaldehyde for 1 h. Embed in Paraplast (paraffin wax) and cut serial sections. Gelatin or polylysine-coated slides will be adequate to ensure adhesion of the sections. Dewax the sections in xylene and mount in Permount (Fisher) or glycerol/carbonate buffer with a pinch of PPDA (Sigma) to retard bleaching. Screen sections using a microscope equipped with epifluorescence illumination for the labelled cells if you have used a fluorochrome-coupled lineage tracer. Note that the clone size will vary depending on the times of implantation and analysis.

[a] For composition of MMR medium, see Chapter 5, *Protocol 3*.
[b] To make a hair knife simply attach a short strong hair (such as those from eyebrows) to the tip of a wooden or metal stick with instant glue.

3. Transplantation of small groups of cells

Changing the position of small groups of embryonic cells has allowed the identification of regional differences in the early embryo. This kind of manipulation can reveal the potential of cells in one region to influence the fate of other cells in distant positions, revealing the presence of inductive signals. It has also been used to reveal the capacity of host cells to respond to a signal, thus revealing the competence of responding cells. However, it should be noted that there is no such thing as a 'neutral' environment in the embryo and results should always be interpreted with caution. As mentioned in Chapter 5, classical embryological experiments in amphibians revealed the existence of the organizer, a group of dorsal mesodermal cells of early gastrula embryos that can influence the fate of neighbouring cells (3, 4). Here procedures are given to perform both a classical organizer transplant into the ventral marginal zone (5, 6) and a blastocoele implantation or '*Einsteck*' assay (7). Although these procedures were originally performed in amphibians other than *X. laevis* they can easily be done in this species (6, 8).

Protocol 2. Organizer grafts into the ventral marginal zone

  1. Locate the dorsal marginal zone of embryos at about stage 10. Stage 10 embryos display a pigmented arc in the dorsal subequatorial region (see Appendix 1). This is the dorsal lip of the blastopore and indicates that gastrulation has started. Use the dorsal lip as an indication of the position of the dorsal side.

  2. Place embryos in agarose-laid dishes containing 0.5 × MMR plus antibiotics. With fine forceps, free the donor embryos from their vitelline membranes.

  3. Turn the embryos vegetal side up and cut the dorsal lip region with hair needles as if the embryo were a pie. Turn the excised dorsal region on its side and cut through the lip. Make sure vegetal tissue is not taken along with the mesoderm.

  4. Cut the animal cap cells (pigmented cells) from the dorsal region. You want the early ingressing mesoderm as well as the mesoderm in the dorsal-most equatorial region. Vegetal cells are large and yolky. Mesodermal cells are small, fairly round and white. Animal cap cells display epithelial morphology and the outer cells are pigmented. Throughout the operation do not lose sight of the dorsal lip. Discard unwanted pieces at every step. This will ease the task of keeping track of the dorsal cells. For grafting in the ventral marginal zone you want to keep the superficial layers (non-pigmented prospective archenteron roof endoderm and some dorsal ectoderm) together with the mesoderm. The explants to be grafted should be approximately square.

  5. Make sure that isolated pieces are free of dead or broken cells by pipetting them up and down very gently. Transfer isolated dorsal cell groups (organizers) to an agarose dish containing fresh 0.5 × MMR plus antibiotics.

6. In a separate dish remove the vitelline membrane from host embryos.

7. With hair needles and/or fine forceps excise a portion of tissue from the ventral marginal zone equivalent in size to the dorsal explant to be grafted. Make side cuts first, as if cutting a square out of the ventral maginal zone. Push the piece to be excised towards one side and then towards another. Here the whole embryo is important and not the tissue that is being dissected out, so be very careful.

8. Gently, transfer the organizer explant next to the operated host and graft it into the 'hole' left by the dissected ventral tissue. Push it with forceps and remove any cells that may extrude. Make sure the surfaces of the explants are clean of loose or dead cells, otherwise the host embryo will not heal.

9. Allow the operated embryo to rest for about 20–30 min on the graft to help keep this in place. Transfer embryos to a clean agarose dish containing 0.1 × MMR plus antibiotics and incubate overnight. Placing the embryos at 18–22 °C can improve healing.

10. If successful, the grafted host embryos will form a secondary dorsal lip in the ventral region and will display a double axis visible first at the early neurula stage. Even though a perfect graft will result in an embryo with perfectly symmetric axes at 180° to each other, this procedure usually gives partial axes, mostly lacking head structures, positioned less than 180° to the host axis. This is likely to be due to the time of grafting and to the effectiveness of grafted organizer cells to recruit host ventral cells towards dorsal fates through the process of dorsalization (6). Try it a few times before even thinking of giving up!

Protocol 3. Blastocoele implants

1. Prepare dorsal organizer explants (see above) and dissect out superficial layers by slicing the tissue next to the outer cell layers with a hair needle. Only whitish, small cells should remain. All operations are done in agarose-laid dishes containing 0.5 × MMR plus antibiotics.

2. If the explants are large in relation to the animal cap hole (see below), split these into two to three fragments using a hair needle.

3. Remove the vitelline membrane from host embryos by pinching it at the animal pole and place the hosts with their animal (pigmented) side up.

4. If you have pinched off the animal cap during removal of the vitelline membrane, use this incision to make a small opening. Otherwise gently make a small slit in the animal pole. This incision should ideally be as small as possible without loss of cells.

Protocol 3. *Continued*

5. Carry the dissected organizer fragments and introduce them into the blastocoele by gently pushing them through the hole, one fragment per host embryo. If the hole is too large, start again.

6. Bring the sides of the hole together after implanting the organizer fragments. Contact will speed up closure.

7. Let the host embryos heal for 10–20 min. The hole should disappear. In some cases, the embryos will not heal and this will become obvious. Discard these embryos immediately.

8. Transfer the host embryos very carefully to a clean dish containing 0.1 × MMR plus antibiotics and incubate overnight.

9. If the experiments are successful, the host embryos are likely to develop a secondary head appearing in the ventral side. The mesoderm implanted into the blastocoele ends up in the antero-ventral region of the embryo due to the constriction of the blastocoele during gastrulation. It will then induce competent ventral ectoderm to become neural tissue. Only in a few cases will host embryos develop full secondary axes.

4. Notes

Note that the same group of cells will give rather different results depending on the mode of transplantation: mostly posterior axes if grafted into the ventral marginal zone, or mainly anterior, head structures if grafted into the blastocoele. These experiments can be carried out with labelled donor embryos to follow the fate of the grafts. When the organizer graft induces the formation of a second axis, labelled cells will mainly populate the midline axial structures, the notochord, and floor plate of the neural tube (3, 6). In this case, embryos can be fixed and sectioned to reveal the location of the descendants of the grafted cells. Alternatively, the embryos can be viewed as whole mounts after dehydration in methanol and clearing in 1:2 benzyl alcohol/benzyl benzoate (Sigma).

References

1. Gimlich, R. L. and Gerhart, J. C. (1984). *Dev. Biol.*, **104**, 117.
2. Heasman, J., Wylie, C. C., Hausen, P., and Smith, J.C. (1984). *Cell*, **37**, 185.
3. Spemann, H. (1938). *Embryonic development and induction*. Yale University Press, New Haven.
4. Nieuwkoop, P. D., Johnen, A. G., and Albers, B. (1985). *The epigenetic nature of early chordate development. Inductive interactions and competence*. Cambridge University Press, Cambridge.

5. Spemann, H. and Mangold, H. (1924). *Wilhelm Roux Arch. EntwMech. Org.*, **100**, 599.
6. Smith, J. C. and Slack, J. M. W. (1983). *J. Embryol. Exp. Morphol.*, **78**, 299.
7. Mangold, O. (1933). *Naturwissenschaften*, **21**, 761.
8. Ruiz i Altaba, A. and Melton, D. A. (1990). *Trends Genet.*, **6**, 57.

12

Transplantation in avian embryos

CLAUDIO D. STERN

1. Introduction

This chapter briefly introduces two techniques for microsurgery. In the first, embryos are operated whilst in modified New culture (Chapter 6, *Protocol 3*). This is suitable for operations on embryos of up to 1 day's incubation, and embryos can be cultured for up to 36 h. As an example, an operation in which the anterior tip of the primitive streak (Hensen's node, the organizer of the amniote embryo) is grafted into an ectopic site in a host embryo is described (see ref. 1). The second technique is for operations *in ovo*, through a window in the shell, suitable for embryos of 1–3 days' incubation. The example given is that of a graft of segmental plate (see ref. 2). With some care and a few precautions, embryos operated in this way can be grown to hatching. Older embryos can be operated as described in Chapter 13. A histochemical method for revealing the quail nucleolar marker in chimaeras is also given (adapted from ref. 3).

2. Operation on young embryos in New culture

Here the host embryos are set up in New culture (Chapter 6, *Protocol 3 or 4*), and the donor (which can be a quail embryo so that donor and host cells can be distinguished) can be explanted by the rapid method given in Chapter 6, *Protocol 2* or in New culture.

Protocol 1. Grafting into embryos in New culture

The materials required are as listed in Chapter 6, Section 3.1.1. A pair of mounted needles (either tungsten or steel, prepared as described in Chapter 6, Section 2.2) will also be required, and a Gilson micropipette that can be set to 3 μl.

1. Prepare host embryo(s) from eggs incubated 11–14 h at 38 °C as described in steps **1–7** in *Protocol 3*, Chapter 6.

Protocol 1. *Continued*

2. Prepare donor embryo(s) at the same stage of development (which may be quail embryos) in the same way or as described in steps **1–5** in *Protocol 2*, Chapter 6.

3. Identify the primitive streak, the anterior (cranial) tip of which is Hensen's node (see pp. 299, 305, 306).

4. Place the donor embryo under the microscope. Making sure that it is well submerged in saline, use a pair of mounted needles to cut out Hensen's node. Cut close to the visible edges of the primitive streak to obtain a rectangular piece about 100 μm in lateral extent and 120 μm in anteroposterior dimension. Cut through the whole thickness of the embryo. Place the node gently on the surface of the same embryo.

5. Use a Gilson micropipette set to 3 μl to pick up the graft. Be careful not to allow the graft to attach to the end of the plastic tip or to be caught up in the meniscus at the top of the fluid taken up. Place the host embryo under the microscope. Transfer the graft to the host embryo whilst observing under the microscope at low magnification.

6. Cut a small, superficial nick in the endodermal layer (facing upwards) at the margin between the *area pellucida* and the more yolky, peripheral *area opaca*. The nick should only penetrate the endodermal layer so as to create a pocket.

7. Using mounted needles, push the graft well under the endoderm if possible.

8. Remove all saline inside and outside the ring and transfer the ring with the operated embryo to a plastic dish containing thin egg albumen as described in steps **8–11**, *Protocol 3*, Chapter 6.

9. Seal the dish and place it in a humidified box to incubate at 38 °C overnight as described in steps **11–13**, *Protocol 3*, Chapter 6.

---

3. Operation on older embryos *in ovo*

Here the host embryo is operated *in ovo*, which is only suitable with ease for embryos older than about 36–48 h incubation. The donor embryo may be prepared in the same way or explanted by the quick method described in *Protocol 1*, Chapter 6. As an example, the operation described here consists of a graft of presomitic mesoderm (segmental plate) with a quail embryo as the donor (see ref. 2).

The materials required are as follows:

● dissecting kit: 2 pairs small forceps (watchmakers', No. 4 or 5), 1 pair small scissors (about 2 cm straight blades), 1 scalpel (No. 3 handle, No. 11 blade), a Gilson micropipette for 3 μl, yellow tip(s)

- eye surgeon's micro-knife, 15° angle[a]
- 2 mounted tungsten needles or steel pins as described in Chapter 6, Section 2.2
- plasticine to make ring for resting egg on its side
- 100 ml calcium-/magnesium-free Tyrode's saline (CMF) (Chapter 6, Section 4)
- antibiotic/antimycotic solution, 100 × concentrated (Sigma)
- 70% ethanol to wipe shell
- PVC tape to seal egg
- 2 Pasteur pipettes and rubber teats
- container for egg waste
- 10 ml plastic syringe with Gilson yellow tip stuck into the end, filled with high vacuum silicon grease
- 1 ml syringe, 27Gx3/4" needle (for ink injection)
- 1 ml syringe, 21G needle (for antibiotics)
- 5 ml syringe, 21G needle (for withdrawing albumen)
- paper tissues
- 35 mm plastic dish coated with Sylgard (Chapter 6, Section 2.1) and steel insect pins, size A1 or D1
- Indian ink (Pelikan Fount India is best; most other makes are toxic), diluted 1:10 with CMF and loaded in the ink injection syringe
- 50 ml trypsin (Difco, 1:250), freshly made up to 0.12% (w/v) in CMF
- hens' and quails' eggs incubated 40–44 h so that they are at stage 11. The hens' eggs (hosts) should have been resting on their sides at least 20 min
- dissecting microscope, fibre optic incident illumination

[a] Suitable microknives are: micro-feather microsurgery scalpels for eye surgery, 15° blade angle, catalogue No. 715, manufactured by Feather (Japan) and marketed by pfm GmbH. Sold in boxes of five, not cheap!

Protocol 2. Operation of embryos *in ovo*

1. Prepare a donor quail embryo by explanting it as described in Chapter 6, *Protocol 1*. Pin it through the extraembryonic membranes to a Sylgard dish filled with CMF.
2. Shape plasticine into a ring about 2 in (5 cm) in diameter and place it on the stage of the microscope. Place a hens' egg (host) onto the plasticine

Protocol 2. *Continued*

ring, being careful that it is not rotated with respect to its resting position.

3. Using the 5 ml syringe with 21G needle, held nearly vertical, insert needle into blunt end of egg until the shell is felt at the bottom surface. Withdraw 0.5 ml egg albumen, which should come up easily.

4. Score a shallow 1 cm × 1 cm square on the top of the shell with the scalpel and lift up the square of shell.

5. With a pair of watchmakers' forceps, pierce and remove the underlying shell membrane, after wetting it with CMF. Avoid damage to the embryo underneath: before air is allowed into the egg, the embryo will lie very close to the membrane.

6. Fill the cavity with CMF so that the embryo floats up to the level of the window.

7. After ensuring that there are no air bubbles in the syringe with Indian ink or in the needle, insert the needle under the vitelline membrane, tangentially, at a position as far away from the embryo proper as possible. Point towards and slightly below the embryo, and inject about 50 μl. It is important to minimize movement of the needle after penetrating the vitelline membrane, or the hole will be very large and yolk/ink will leak out. Introduce and withdraw the needle with one clean, decisive movement and do not stir the needle inside the yolk; only one attempt per egg!

8. Draw a shallow, continuous border of silicon grease around the window. This will contain a standing drop in which the operation will be done. Now fill this chamber with CMF saline until there is a standing drop, and adjust the fibre optic light to shine tangentially to the surface of the egg so that the embryo can be seen very clearly with minimal light intensity from the light source.

9. Break the vitelline membrane just over the region to be operated with a needle. The hole should be as small as possible. The segmental plates are the rod-like structures lying on either side of the neural tube at the tail end of the embryo, just behind the last somite. The portion to be rotated is the most anterior half of the plate (see p. 307).

10. Replace the bubble of CMF with trypsin/CMF. Increase the magnification of the microscope as much as possible.

11. Operating in the drop of trypsin, use the micro-knife to make initially very shallow cuts in the ectoderm next to the neural tube in the region of the operation.

12. Gradually deepen the cuts by repeating the cutting movement with the knife. Find the lateral border of the segmental plate, and do the same

there: a shallow cut in the ectoderm first, then repeat the cut once or twice more to deepen. In both cases, make sure you do not penetrate the endoderm (which is one-cell thick) or ink will pour out. Finally, free the posterior end of the piece and loosen the graft.

13. Remove the piece of segmental plate with a Gilson micropipette set to 3 µl. Replace the bubble over the embryo with fresh CMF twice to remove the trypsin solution. Make a new bubble of CMF whilst obtaining the graft from the donor.

14. Turn to the donor embryo in the Sylgard dish. Replace almost all of the CMF in which it was submerged by trypsin solution. Repeat the trypsin wash and perform the dissection in this solution at room temperature.

15. Cut out an equivalent piece of segmental plate as the one removed from the host.

16. Pick up the graft with the Gilson. With the other hand, place the host under the microscope and, observing under low magnification, carefully place the graft into the CMF bubble over the embryo.

17. Use the knife tip and work at low magnification to manipulate the graft into the gap made by removal of the host piece of segmental plate.

15. When the graft is in position (approximately), very carefully remove the most of the fluid from above the embryo with a Pasteur pipette while watching under low power. If necessary, reposition the graft with a mounted needle.

16. Insert the 5 ml syringe with 21G needle into the original hole in the blunt part of the eggshell, vertically, and carefully withdraw 2–2.5 ml thin egg albumen. This will lower the operated embryo back to its original position. Be careful when you insert the needle, since the pressure could make the graft come out of its site.

18. Add 100 µl of antibiotic/antimycotic concentrate (away from the graft site).

19. Wipe the edges of the shell with tissue paper moistened lightly in 70% alcohol to remove the silicon grease.

20. Cut a piece of PVC tape about 8 cm long. Stretch it slightly, and then let it relax. Place it over the window, smoothing out any unevenness carefully so as to avoid breaking the shell or applying too much pressure on the window.

21. Keeping the egg on its side, place it (window up) into an egg tray in a humidified incubator at 38 °C.

22. Incubate 1–3 days. Embryo survival 2 days after this operation should be 80–100%.

4. Fixation and processing for quail/chick chimaeras

Quail cells can be distinguished from their chick counterparts by having a prominent, single nucleolus highlighted by dense heterochromatin. Chick cells tend to have multiple and more diffuse nucleoli. The difference can be revealed with Feulgen's staining technique. However, we use an alternative technique (3), found to be more reliable and consistent and to produce results that can be photogaphed more easily. For both procedures, the best fixative is Zenker's. Aldehydes interfere with the reaction.

4.1 Fixation

Zenker's fixative: mix together 5 g mercuric chloride, 2.5 g potassium dichromate, 1 g sodium sulphate, and dissolve in 100 ml distilled water. Keep at room temperature as a stock. Before use, add 5 ml glacial acetic acid to every 100 ml. Fix for 30 min to 3 h, with embryos pinned in a Sylgard dish (see Chapter 3). Wash embryos three times (10 min each) in tap water and place into 70% alcohol. Change the alcohol several times over 3 h to 1 day (depending on embryo size) until all the yellowish colour disappears.

4.2 Processing and staining to reveal quail nucleolar marker

The method given here is adapted from the technique described in ref. 3.

Protocol 3. Staining wax sections for quail nucleolar marker

1. After fixation and washing, embed in paraffin wax and section conventionally.
2. Mount 10 μm paraffin sections onto slides smeared with gelatin/albumen (BDH/Gurr).
3. Dewax sections in xylene or Histoclear, 2 × 30 min.
4. 100% ethanol twice for 10 min.
5. 70% ethanol, 30 min.
6. Remove mercury precipitate in 0.5% iodine in 70% ethanol, 5 min.
7. Rinse in tap water.
8. Remove iodine in 2.5% hypo (sodium thiosulphate), 5 min.
9. Rinse in running tap water, 5 min.
10. Place in a container of distilled water pre-warmed to 60 °C for 30 min.
11. Hydrolyse in pre-warmed 3.5 M HCl at 37 °C for 45 min.
12. Rinse three times in distilled water at room temperature.

13. Stain in Harris's haematoxylin[a] for 10 min.

14. Differentiate in acid alcohol[b] for 10 sec.

15. Immediately place into running tap water and leave 30 min.

16. Rinse in distilled water.

17. Dehydrate up series of alcohols (70%, absolute twice, 10 min each)

18. Clear in xylene or Histoclear.

19. Mount in Canada balsam or DPX.

[a] Harris's haematoxylin. Dissolve 2.5 g haematoxylin powder in 25 ml absolute ethanol. Dissolve 50 g ammonium-potassium alum in 500 ml distilled water. Mix the two solutions together, and boil. Add 1.25 g mercuric oxide *slowly* (or it will boil over). Allow to cool and add 20 ml glacial acetic acid. Filter before each use through Whatman No. 1 paper.
[b] Acid alcohol: 1% concentrated HCl in 70% ethanol.

References

1. Storey, K. G., Crossley, J. M., De Robertis, E. M., Norris, W. E., and Stern, C. D. (1992). *Development*, **114**, 729.
2. Keynes, R. J. and Stern, C. D. (1984). *Nature*, **310**, 786.
3. Hutson, J. M. and Donahoe, P. K. (1984). *Stain Technol.*, **59**, 105.

13

Chick limb buds

CHERYLL TICKLE

1. Introduction

Since the pioneering work of Harrison, Zwilling, Saunders, and others, the limb has become one of the classical model systems in which to study cell interactions in development. A considerable body of work in the last few decades has established the basic principles of limb development and regeneration. Amphibians are often used for studying limb regeneration, while the chick limb has been the main source of information about limb development. Experiments on chick limb buds have led to the identification of regions such as the apical ectodermal ridge (AER), zone of polarizing activity (ZPA) and progress zone as key players (for reviews see refs. 1–6). Recently, molecular approaches have started to be combined successfully with more classical experiments (e.g. ref. 7), and localized misexpression of genes achieved through the use of retroviral vectors (8) (see Chapter 20).

This chapter describes the main technique used for grafting a polarizing region (ZPA) between a donor and a host limb bud, which has formed the basis for much of the experimental work in the field. The polarizing region resides in the mesenchyme at the posterior margin of the limb. When grafted anteriorly to a host limb, it induces the duplication of structural elements, with mirror-symmetry with respect to the normal structures. The extent of the duplication depends mainly on the stage of the host embryo: if done late, only the digits are duplicated; if early, more proximal structures are also duplicated. This experiment suggested that the posterior margin of the limb bud has an organizing role and that it emits positional cues, interpreted by other cells in the bud.

The techniques described here can also be modified slightly for other operations on chick limbs. For example, it will produce suitable embryos for fate mapping or cell lineage analysis in the limb (Chapter 9), for introducing viruses for lineage analysis or misexpression of genes (see ref. 8 and Chapter 20), or for implanting beads soaked in putative morphogens like retinoic acid into known positions in the limb (see ref. 7).

2. Preparation for operations on chick limb buds

Operations on the limb buds of developing chick embryos can be performed through a window made in the shell. The window is usually made about 2–3 days after the start of incubation, and can be larger than those described in Chapter 12. The embryo can then be staged and reincubated so that it reaches the correct stage for carrying out the operations. The limb buds start developing around 2½ days after the start of incubation.

Protocol 1. Windowing eggs for limb bud operations

1. The egg is turned so that the embryo floats freely to the top of it. This can be checked by candling[a].

2. The blunt end of the egg is swabbed with 70% alcohol and a hole is made into the air sac using a blunt pair of forceps. The forceps and all the instruments should be kept clean by wiping them with 70% alcohol each time before they are used.

3. The uppermost part of the egg is then swabbed with 70% alcohol and a small hole is made in the shell. A small fragment of the shell can then be picked away. The white membrane that is now exposed is the shell membrane and the embryo lies immediately beneath it. One prong of the forceps is then used to make a hole in the shell membrane and the edge of the membrane is lifted carefully so that air can now enter beneath it. Because there is a hole at the blunt end, the embryo drops away from the shell membrane as air escapes from the air sac (*Figure 1*). It should be possible to see a change in the whiteness of the shell membrane which indicates that the embryo has dropped. Now it is safe to enlarge the hole in the shell and the shell membrane.

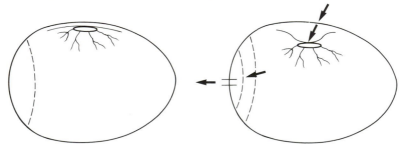

Figure 1. Diagram showing the position of the chick embryo on top of the yolk inside the egg. When a hole has been made in the air sac, tearing the shell membrane above the embryo allows air to enter and the embryo drops.

4. A rapid way to enlarge the hole without dropping too much shell on the embryo is to place a piece of adhesive tape (e.g. Sellotape or Scotch tape)

over the hole and then enlarge the hole by cutting into the sellotape and the shell with a pair of small scissors.

5. The embryos can now be staged according to various features including the shape of the limb buds, using the Hamburger and Hamilton staging system (9) (see Appendix 1).

a Candling consists of holding the egg over a light source strong enough to give a shadow of the image of the embryo through the unopened shell. A simple candling device can be made by cutting a circular hole, about 5 cm in diameter, into the bottom of an old cardboard box containing a 40–100 W tungsten bulb. Place an egg in the opening and observe. After the eye has become pigmented at stage 22, this should be visible through the shell. Before this only the shape of the embryo will be visible.

Operations on the limb buds are most commonly carried out on the right wing bud, which lies uppermost. The left wing bud acts as a control. Operations can be carried out on the right leg bud but more care is needed because this bud lies close to the allantois. Most operations involve cutting and pinning tissue. The special instruments used for these operations are small needles and miniature spatulas. The needles can be made from lengths of 300 μm diameter tungsten wire, which is sharpened electrolytically by applying a DC potential of 10 V across the wire in 1 M NaOH; during this, repeatedly immerse and withdraw the wire to etch a tapered tip (see also Chapter 6, Section 2.2) The spatulas are made from 500 μm diameter nickel-chrome wire which is beaten into shape. In both cases, the wire is fixed with Araldite into glass tubing which forms a handle. Wire for these instruments can be obtained from Goodfellow metals.

3. Grafting a polarizing region to the anterior margin of a wing bud

3.1 Preparation of the host

It is better to do this first, before preparing the graft. The slight delay will give time for any bleeding to stop.

Protocol 2. Host preparation

1. Before any operations can be carried out, there are two membranes that lie over the embryo that must be torn away. This is done using a fine pair of forceps. The position of the upper (vitelline) membrane may be easy to judge if some small pieces of shell have dropped onto it. The lower membrane is the amnion and is closely wrapped around the embryo like a sleeping bag.

Protocol 2. *Continued*

2. When both membranes are cleared away, it should be possible gently to pass the miniature spatula beneath the wing bud. It will now be possible to make out a clear rim around the tip of the bud, which is the apical ectodermal ridge.

3. The miniature spatula is used to support the wing bud during cutting. Clean cuts are made with the fine needles so that a cube of tissue is removed from the anterior margin of the wing bud (*Figure 2*). The piece of tissue that has been cut out can be removed from the egg, carried on the spatula.

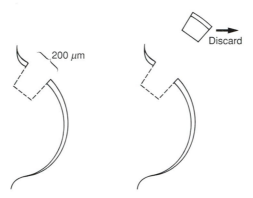

Figure 2. Diagrams illustrating how a cube of tissue is cut out from the anterior margin of a wing bud to make a site for a graft of polarizing region.

3.2 Preparation of the graft

The graft will be held in place by a platinum pin. The pins are made by cutting platinum wire (0.025 mm in diameter; Goodfellow metals) into lengths of about 200–300 μm in 70% alcohol. One end can then be bent to form a hook. The pins are then transferred to a dish containing tissue culture medium or PBS to wash off the 70% alcohol before being used.

Protocol 3. Cutting out the polarizing region

This is most easily done by cutting out tissue from a wing bud in a dish.

1. An embryo is removed from the egg by grasping it from behind the waist region with a fine pair of forceps.

2. Place the embryo in a Petri dish containing tissue culture medium or PBS (equilibrate with air).

3. Remove the wing buds by pinching through the tissue at the base of the wing bud with a pair of fine forceps. It should still be possible to identify

the posterior margin by the shape of the bud. The bud is also slightly convex dorsally and the tip tilts ventrally. It' is even. easier to orientate the buds if left and right are kept separate.

4. Cut the polarizing region from the posterior margin of the bud. One needle is used to impale the bud and the other for cutting the tissue. Two cuts are made (*Figure 3*).

5. After making the cuts, insert a pin, which will hold the graft in place after transplantation, and also serves at this time to mark the orientation of the tissue.

6. Use the pin to secure the polarizing region before making the final cut. The hooked end marks the edge of the tissue covered by ectoderm. The polarizing region impaled on the pin is now ready to be grafted.

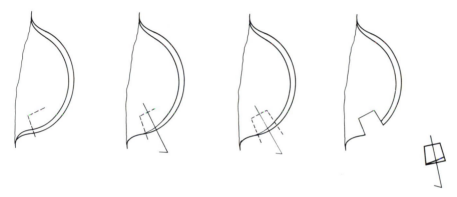

Figure 3. Diagrams showing how the polarizing region is cut out from the posterior margin of a wing bud.

3.3 Grafting the polarizing region

Protocol 4. Grafting the polarizing region

1. The polarizing region can be transferred directly to the host embryo on the pin by gripping the pin at the hooked end with a pair of forceps. Alternatively, the graft can be transferred on the miniature spatula. Grafts of tissue that is not very compact, such as a pellet of cells, can be transferred using a Pasteur pipette.

2. It is best first to transfer the polarizing region to some safe place in the host egg, for example, rest it on the body wall next to the wing bud, and then to grip the pin again and manoeuvre the graft into the site that has been prepared (*Figure 4*). Press the pin well down into the host tissue. The pins can be removed later, e.g. next morning, if the bud is to be sectioned.

Protocol 4. *Continued*

3. Reseal the egg of the operated embryo with clear adhesive tape (Sellotape or Scotch tape). It is usually a good idea to check the embryos the next day to see whether the graft is still in place. An effect on pattern can be seen a minimum of 4–5 days later.

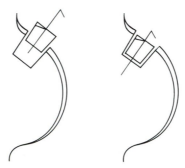

Figure 4. Diagrams to show a polarizing region graft, positioned into a hole cut in the anterior margin of a host wing bud.

4. Preparation of whole mounts of wings to show pattern of skeletal elements

Here, operated wings are collected, fixed and stained with Alcian Green or Alcian Blue to show cartilaginous skeletal elements.

Protocol 5. Preparation of whole mounts of limbs and staining for cartilage

1. To remove the embryo from the egg, first enlarge the window using scissors. Then pick away the membranes that have grown back over the embryo using forceps until the head is clearly seen. Then pass a small spatula under the neck and gently ease the embryo out of the egg.
2. Place the embryo in a dish containing tissue culture medium and snip through the neck with forceps.
3. Transfer the body into a dish containing 5% trichloroacetic acid (TCA) and trim off the limb buds by cutting through the body wall with forceps.
4. Fix the limbs 3 h or longer in TCA.
5. Wash the limbs in water and then place in Alcian Green or Alcian Blue (0.1% in acid alcohol[a]) for at least 3 h (can be overnight).
6. Differentiate in acid alcohol overnight, dehydrate with two changes (1 h each) of absolute alcohol and clear in methyl salicylate.

7. The cleared limbs can be trimmed and flattened, and then viewed while submerged in methyl salicylate in a *glass* dish.

a Acid alcohol: 1% HCl in 70% ethanol.

References

1. Wolpert, L. (1989). *Development*, **107** (Supplement), 3.
2. Brickell, P. M. and Tickle, C. (1989). *Bioessays*, **11**, 145.
3. Wolpert, L. (1990). *J. Cell Sci.*, **13** (Supplement), 199.
4. Tickle, C. (1991). *Development*, **1** (Supplement), 113.
5. Tabin, C. J. (1991). *Cell*, **66**, 199.
6. Tabin, C. J. (1992). *Development*, **116**, 289.
7. Oliver, G., De Robertis, E. M., Wolpert, L. and Tickle, C. (1990). *EMBO J.*, **9**, 3093.
8. Morgan, B. A., Izpisúa-Belmonte, J.-C., Duboule, D., and Tabin, C. J. (1992). *Nature*, **358**, 236.
9. Hamburger-Hamilton, V. and Hamilton, H. L. (1951). *J. Morphol.*, **88**, 49.

14

Nuclear transplantation

JOHN B. GURDON

1. Introduction

Nuclear transplantation is a procedure by which the nuclear material of a fertilized egg is replaced by that of another cell. The purpose of this procedure is twofold. The primary purpose, and that for which the technique was devised, is to ask whether the genetic material in a somatic cell is functionally equivalent to that of the zygote nucleus. If a somatic nucleus can replace the zygote nucleus in an egg by supporting normal development, this shows that there has been no loss or permanent inactivation of genes in the course of development leading to the formation of the somatic cell in question. Once this result has been achieved, a related question is how the nucleus of a specialized cell can be switched back to an embryonic state so as to express all genes required for normal development. The second and more recent purpose of nuclear transplantation is to create substantial numbers of genetically identical individuals from the nuclei of one embryo. This second aim has possible practical value in propagating animals with a valuable genetic constitution.

Several methods have been used to carry out nuclear transplantation in animals, and the methods differ substantially according to the species concerned. Most nuclear transplantation experiments have been carried out so far on *Xenopus*, and these are described in some detail. In insects, nuclear transplantation was among the techniques that led to the identification of the germ plasm as a determinant of cell fate. Considerable efforts have recently been made to achieve satisfactory results in mammals; surprisingly certain kinds of mammals, such as sheep, are far more suitable for this technique than others, such as mice.

The principle of the nuclear transplantation procedure in amphibia was worked out by Briggs and King (1) who obtained some normal embryos by transplanting blastula nuclei in *Rana pipiens*. The method used was to eliminate the genetic material of an unfertilized egg by physically removing the tiny amount of cytoplasm that contains the second meiotic spindle and attached chromosomes. In *Xenopus*, the enucleation of the egg is best achieved by UV irradiation which kills the second metaphase chromosomes located on the surface of the egg, but does not penetrate far enough to damage the egg cytoplasm. In some species of animals the enucleated egg needs to be activated, for example by an electric shock. In other species this requirement is fulfilled by penetrating the egg with a micropipette in such a way as to break the cell wall but not to displace all of the cytoplasm from around the nucleus which is protected by it. This broken cell is then introduced into the egg by a specially shaped micropipette. The amount of donor cell cytoplasm is negligible in relation to the large volume of the egg. The egg then commences cleavage and further development forcing the introduced nucleus to divide on the normal time schedule for the egg. When successful, the enucleated egg and introduced nucleus will develop normally.

In all nuclear transplantation experiments it is very desirable, if not essential, to use some kind of genetic marker to prove that the resulting embryos or individuals are wholly dependent on the introduced genetic material and do not represent a failure of enucleation.

A refinement in the technique of nuclear transplantation first introduced by King and Briggs (2) is to carry out serial nuclear transfers. The idea here is to use an embryo obtained by the transplantation of a single somatic nucleus to provide many more nuclei of the same kind for subsequent transfers. By this means it is possible to select the most normal embryos resulting from several initial single nuclear transfers and to use the cells of such a normal embryo for further transfers of these nuclei to a new set of enucleated recipient eggs. This procedure can, in principle, be repeated indefinitely so long as there is a continuing supply of recipient eggs. In this way it is possible to create an indefinite number of genetically identical individuals.

2. Results obtainable with nuclei of different cell-types

In all species in which nuclear transplantation has been successful, it has been found that the earlier the developmental stage from which a nucleus is taken, the more normal is the development which it will sustain after transplantation. With *Xenopus*, the best results are obtained from the nuclei of blastulae and from the endoderm of gastrula and neurula stages (see ref. 3 for review). In these cases, about one-third of all nuclear transplantations will develop to a swimming tadpole stage, and more than one-half of these will subsequently from sexually mature adults. However when nuclei from differentiating or differentiated cells are used as donors, the normality of development greatly declines. When differentiated cells such as those of adult skin are used, about 20% of the nuclear transfers can reach the stage at which muscular contractions occur, indicating the presence of functional muscle and nerve (4). Results with *Rana pipiens* are broadly comparable, although, in this species, it has been found helpful to transplant nuclei into maturing oocytes, thereby giving them longer in the environment of an oocyte or egg cytoplasm before they are required to commence division (5).

Primary cultured cells have been used for nuclear transfer. The cells which grow out from different tissues, presumably fibroblasts, all give similar results, comparable to those mentioned above for adult skin cells (6).

In order to obtain the best results from differentiated or cultured cells, it is necessary to carry out serial nuclear transplantation. The reason why this seems to give better results than first transfers is believed to relate to the rapid switch in replication rate required of the transplanted nucleus, which, once it is in egg cytoplasm, must conform to the rapid cell cycle of an embryonic cell. It often happens that the nucleus of a slow dividing cell enters only one of the first two blastomeres after the nuclear transplant egg has divided and commences its own division only at the second cell cycle of the egg (4). In this way it is believed that the transplanted nucleus has a second S phase in which to complete its chromosome replication, and is therefore more likely to achieve complete replication and normal chromosome segregation at the second division of the egg than it would at the first division so soon after transplantation.

3. The procedure as applied to *Xenopus*

This has been reviewed in detail elsewhere (7, 8), and the reader is recommended to refer to these articles.

3.1 Description of equipment

By far the most critical part of the necessary equipment is a suitably shaped micropipette or injection needle. This needs to have a smooth-edged opening just smaller than the diameter of the donor cell, i.e. from 5–20 μm, according to cell type. In addition there needs to be at least 2 mm of parallel-sided glass tubing of the same diameter and immediately behind the opening. Lastly, the smooth-edged opening needs to have the general shape of a hypodermic syringe needle, i.e. there must be a spike on the periphery of the opening to enable the egg to be penetrated harmlessly. There are several commercial instruments which are capable of pulling the appropriately sized pipette. There are also several commercial instruments which are capable of sharpening the edge of an open glass needle.

The needle needs to be held in a simple micromanipulator. Any one which gives a 5–10-fold reduction in hand movement will be suitable. One of the best is that made by the Singer Company. The shaped glass needle is filled with light paraffin oil and this is also used to fill the plastic tubing which connects the micropipette to a syringe which itself should be filled with the same oil. Very suitable is the Agla syringe (Burroughs Wellcome). The best control is obtained by having the whole system air-free and filled with light paraffin oil (0.85 g/ml) except for the very tip of the injection pipette. A stereomicroscope giving about 25× magnification is suitable for amphibian nuclear transplantation.

3.2 Production and enucleation of recipient eggs

Xenopus unfertilized eggs are obtained by the standard hormone treatment. If eggs are laid in a high salt solution, they need to be transferred to dilute salt such as 0.1 × MBS[a] to enable the jelly to swell. It is best to use unfertilized eggs within 20 min of laying or of being transferred to low salt solution. Each egg is then transferred to a glass slide with the animal pole uppermost. All excess water is removed and the glass slide carrying the eggs (usually four to six) is placed under the appropriate UV light beam.

Enucleation by UV light can be achieved with either or both of two sources. One is the UVS Mineralite (Ultraviolet Products Inc., UVSL-15). Eggs should be placed about 5 cm from the bulb of this lamp and exposed for 30 sec, thereby receiving about 16 000 ergs/sec/cm^2 (1 erg = 10^{-7} J). The second source of UV light that can be used is a medium pressure mercury arc lamp, originally made by Hanovia. This source is particularly valuable for softening the jelly on the animal pole of the egg so as to make it more easily penetrable by a pipette. The normal dose of this second UV light is 10 sec at 10 cm distance, the exact amount being judged by that necessary to make the jelly easily penetrable by micropipette. We use both UV lights, the Hanovia source after the Mineralite.

An alternative procedure is to 'dejelly' the eggs by the standard cysteine hydrochloride method (2% cysteine hydrochloride in 0.1 × MBS at pH 8.0 for 6 min). This makes penetration by the micropipette very easy, if the eggs are injected and maintained in a 2% Ficoll in 1 × MBS[a] solution. However, the removal of jelly makes it harder to orientate the eggs correctly under an UV light. It should be remembered that water strongly absorbs UV light and enucleation will not work well if the eggs are overlaid by more than about 300 μm of water or jelly. As soon as the recipient eggs have been exposed to UV light, they may be injected with a nucleus.

[a] 1 × MBS (modified Barth's saline; ref. 7): 88 mM NaCl, 1 mM KCl, 0.41 mM CaCl$_2$, 0.33 mM Ca(NO$_3$)$_2$, 0.82 mM MgSO$_4$, 2.4 mM NaHCO$_3$, 10 mM HEPES pH 7.4, 0.001% (w/v) benzyl penicillin, 0.001% (w/v) streptomycin sulphate.

3.3 Preparation of donor cells

The appropriate part of an embryo may be dissected in 1 × MBS, or cultured cells may be trypsinized as for subculturing. Cells of an embryonic tissue readily dissociate if incubated in calcium-free 1 × MBS solution. If necessary, this can be made 0.1 mM EDTA at pH 8 to accelerate dissociation.

As soon as cells are dissociated, they should be placed in calcium-free MBS containing 0.1% bovine serum albumin. The dissociated cells are best maintained on an agar-covered slide in this medium.

3.4 Donor cell markers

The two most commonly used forms of genetic markers are polyploidy and an anucleolate mutant. Triploid or tetraploid embryos have a maximum of 3 or 4 nucleoli (representing nucleolar organizers) whereas haploid or diploid cells have a maximum of 1 or 2 nucleoli per nucleus). It is therefore possible to take a few cells from a nuclear transplant embryo to determine the number of nucleoli per nucleus and hence to confirm that the embryos resulting from the transplantation of a triploid or tetraploid nucleus are truly triploid or tetraploid. The same point can be checked by counting chromosomes. Alternatively, the *Xenopus* 1-nucleolate mutant (9) can be used, in which each diploid nucleus has a maximum of 1 nucleolus, due to the deletion of the ribosomal genes (and hence nucleolus) on 1 set of chromosomes. Thus 1-nucleolated diploid nuclei when transplanted should yield diploid nuclear transplant embyros having only 1 nucleolus in all their nuclei.

If nuclear transplant embryo survival is expected to be sufficiently good, it is also possible to use nuclei of albino embryos (10). However the pigmentation difference between an albino and a wild-type embryo does not show until around the heartbeat stage.

3.5 Nuclear transfer

To transfer a nucleus, a single isolated donor cell is sucked into the micropipette in such a way that it has its cell wall broken but not the cytoplasm removed from around the nucleus. This involves a very small distortion of the shape of the donor cell so that its length in the micropipette is only two to three times greater than its diameter. Under these conditions the cytoplasm does not appear to be dislodged from around the nucleus which is not broken. It is better to err on the side of not breaking the cell wall than in the direction of disturbing cytoplasm too much from around the donor cell nucleus. This part of the technique is critical and requires a very clean and smoothed-edged pipette. Donor cells can often be torn as they enter the pipette and the contents dispersed. Nuclei of cells to which this has happened yield poor results. Since the tip of the pipette has entered the egg, it is usual to include a small air bubble in the pipette shaft which can be followed to monitor the expulsion of the donor nucleus from the end of the pipette. The amount of medium injected with the donor nucleus in the recipient cell is not critical; so long as the nucleus in the recipient cell is deposited somewhere in the animal half of the egg, the mechanism which brings a sperm nucleus to the appropriate part of the cytoplasm seems also to apply to the injected nucleus.

3.6 Culture of nuclear transfer embryos

If the enucleated recipient eggs are to be injected with a nucleus while they are still uncovered with medium, it is essential to ensure that they do not become dried at the surface. The advantage of injecting recipient cells dry rather than under water is that they are held in position more conveniently and can be injected in quick succession. Once this has happened, the slide containing the recipient eggs should be transferred to 1 × MBS; this may contain 2% Ficoll if there is any tendency for the cytoplasm of the injected egg to leak out. When the nuclear transplant eggs have reached the mid-blastula stage, they should be transferred to 0.1 × MBS. This is because full strength MBS will often cause abnormal gastrulation. Embryos are then cultured under standard conditions as long as is required.

4. Amphibia other than *Xenopus*

All amphibian species can be used for nuclear transplantation, but each species has certain technical requirements. These primarily concern methods of enucleation and activation.

In *Rana pipiens*, enucleation is carried out by a laser beam (see Chapter 15 for details of the use of a laser for cell ablation; the procedure is similar) and this also provides the activation stimulus required. In this species, the most recent work (5) uses the injection of several donor nuclei into oocytes which are then matured into eggs and cultured. This procedure works better, in *Rana pipiens*, than does the injection of single nuclei into unfertilized eggs.

An ancillary technique which has been employed in work with *Rana pipiens* (11) is to graft tissues from an advanced, but not completely viable, nuclear transplant larva on to normal host embryos. In this way, the differentiation of the cells containing transplanted nuclei can be allowed to proceed further than would be the case if they had been left as part of the nuclear transplant embryo. Although the conclusions from such experiments are affected by the unknown benefits which the transplanted tissues receive from their host embryos, this has been a useful approach in work on the transplantation of Lucke renal carcinoma nuclei.

Another amphibian species used for nuclear transplantation is *Pleurodeles*. The techniques required for nuclear transplantation differ in detail from those described for *Xenopus* and may be found in ref. 12.

5. Mammals

By comparison with amphibia, nuclear transplantation in the mouse has yielded very poor results, and it does not appear to have been possible to obtain normal live born young from the nuclei of any stage beyond the advanced two-cell, even though it is known that mouse cells are totipotent at the four-cell stage. Apparently normal blastocysts have been obtained in the mouse from eight-cell stage nuclei and, in a low percentage, from the nuclei of the inner cell mass. In addition, primordial germ cell nuclei from later stages have also yielded blastocysts. Recent work involving mouse nuclear transfers is described in refs. 13–17. Very limited development has so far been reported with teratocarcinoma cells (18).

Most work with nuclear transplantation in the mouse make use of micromanipulation to remove chromosomes from an unfertilized egg at second meiotic metaphase (in mammals, called an oocyte). It is not easy to see the chromosomes to be removed, and an alternative procedure is to suck out both, clearly visible, pronuclei from a fertilized egg. The former procedure is preferred since it is thought preferable to let the donor nucleus 'programme' the egg from its unfertilized state.

Since a mouse egg, like most mammalian eggs, is only 70 μm in diameter, and a two-cell nucleus can be 20 μm in diameter, it is found best to inject a complete donor cell under the zona, so that it is now in contact with, but not in, the recipient egg. It is then fused to the egg using Sendai virus, electrofusion or other means.

More substantial success has been with other mammals, notably in the sheep (19), pig, and rabbit. Thus, live born young have been obtained from the nuclei of the inner cell mass in the sheep (20), from the four- and eight-cell stage of the pig (21), and cow (22), and also from the 8–16-cell stage of the rabbit (23, 24). In many of these cases adults have been obtained from live born young; this has been achieved from as late as a 120-cell stage in the sheep (25, 26).

The methods used for nuclear transplantation in mammals differ considerably from species to species and from author to author. In some cases, nuclei are injected directly into recipient eggs, whereas in other cases a cell fusion technique is used to cause a cell injected under zona to fuse with the egg. Similarly the time when enucleation is carried out differs according to species and author. The reader is referred to the references to obtain details of the procedures in each case.

6. Insects

A protocol for nuclear transplantation in *Drosophila*, also applicable to some other insects, is described in Chapter 10 (Section 4 and *Protocol 4*). It is best carried out between embryos at the syncytial blastoderm stage, when no cell membranes intervene between the nuclei.

Acknowledgement

J. B. Gurdon gratefully acknowledges support of the Cancer Research Campaign, and is a member of the Department of Zoology, Cambridge.

References

1. Briggs, R. and King, T. J. (1952). *Proc. Natl. Acad. Sci. USA*, **38**, 455.
2. King, T. J. and Briggs, R. (1956). *Cold Spring Harbor Symp. Quant. Biol.*, **21**, 271.
3. Gurdon, J. B. (1986). *J. Cell Sci. (Suppl.)*, **4**, 287.
4. Gurdon, J. B., Laskey, R. A., and Reeves, O. R. (1975). *J. Embryol. Exp. Morphol.*, **34**, 93.
5. Di Berardino, M. A. and Orr, N. H. (1992). *Differentiation*, **50**, 1.
6. Laskey, R. A. and Gurdon, J. B. (1970). *Nature*, **228**, 1332.
7. Gurdon, J.B. (1977). *Methods Cell Biol.*, **16**, 125.
8. Gurdon, J. B. (1991). *Methods Cell Biol.*, **36**, 299.
9. Elsdale, T. R. Gurdon, J. B., and Fischberg, M. (1960). *J. Embryol. Exp. Morphol.*, **8**, 437.
10. Hoperskaya, O. A. (1975). *J. Embryol. Exp. Morphol.*, **34**, 253.
11. Lust, J. M., Carlson, D. L., Kowles, R., Rollins-Smith, L., Williams, J. W., and McKinnel, R. G. (1991). *Proc. Natl. Acad. Sci. USA*, **88**, 6883.
12. Lesimple, M., David, J. C., Dournon, C., Lefresne, J., Houillon, C., and Signoret, J. (1989). *Dev. Biol.*, **135**, 241.
13. Landa, V. (1989). *Folia Biol. Praha.*, **35**, 353.
14. Tsunoda, Y., Tokunaga, T., Imai, H., and Uchida, T. (1989). *Development*, **107**, 407.
15. Kono, T., Tsunoda, Y., Watanabe, T., and Nakahara, T. (1989). *Gamete Res.*, **24**, 375.
16. Kono, T., Kwon, O. Y., and Nakahara, T. (1991). *J. Reprod. Fertil.*, **93**, 165.
17. Kono, T., Tsunoda, Y., and Nakahara, T. (1991). *J. Exp. Zool.*, **257**, 214.
18. Modlinski, J. A., Gerhauser, D., Lioi, B., Winking, H., and Illmensee, K. (1990). *Development*, **108**, 337.
19. McLaughlin, K.J., Davies, L., and Seamark, R. F. (1990). *Reprod. Fertil. Dev.*, **2**, 619.
20. Smith, L. C. and Wilmut, I. (1989). *Biol. Reprod.*, **40**, 1027.
21. Prather, R. S., Sims, M. M., and First, N. L. (1989). *Biol. Reprod.*, **41**, 414.
22. Willadsen, S. M. (1989). *Genome*, **31**, 956.
23. Stice, S. L. and Robl, J. M. (1988). *Biol. Reprod.*, **39**, 657.
24. Collas, P. and Robl, J. M. (1990). *Biol. Reprod.*, **43**, 877.
25. First, N. L. (1990). *J. Reprod. Fertil. Suppl.*, **41**, 3.
26. Prather, R. S. and First, N. L. (1990). *Int. Rev. Cytol.*, **120**, 169.

15

Laser ablation of cells

JUDITH EISEN, RACHEL M. WARGA, LOIS G. EDGAR,
and WILLIAM B. WOOD

1. Introduction

In principle, almost any cell that can be visualized with Nomarski DIC optics
can be ablated by laser-irradiation. But such optical clarity is a property of the
embryos of only a few animal species, such as the nematode, *Caenorhabditis
elegans*, some other invertebrates (sea urchin, annelids, etc.), and fish
embryos. In this chapter, we give some basic methods for laser ablation of
individual cells in fish and nematode embryos, but the methods are broadly
applicable to embryos of other species where cells can be resolved sufficiently
clearly in the intact embryo by interference optics.

2. Laser ablation of cells in zebrafish embryos

The power output of the laser and the type of optical system used to focus the
beam onto the cell will determine whether, in practice, a particular cell can be
ablated. The following protocol has been used to ablate single identified
neurons in the spinal cord, individual floor plate, muscle, notochord, and
neural crest cells, and entire ganglia of zebrafish embryos.

First, mount embryos of the desired stage, either on microslides in 1.2%
agar in embryo medium (see Chapter 4) or between bridged coverslips in
embryo medium in such a way that the cell of interest is easily visualized with
Nomarski DIC optics (see Chapter 9, *Protocol 5*). A high numerical aperture
lens, such as the Leitz 50× fluorescence water-immersion objective, is
necessary for focusing the laser beam within a narrow field. This objective
may be currently unavailable since it is no longer made, however a Zeiss 40×
water-immersion objective should work, although the numerical aperture is
lower than that of the Leitz objective. Although there are other ways to do it,
we focus the laser beam into a Zeiss standard microscope through the port of
the vertical illuminator. This set-up is described in ref. 1. More details on the
design of such systems can be found in ref. 2.

We use a pulsed laser for ablations. Currently at least two types are available: dye-pumped pulse lasers, such as those made by Phase-R and Candela, and pulsed nitrogen lasers with dye modules, such as those made by Laser Sciences. The advantage of the first type is that it contains a helium–neon spotting laser for relatively easy alignment. The advantage of the second type is that it is cheaper, smaller, and quieter. In both cases, the dye is Coumarin 450, (made by Exciton Corp.) which is dissolved in high grade methanol (1 g/4 litres) for the dye-pumped laser and 17.5 mg/20 ml for the nitrogen laser. Depending on the cell, the condition of the dye, and the type of laser, 1–100 pulses are needed to ablate a cell.

A cell can be ablated by focusing the laser microbeam onto the cell nucleus; however, it can also be focused onto the membrane, creating a large hole. Cells may die instantly after irradiation, or they may take up to about 15 min. It is important to watch the cell die, or to have an independent measure, such as the loss of cell-specific antibody labelling. As cells die, their nuclei typically become granular in appearance, and they often swell. Once the cell of interest has been ablated, the embryo should be removed from the agar or the bridged coverslip and placed in a dish of embryo medium to continue development.

3. Laser ablation of cells in nematode embryos

In the nematode, a focussed laser beam may be used to ablate individual cells (3), to bring about the fusion of cell membranes (3), and even to permeabilize individual embryos for fixation at precise stages (4). As for fish embryos (Section 2), the best laser system is currently a pulsed dye laser aligned by deflection with a semi-silvered mirror through the 100× objective of a compound microscope. A coumarin dye emitting at 450 nm, and a peak laser power of approximately 100 kW, will produce a suitable range of cell damage in a focus of 1 μm spot diameter. The laser is aligned to a cross-hair in the eyepiece by making hits on the coverslip, or by focussing an additional gas laser beam following the same light path (1). For cell ablations, nuclei are targeted and several subthreshold shots are delivered in the target area until the cells become refractile. A cell should be observed for some time to be sure it does not recover. For cell fusions the desired intercellular membranes are punctured, using a dye emitting at 386 nm (3).

References

1. Sulston, J. E. and White, J. G. (1980) *Dev. Biol.*, **78**, 577.
2. Berns, M. W. (1972) *Nature*, **240**, 483.
3. Schierenberg, E. (1984). *Dev. Biol.*, **101**, 240.
4. Priess, J. and Hirsch, D. (1985). *Dev. Biol.*, **107**, 337.

III
Cellular
techniques

16

Studying cell movements *in vivo* by time-lapse video microscopy

PETER V. THOROGOOD and DEE AMANZE

1. Introduction

Cell movement plays an important role during embryonic development (1). Much of what is presently known about cell movement is based on the behaviour of cultured cells. Despite refinements of culture conditions, however, extrapolation from events *in vitro* to behaviour in the embryo is difficult. Direct observation of living cells *in vivo*, when this can be done, is clearly preferable. Naturally transparent tissues such as the cornea (2), nematode embryos (3) (Chapter 2) and teleost fish embryos (4) (Chapters 4 and 9) (*Figure 1*) lend themselves particularly well to *in vivo* analyses of cell movement. Techniques such as Nomarski DIC (Chapter 8), confocal scanning microscopy, and the labelling of living cells with fluorescent dyes coupled with fluorescence microscopy (Chapter 9), have begun to give us a much better appreciation of how cells normally behave *in vivo*.

Most embryonic cells move slowly, often at only 25–50 µm/h. *Time-lapse* is therefore required: cell behaviour is recorded in 'real time' and subsequently accelerated for viewing. This chapter gives guidelines on different strategies, ranging from cheap improvised systems with manual analysis, to more sophisticated electronic systems with computer-based analysis.

2. Choosing a time-lapse system

2.1 Video versus cine

The time-lapse medium chosen will depend upon the investment of time and the budget available. Two media are generally appropriate, each with advantages and disadvantages:

(a) Time-lapse cine filming (16 mm) provides a final image of high resolution but is expensive, not instant (film has to be developed), there is no opportunity for live display whilst filming and the information cannot be transmitted electronically, limiting subsequent analysis. On the other hand, publication quality photographs can be produced from individual 16 mm frames.

(b) Time-lapse video recording lacks such a high quality image, yet allows instant playback, electronic transmission, and video tape is re-usable. It also allows the time and date to be placed onto each frame, which is invaluable for later analysis. Whenever quantitative data have to be generated and/or an image processed, time-lapse video recording is the preferred medium.

For both cine and video cameras, the microscope needs to have a 'C-mount' adapter on the phototube. The microscope provides the optics; camera lenses are not used. A well supported video camera can simply be juxtaposed to the phototube and an acceptable image recorded even without an adapter, provided that extraneous light is excluded with black paper.

2.2 Cameras, video formats, and recording media

For video microscopy, a variety of cameras of different types is available. They vary according to whether they operate in analogue or digital fashion and in their sensitivity to light. Analogue cameras consist of a cathode ray tube (CRT) and amplifiers, digital cameras (called charge couple devices, CCD) have arrays of light-sensitive silicon elements. The former are typically more sensitive but have a worse signal-to-noise ratio. However, it is possible to increase the sensitivity and the stability of both. Analogue tube cameras vary in price and sensitivity from the cheap and nasty Vidicon (mainly for home use and surveillance), the more stable and sensitive Newvicon, to the very expensive and sensitive silicon intensifier target (SIT) and intensified-SIT (ISIT) cameras, which are *only* suitable for very low light applications. Digital cameras also vary; standard and high-resolution CCDs are available, and more recently integrating and/or Peltier-cooled CCDs have become available for extremely low light applications. Integrating CCDs compose an image from photons hitting the array over a long period, and Peltier cooling reduces noise even further. For biological applications, the most commonly used tube cameras are Newvicons, Plumbicons, and SITs. High-resolution and Peltier-cooled, integrating CCDs are also used. Fluorescence requires the more sensitive cameras but, since bright light affects many embryos, these cameras are also useful for imaging under low level transmitted light conditions. It is also worth mentioning that an alternative way of increasing the sensitivity of any camera, be it tube or digital, is to interpose an electronic front-end intensifier or multi-channel plate between the microscope and the camera. This is widely used for biological applications despite some loss of resolution.

In addition, North American ('NTSC', 525 lines) video/TV systems differ from the European ('PAL', 625 lines) counterparts in image resolution.

Either is suitable for biological applications, but equipment cannot usually be mixed between the two systems unless there is a specific switch.

There are also different types of recording media for video use. In cassette format, the one in most common use is VHS, recently improved to the higher resolution, super-VHS format (which requires special equipment). Professional video engineers often use U-matic cassette tapes, which also require special equipment, are wider and move faster across the recording heads, thus increasing image quality. They also allow more accurate positioning of the tape for selecting particular frames. For biological applications, there is an additional requirement: the video cassette recorder (VCR) should be able to record at a slower speed than used for playback. This requires a specialized (time-lapse) VCR[a].

The recent introduction of video discs has represented a significant advance over video tapes because of they store images in digital (therefore more accurate and highly reproducible) form. Furthermore, discs do not wear despite repeated viewing. However, they are relatively expensive and the discs cannot be reused.

[a] Note that tapes recorded on most (not all; see below) time-lapse video recorders cannot be played back on a domestic VCR or *vice versa*, because of different methods of internal synch pulse generation.

2.3 Standard laboratory set-ups

The standard laboratory set-up[a] requires a Nomarski DIC microscope, video camera (the best choice currently being a high-resolution CCD), time-lapse video recorder and monitor for display, with a computer-based image analysis system (sometimes a time-base corrector is also needed for synchronization between various modules). However, in some cases, particularly with fast-moving cells (e.g. ciliate and amoeboid protozoa, leukocyte chemotaxis, and the 'swimming' of spermatozoa), a domestic video recorder will be perfectly adequate. Of course, a good microscope and appropriate video camera will still be needed, as will much time and patience for manual analysis! Useful data can sometimes be extracted from a sequential series of still 35 mm photographs, taken at suitable intervals. This requires a stopwatch and stamina, although some photomicroscopes will have an intervalometer capable of controlling 16 mm or 35 mm camera backs and activate a light shutter. With ingenuity, simple set-ups such as this can be used effectively.

[a] *Figure 1* was produced with the following equipment: JVC high resolution (330 lines) CCD camera model TK-87OU, JVC VHS time-lapse VCR model BR-9000 UEK and time-base corrector (Cel Electronic Ltd, model P147-30), through a Zeiss Axiophot microscope with Nomarski DIC optics. Hamamatsu manufacture a wide range of low- and intermediate-light level cameras and accessories, as well as multi-channel plates, which are all suitable for biological photomicrography. Time-lapse video recorders in VHS format that are compatible with domestic VCRs are manufactured by Mitsubishi and by Hitachi.

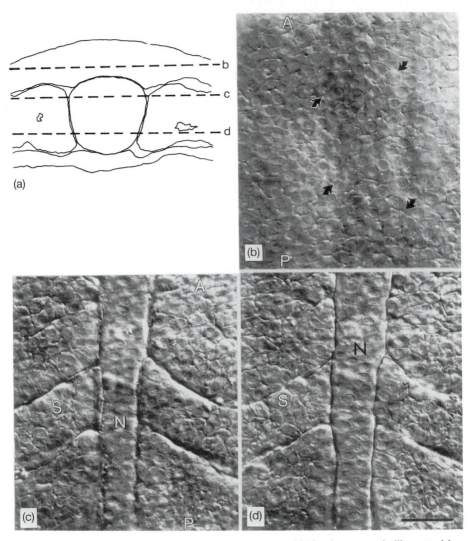

Figure 1. The non-invasive, optical sectioning ability of DIC microscopy is illustrated by various depths of focus in a cultured two-somite rosy barb embryo. (A) Diagram of a transverse section through the embryo at the level at which the somites are forming. The notochord is flanked by paraxial mesoderm; these tissues are overlain by the thickened epithelium of the neural keel (presumptive neural tube). The dashed lines indicate the optical sectioning levels in the other panels. (B) Optical section focused within the neural keel. The arrowheads indicate the faint shadow of the subjacent notochord. (C) Optical section focused dorsally within the notochord and somites; the first two somitic clefts have formed. (D) Optical section focused at the level of the ventral notochord and sclerotomal portion of the somites. A, anterior; P, posterior; N, notochord; S, somite (from Wood and Thorogood, in preparation).

2.4 Video-enhanced microscopy

This is a recent application combining computing and video microscopy which greatly enhances the true resolution of images by electronic means (this is not the same as contrast enhancement, which is a computer-generated stretching of the grey levels of the image). This technique is beyond the scope of this chapter but has been reviewed in ref. 5.

3. Recording/filming

Five factors need to be considered at the start of a new set of experiments:

(i) Stability
The microscope stage must be stable. It is also worth adjusting the tension on the focusing mechanism to keep it from drifting out of focus over long periods and using a locking device on the coarse focus if available.

(ii) Temperature
Clearly, the embryo will need to be maintained at a temperature compatible with a normal rate of development. Heating can be supplied through a *heated stage*, which incorporates a thermostatically controlled heating coil but can generate temperature gradients. Alternatively, a *heated chamber* can be fitted around the stage or microscope: air at a controlled temperature is blown into the chamber. It is possible to construct such a chamber using a simple electronic circuit with a thermistor probe, controlling the current through the heating element of a domestic hair dryer. It is important to position the thermistor probe as near as possible to the specimen rather than near the fan. A third solution is to house all of the equipment in a *hot room*.

(iii) Landmarks
Fixed reference points are crucial for assessing relative movement. With cultured cells there is usually some fortuitously located static feature (e.g. debris on the substrate). However, in embryos the whole field may be changing and contain no fixed points of reference. Unfortunately, there is no simple or general solution to this problem.

(iv) Filming/culture chambers
A standard plastic flask may suffice for some purposes, but tissue culture plastic is optically inferior to glass. The best chambers contain a glass coverslip on each side of the specimen, held apart by a gasket, and may incorporate a device for perfusion. An example is shown in *Figure 2*.

(v) Calibration
Calibration of the field size is important if quantitative data are required. A convenient way for doing this is to include some shots of a micrometer slide at the start of each sequence.

During filming, it is important to check the focus, field of view and temperature at regular intervals.

perfusion

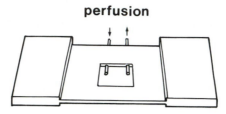

Figure 2. The main component of the embryo culture chamber used by the authors is a gasket machined out of Perspex (shown in figure). To form the chamber, coverslips with their edges lightly coated with vacuum grease are positioned on the upper and lower surfaces of the gasket over the central hole. A perfusion facility can be incorporated.

4. Analysis

Analysis of the behaviour of single cells may be necessary to assess the heterogeneity of embryonic cell populations. Many aspects of cell behaviour can be analysed and statistical means of doing this are reviewed in ref. 6. Since cell behaviour can be somewhat stochastic, it is important to continue the analysis for long enough to collect representative data about the population. This is illustrated by considering an example of directionality of movement: *Figure 3* shows that the use of the shortest time window can give a misleading impression about the trajectory of the cell.

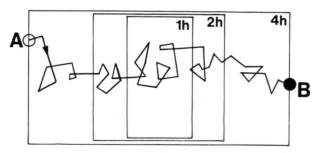

Figure 3. Traces of the path followed by a migrating avian neural crest cell in culture, moving from A to B, made by tracing its movement on a cellulose acetate sheet placed over the monitor screen. Within a 1 h window of recording, the trajectory appears random; the overall directionality only becomes apparent with the 4 h window.

For manual analysis, transparent acetate sheets can be taped over a video monitor or a cine film projected onto sheets of paper. In both case, cells or their trajectories can be traced. Computer-based image analysers simplify some aspects of the analysis, particularly the measurement of lengths of trajectories, surface areas, and optical density, but they are usually less good at object discrimination or pattern recognition. Even with an image analyser, a certain amount of manual analysis is necessary.

5. An example: observing cell movement in fish embryos

As described in Chapter 4, embryos of the zebrafish, *Brachydanio rerio*, are easy to culture and are optically transparent (properties shared with some other teleosts, e.g. the rosy barb, *Barbus conchonius*). These properties make zebrafish an ideal system for observing cell movements *in vivo* in a vertebrate embryo. Here we describe a method for setting up fish embryos for Nomarski DIC observation and, if required, recording/filming (*Protocol 1*). The basic steps are enzymatic dechorionation of the eggs/embryos and assembly of a culture chamber in which the embryos are held stationary, in a semi-solid medium. The enzymatic dechorionation step may be replaced by manual dechorionation using watchmakers' forceps, particularly for older embryos (as described in Chapter 4).

Protocol 1. Observing cell movement *in vivo* in fish embryos

1. Using a wide-mouthed pipette, transfer the fish eggs in a minimum volume of liquid to a solid watch glass containing 0.5% pronase (Sigma) in BSS[a].

2. Carefully monitor digestion of the chorion under a dissecting microscope, until just complete[b]. Zebrafish embryos of less than 24 h of development will take 10–20 min; rosy barb embryos need only 2–3 min.

3. When the outer chorion layer has been digested (leaving an indistinct reticulate inner layer), remove the pronase solution and replace with fresh BSS; gently swirl the watch glass to wash the embryos. Repeat with fresh BSS several times.

4. Assemble a Perspex (Plexiglass) culture chamber (*Figure 2*), with a single cleaned coverslip held in place on one side with the minimum of vacuum grease. Place the chamber onto the stage of a dissecting microscope, with the open side of the chamber uppermost.

5. Add 2.5 ml BSS to 12.5 mg low-gelling temperature agarose (Sigma, type VII), warm gently over a bunsen to dissolve the agarose (but do not boil). Cool to about 25 °C.

6. Place the dechorionated embryos into the agarose solution and transfer to the prepared culture chamber. Whilst the solution is still liquid, quickly and carefully orient the embryos using a nylon thread or hair loop.

Protocol 1. *Continued*

7. Leave for about 15 min, then top up the chamber with BSS; do not overfill to the extent that the meniscus is prominent.

8. Place a second coverslip (with greased edges) on top of the chamber to seal it, and check the preparation with the dissecting microscope.

9. Place onto the stage of the compound microscope for observation with Nomarski DIC optics. For zebrafish embryos, maintain at 27–28 °C; the embryos will continue developing normally for up to 20 h of observation. For prolonged periods of recording/filming, place a glass heat filter in the path of the light source, to prevent the embryo from overheating.

10. After observation and recording, liberate the embryos and monitor subsequent development, to confirm that normal development was not perturbed.

---

^{a} BSS. 137 mM NaCl, 5.4 mM KCl, 1.3 mM $CaCl_2$, 1.0 mM $MgSO_4$, 0.44 mM KH_2PO_4, 0.25 mM Na_2HPO_4, 4.2 mM $NaHCO_3$ (made freshly).
^{b} Pronase activity can vary between batches and suppliers. Do not leave embryos in pronase longer than necessary; the earliest stages are particularly sensitive to 'over-dechorionation'.

References

1. Trinkaus, J. P. (1984). *Cells into organs; the forces that shape the embyro*. Second edition. Prentice-Hall, New Jersey.
2. Bard, J. B. L. and Higginson, K. (1977). *J. Cell Biol.*, **74**, 816.
3. Sulston, J. E., Schierenberg, E., White. J. G., and Thomson, J. N. (1983). *Dev. Biol.*, **100**, 64.
4. Thorogood, P. and Wood, A. (1987). *J. Cell Sci. Suppl.*, **8**, 395.
5. Shotton, D. M. (1988). *J. Cell Sci.*, **89**, 129.
6. Lackie, J. M. (1986). *Cell movement and behaviour*. Allen and Unwin, London.

17

Induction assays in *Xenopus*

ARIEL RUIZ i ALTABA

1. Introduction

The study of inductive interactions requires the identification of tissues that produce the inductive signal and tissues that are competent to respond (1). The identification of the signal as well as the study of the molecular mechanisms underlying induction requires the development of an assay system with which to test different preparations and factors. Amphibian embryos have been used to identify and to define a series of embryonic inductions. Two of these, the induction of mesoderm and the induction of neural tissue, are crucial in the formation of the vertebrate body plan and were first described in amphibian embryos (2–4). In the first case, vegetal cells signal overlying animal cap cells at the equator of the blastula-stage embryo to change fate from ectodermal to mesodermal. In the second case, mesodermal cells signal to adjacent animal cap cells to change fate from epidermal to neural during gastrulation.

These inductive interactions can be recapitulated *in vitro* by apposing reactive tissues in recombinates. For example, vegetal cells can instruct animal cap cells to differentiate as muscle or notochord and mesodermal cells can in turn instruct animal cap cells to differentiate as neurons. In the embryo, not all competent cells are induced. Indeed, mesodermal development occurs only at the equator, where the two cell types are apposed. The distribution of induced cell types results in part from the local action of inductive signals and from the timing of competence of responding cells. Ectodermal cells of midgastrula embryos have already lost the competence to respond to a mesoderm-inducing signal while still retaining the ability to respond to a neural-inducing signal. Here a protocol is described for an induction assay in which animal cap cells of late blastula-stage embryos are induced to form mesoderm by vegetal cells as well as by purified peptide growth factors. This assay, developed by P. D. Nieuwkoop (5) was used to identify the first molecule able to induce mesodermal development (6).

2. Animal cap assays

Protocol 1. Animal cap assay: preparation of animal cap explants

1. 'De-jelly' embryos at early cleavage stages and place in agarose dishes containing $0.5 \times MMR^a$ with antibiotics.

2. Remove the vitelline membrane enclosing the embryo with a pair of fine forceps. It is recommended to pinch the vitelline membrane in the vegetal region for animal cap explants or in the animal region for vegetal explants.

3. Using a hair needle and a pair of forceps, dissect the animal-most region of the darkly pigmented animal cap, staying some 30° from the animal pole (*Figure 1*). The animal cap explant can be cut either by pressing laterally on the animal region of an embryo lying on its side with a hair needle and then trimming the explant, or by cutting it from the top with incisions parallel to the animal–vegetal axis. Try not to break open too many cells and do not take caps that are too big or you run the risk of taking some marginal zone (mesodermal) cells.

4. Rinse the animal cap explants gently by pipetting them up and down with a wide-bored pipette. Make sure explants do not reach the surface at any point or they will burst due to the surface tension of the liquid.

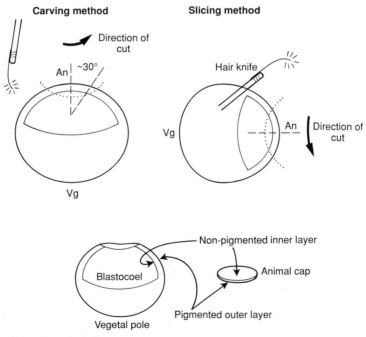

Figure 1. Animal cap isolation.

5. Transfer the animal cap explants to a clean agarose dish with $0.5 \times$ MMR containing antibodies.

6. Keep animal cap explants with their inner side facing upwards and separate them from each other as they tend to fuse.

7. To treat animal caps with mesoderm-inducing peptide growth factors, follow *Protocol 2*. To make animal-vegetal recombinates or sandwiches, follow *Protocol 3*.

   [a] For composition of MMR medium, see Chapter 5, *Protocol 3*.

Protocol 2. Treating animal cap explants with mesoderm-inducing factors

1. Treat animal caps with mesoderm-inducing peptide growth factors such as basic fibroblast growth factor and activin A at concentrations of 0.1–200 ng/ml. These treatments can be done in the wells of six- to 96-well tissue culture plates coated with agarose.

2. The animal cap explants will roll up with the outer pigmented epidermal layer outside. Only the inner deep ectodermal layer is resposive to mesoderm induction. After the explants roll up they can be transferred to a clean dish with fresh $0.5 \times$ MMR plus antibiotics.

3. Culture a few animal cap explants alone. This is very important as a negative control for induction and will reveal how cleanly the dissection was performed.

4. Culture the explants for 3–6 h in $0.5 \times$ MMR plus antibiotics for elongation of the explant and for 1–3 days for overt tissue differentiation.

Protocol 3. Making animal–vegetal recombinates and sandwiches

1. To make animal–vegetal recombinates, remove the vitelline membrane of similarly staged embryos as those used to isolate the animal caps (*Protocol 1*).

2. Carefully dissect a portion of the vegetal pole, avoiding the equatorial region. Be careful with the large vegetal blastomeres, as they are much more fragile than the smaller animal cap cells. Rinse the vegetal explant and transfer to a clean dish.

3. Take one animal and one vegetal explant and juxtapose them so that their original inner surfaces are touching as much as possible. Hold them in place for a few seconds to allow them to heal. It is convenient to place the animal cap explant inner side up on the agarose and place the vegetal

149

Protocol 3. *Continued*

explant on top of it. Two animal caps can also be used to sandwich the vegetal cells. This is the animal/vegetal recombinate.

4. Culture a few animal cap explants alone. This is very important as a negative control for induction and will reveal how clean the dissection was performed.

5. Culture the explants for 3–6 h in 0.5 × MMR plus antibiotics for elongation of the explant and for 1–3 days for overt tissue differentiation.

The age of the animal cap cells is important. As mentioned in the Introduction, animal cap cells lose their competence to respond to mesodermal induction by the mid-gastrula stage (stage 11–11.5). Late blastula stage embryos are routinely used for these assays. To test for potential neural-inducing factors, midgastrula embryos are used since their ectodermal cells can respond to neural but not mesodermal induction. This is important because not all cells in the animal cap respond to induction and secondary interactions, such as the induction of some neural tissue, usually occur in cultured caps in which mesodermal differentiation has been induced.

Notwithstanding the difference between deep and superficial cells in the animal cap, these have been considered homogeneous for a long time. Indeed, many experiments have been performed with dissociated single animal cap cells (7). However, the results of these induction assays should be interpreted with caution since some degree of heterogeneity has been observed in different regions of the animal cap (8, 9). It is recommended always to dissect small caps (comprising tissue within 30° of the animal pole), as these provide a stringent test for induction. Larger caps may reveal a predisposition of cells adjacent to the mesoderm at the equator.

Different concentrations of mesoderm-inducing peptide growth factors will elicit the differentiation of various types of mesodermal cells (10, 11). These include members of the transforming growth factor family (12) such as activin A (13–15) and of the fibroblast growth factor family (16, 17). High concentrations of activin will induce dorsal mesodermal differentiation and tissues such as notochord and somites will be present in the explants. Neural tissue will also appear but this is likely to result from a secondary induction. Low concentrations of activin will induce ventral type mesoderm and blood cells and mesenchyme will differentiate in the induced explants. Inductions with fibroblast growth factor (acidic or basic) will not result in the differentiation of notochord cells. However, high concentrations of FGF will induce muscle while low concentration will induce mainly mesenchyme.

Mesodermal development can be assayed by looking for the elongation of the explant. The animal cells induced to become mesoderm undergo movements that mimic those of mesodermal cells during normal gastrulation. Because early embryonic cells have an internal clock that is not disturbed by these manipulations, the onset of movements of induced cells in an explant is an indication of the type of mesoderm they will form (18). Dorsal mesodermal cells begin their gastrulation before their ventral counterparts do (19, 20).

Identification of differentiated mesodermal cells can be done by looking for the appearance of:

(a) typical morphological and histological features such as striated muscle fibres and vacuolated notochord cells

(b) induced secondary cell types such as neural tissue, pigmented melanocytes and fin tissue

(c) by testing for the expression of mesodermal cell type-specific molecules such as those recognized by the monoclonal antibody 12/101 (21) or α-actin (muscle-specific) mRNA (22), both of which identify the presence of somitic mesoderm (muscle).

The type of induced mesoderm has also been tested in blastocoele implantations (Chapter 11). In this case, the type of neural tissue induced is taken as a reflection of the type of mesoderm in the implant: anterior mesoderm can induce anterior neural structures including eyes and posterior mesoderm can induce spinal cord-like structures (11, 23).

References

1. Gurdon, J. B. (1987). *Development*, **99**, 285.
2. Spemann, H. (1938). *Embryonic development and induction*. Yale University Press, New Haven.
3. Saxén, L. and Toivonen, S. (1962). *Primary embryonic induction*. Logos Press, London.
4. Nieuwkoop, P. D., Johnen, A. G., and Albers, B. (1985). *The epigenetic nature of early chordate development. Inductive interactions and competence*. Cambridge Univeristy Press, Cambridge.
5. Nieuwkoop, P. D. (1969). *Wilhelm Roux Arch. EntwMech. Org.*, **163**, 298.
6. Smith, J. C. (1987). *Development*, **99**, 3.
7. Green, J. B. and Smith, J. C. (1990). *Nature*, **347**, 391.
8. Ruiz i Altaba, A. and Jessell, T. M. (1991). *Genes Dev.*, **5**, 175.
9. Sokol, S. and Melton, D. A. (1991). *Nature*, **361**, 409.
10. Green, J. B. and Smith, J. C. (1991). *Trends Genet.*, **7**, 245.
11. Ruiz i Altaba, A. and Melton, D. A. (1989). *Nature*, **341**, 33.
12. Rosa, F., Roberts, A. B., Danielpour, D., Dart, L. L., Sporn, M. B., and Dawid, I. B. (1988). *Science*, **239**, 783.
13. Asashima, M., Nakano, H., Shimada, K., Kinoshita, K, Ishii, K., Shibai, H., and Ueno, N. (1990). *Roux's Arch. Dev. Biol.*, **198**, 330.

14. Thomsen, G. Woolf, T., Whitman, M., Sokol, S., Vaughan, J., Vale, W., and Melton, D. A. (1990). *Cell*, **63**, 485.
15. Smith, J. C., Price, B. M., Van Nimmen, K., and Huylenbroeck, D. (1990). *Nature*, **345**, 729.
16. Slack, J. M. W., Darlington, B. G., Heath, J. K., and Godsave, S. F. (1988). *Nature*, **326**, 197.
17. Kimelman, D. and Kirschner, M. (1987). *Cell*, **61**, 859.
18. Symes, K. and Smith, J. C. (1987). *Development*, **101**, 339.
19. Gerhart, J. C. and Keller, R. (1986). *Annu. Rev. Cell Biol.*, **2**, 201.
20. Keller, R. and Danilchik, M. (1988). *Development*, **103**, 193.
21. Kintner, C. R. and Brockes, J. P. (1984). *Nature*, **308**, 67.
22. Dworkin-Rastl, E., Kelly, D. B., and Dworkin, M. B. (1986). *J. Embryol. Exp. Morphol.*, **91**, 153.
23. Mangold, O. (1933). *Naturwissenschaften*, **21**, 761.

18

Culture of avian neural crest cells

ELISABETH DUPIN and NICOLE M. LE DOUARIN

1. Introduction

As the unique source of very diverse cell types in the adult (such as peripheral glial and neuronal cells, mesenchymal cells, melanocytes, and endocrine cells), the neural crest (NC) of the vertebrate embryo has been the subject of considerable research effort aimed at elucidating several basic embryological problems related to cell commitment and differentiation, and to the segregation of cell lineages during morphogenesis. Although NC derivatives and migration pathways of crest cells have been mapped with great precision in birds, much remains to be known about the mechanisms underlying cell determination and phenotypic diversification in NC development. Attempts to understand the molecular basis of these fundamental processes have led many laboratories to undertake *in vitro* experiments where the differentiation of defined populations of NC cells can be analysed in controlled environmental conditions.

Since the pioneering experiments of Dorris (1), who first described the production of pigment cells by NC excised from chick embryos, much work has been devoted to the study of the differentiation of cultured avian NC cells, mainly into melanocytic and neuronal phenotypes. As several research groups have devised different methods for culturing the NC, this article will deal with the description and comparison of these techniques, excluding organotypic cultures and *in vivo* cultures on the chorioallantoic membrane.

What does 'culture of NC cells' mean? Because the NC is only a transient structure, giving rise soon after its formation to several cellular flows migrating according to precise spatiotemporal patterns in the developing embryo, there are several answers to this question. Indeed, culturing NC can be achieved in various ways, depending mostly on the stage at which NC cells are removed from donor embryos and on the excision procedure used. Thus, it is possible to isolate NC cells before or during their migratory phase, either as a pure or as a mixed cell population according to the methods chosen; these methods are based on previous knowledge of the migration routes that crest cells follow *in vivo* at the various levels of the neural axis. In addition, several experimental conditions can be applied to the avian NC, such as explant cultures and primary or secondary cultures of dissociated cells. Sophisticated techniques for clonal culture are also available. Finally, different substrates and culture media can be used, the choice of which may be crucial to obtain growth and differentiation of specific NC-derived cell types.

This chapter will be limited to methods applied to NC cells taken before or during their emigration from the neural primordium, and excludes NC-derived cells that have already reached their final target organs. Most of the commonly used procedures will be described, with emphasis on the advantages and goals of each method, with the aim of providing some guidance to new investigators entering the field of avian NC development.

2. Removal of NC from avian embryos

NC can be removed from chick or quail embryos during early embryonic stages ranging from 1–3 days of incubation (E1–E3) according to the anteroposterior level and the stage (premigratory or migratory) concerned. Both species have been used for culture experiments but recent investigations, particularly dissociated and clonal cultures, are commonly done with quails since, for unknown reasons, crest cells from the quail develop *in vitro* more easily than do chick NC cells. *Table 1* gives precise stages for quail embryos according to several criteria, and indicates when different NC removal procedures can be applied.

2.1 Premigratory NC cells

2.1.1 Total neural primordium

Initially devised and described in detail by Cohen (2), this method still remains the most commonly used to obtain NC cells at trunk levels of the neural axis. It consists of the removal of the whole neural primordium (NC + neural tube) prior to emigration of NC cells, so that NC cells will subsequently migrate away from the isolated neural tube *in vitro* (*Protocol 1*).

Isolation of the neural tube + NC can be performed virtually at all levels of the neural axis, thus providing antero-posteriorly regionalized (i.e. cranial, vagal, cardiac, trunk) NC cells. For this purpose, it is crucial that embryonic stages are matched with the desired levels of the crest, so that neural tubes will be isolated only from regions where crest cells have not yet started to migrate *in vivo* (see *Table 1*).

Protocol 1. Removal of trunk neural tubes

1. Incubate Japanese quail (*Coturnix coturnix japonica*) eggs at 38 °C for approximately 48 h to obtain embryos at stages 12–13 of Zacchei (3) (equivalent to Hamburger and Hamilton, stages **14–15**; ref. 4; see Appendix), with 22–26 pairs of somites.

154

Table 1. Methods for removal of avian NC cells according to the stages and axial levels of donor embryos

| Tissue removed | Stages of donor embryos | | | | | Axial levels |
	Incubation time	Somite number	Zacchei stage	H&H stage		
Cephalic neural folds	~ 28 h	5–7	6–7	8^+–9		Mes-metencephalon
Cranial neural tube + NC	~ 30 h	7	7	9		Mes-rhombencephalon
Cephalic migrating NC	~ 32 h	9–12	8–9	10^-–11^-		Mes-metencephalon
Trunk neural folds	~ 38 h	17–20	10^+–11^+	12^+–13^+		posterior to last somite, anterior to Hensen's node
Trunk neural tube + NC	~ 48 h	22–26	12–13	14–15		last 6–10 somites formed ± segmental plate
Somites including trunk migrating NC	~ 48 h	~ 25	13	15		somites 15–20
Sclerotomes and associated trunk NC	~ 3 days	~ 35	14–15	18–19		thoracic level between fore- and hind-limb buds

NC can be removed during premigratory or migratory phases at different axial levels. Corresponding stages of donor embryos are indicated by incubation time, somite number, and according to staging systems of Zacchei (3) and of Hamburger and Hamilton (4) (H&H; see Appendix 1) for quail and chick embryos, respectively.

Protocol 1. *Continued*

2. Explant the embryos as described in Chapter 6, *Protocol 1*, and place them in PBS for precise staging by observation with a dissecting microscope.

3. Dissect out blocks of trunk (including ectoderm, neural tube, somites, notochord, and endoderm) with sharpened tungsten or steel needles (see Chapter 6) at the level of the last 10 somites formed plus the unsegmented mesoderm anterior to Hensen's node.

4. Immerse the trunks into either a crude 0.1% trypsin or a 25% pancreatin solution in Ca^{2+}- and Mg^{2+}-free phosphate-buffered saline (CMF-PBS) for about 15 min to allow the neural tubes to separate from the surrounding tissues.

5. Stop the digestion by adding 10% serum-containing medium.

6. Complete the isolation of tubes by gentle pipetting and, if necessary, using needles.

2.1.2 Neural fold

Neural crest cells before or at the onset of their migration can be mechanically excised when still in the neural fold. The surgical procedure has to be performed before complete closure of the neural tube and has been done either at the mesencephalic level at the 5–7-somite stage (5), or in the trunk region of 17–20-somite embryos (6) (*Table 1*).

The dorsal tips of the forming mesencephalic vesicle or neural tube are first moved apart and then the crest is surgically removed together with overlying ectoderm. This method is the only one to provide early premigratory crest in the absence of the neural tube. However, the resulting NC cell population is small, comprising approximately 600 and 1000 cells per crest from trunk and mesencephalic levels, respectively. In addition, because there is no clear demarcation between the crest and the neural tube epithelium, the excision procedure is somewhat delicate, and must involve only the most dorsal part of the folds.

2.2 Migratory NC cells

Identification of *in vivo* migration pathways of NC cells by means of the construction of quail-chick chimeras and other labelling methods, such as immunostaining with monoclonal antibody HNK-1, has revealed the precise location of migratory crest cell populations during development. This provided the opportunity to devise procedures for removing crest cells during their migration phase. Accordingly, methods for removing cranial versus trunk migratory NC differ largely because the directions of crest cell flows were shown to be different in developing head and trunk.

2.2.1 Migrating cephalic NC

Cephalic crest cells migrate away from the encephalic vesicle along a dorsolateral pathway. At the mes-metencephalic level of 9–12-somite quail embryos, migrating NC cells are clearly visible as a multilayered sheet underneath the superficial ectoderm, and therefore can be dissected out mechanically (5) (*Table 1*). This provides a homogeneous crest cell population (plus the overlying ectoderm) free from contamination by neural tube or mesenchyme, each fragment so obtained containing approximately 1500 NC cells.

2.2.2 Migratory trunk NC

The migration of cervicothoracic NC cells proceeds dorsoventrally in the developing trunk, where most migrating crest cells thus are localized within mesodermal tissues, before the subsequent individualization of PNS ganglia and adrenal medulla. Therefore, it is not possible to excise migrating trunk NC cells as a pure population. However, they can be removed together with their accompanying mesenchymal cells when they have colonized the rostral part of the somites or later, after they have become included in the sclerotomes.

(a) Somites containing migratory crest cells are isolated from surrounding tissues by proteolytic treatment from E2 embryos exactly as described for the removal of trunk neural tubes (*Protocol 1*), except that trunk regions are excised out from somite levels 15–20 (*Table 1*). These levels correspond, in 25-somite embryos, to advanced migration stages of crest cells into the rostral part of the corresponding somites and between adjacent segments.

(b) Sclerotomes: at E3, the somitic mesoderm has differentiated into two morphologically distinct components, dermomyotome, and sclerotome. The latter is the vertebral anlage and includes migrating NC cells that will condense to form dorsal root ganglia and primary sympathetic chains. For removal of quail sclerotomes (6) (*Table 1*), ventral structures in the thoracic region of E3 embryos, including the dorsal aorta, are first removed mechanically. The remaining trunk blocks are then excised between fore- and hind-limb buds and transferred in 25% pancreatin solution. Such proteolytic treatment leads to separation of embryonic tissues within 20–25 min. After peeling off the ectoderm and dermo-myotomes, sclerotomes can be isolated from the neural tube and notochord using sharpened needles.

3. Culture conditions

3.1 Explant cultures

Neural fold, migrating cephalic crest, somites, and sclerotomes (including migrating trunk NC cells; see ref. 7) can be cultured as explants by transferring them with a micropipette into culture dishes or wells, in the presence of standard media (see Section 4). A small volume of medium is used initially to cover the explants and to prevent them from floating; more medium is then added once the explants have attached firmly to the substrate, which usually occurs within 2–3 h. To culture migrating cephalic crest cells, it is important to check that explants are positioned with the NC, and not the ectoderm, facing the substrate.

3.2 Primary and secondary cultures of dissociated NC cells

3.2.1 Primary cultures of dissociated cells from neural fold and migrating crest

NC or NC-containing tissues (see Section 3.1) can be rinsed in CMF-PBS and subsequently enzymatically treated to dissociate the cells, using trypsin or trypsin-EDTA solution in CMF-PBS. Concentrations used and approximate times of treatment are as follows: 0.025% trypsin ± 0.01% EDTA (pH 7.4), less than 5 min for trunk and cephalic neural folds, and 5–7 min for migrating cephalic crests; 0.25% trypsin ± 0.2% EDTA, about 5 min for somites and sclerotomes, which can be also mechanically dissociated. After stopping the reaction by addition of serum-containing medium, cells are separated by gentle pipetting with a micropipette or a glass pipette pulled over a Bunsen flame, and centrifuged at 300 g for 10 min. Cells in the resulting suspension can be counted and plated as equivalent cultures (containing an identical number of cells from a uniform, pooled population), which provides the opportunity of quantitative evaluations, such as testing the effect of hormones or growth factors (8–10).

3.2.2 Cultures of total neural primordium

The purpose of primary culture of the whole neural primordium is to allow crest cells to migrate away from the explant, after which the tubes are discarded. Depending on the conditions and mainly on the culture substrata used, emigrating NC cells can develop either attached onto the floor of the dish (neural tube outgrowth) or remaining close to the tubes when forming non-adherent clusters.

i. Neural tube outgrowth

This procedure was first described by Cohen and Königsberg (11), and is still the most commonly used to culture NC cells. Isolated neural primordia (see Section 2.1.1 and *Protocol 1*) are put in Petri culture dishes with standard serum-containing medium, where they firmly attach to the plate within 1 h at 37 °C. Emigrating crest cells adhering to the substratum form primary outgrowth within 3–6 h, with approximately 500–1000 cells left behind. In order to avoid contamination by neural tube cells and to obtain a pure NC cell population, it is usually recommended carefully to remove the neural tubes (which still remain as epithelial structures) from the culture dishes after 24–48 h. It is interesting to note that 48 h, compared to 24 h, in the presence of the neural tube provide a larger crest cell population and enhance melanogenesis, but decrease the proportion of cells undergoing adrenergic differentiation. Crest cells from the primary outgrowth can be detached and subcultured (*Protocol 2*). This offers the unique possibility quantitatively to evaluate the growth or differentiation of equivalent cell populations plated under identical or altered culture conditions.

ii. Neural crest clusters

A second method of culturing crest cells from the whole neural primordium takes advantage of the observation that, under particular medium and substrate conditions, small spherical clusters of cells form *in vitro* on explanted neural tubes (12). These NC clusters can be isolated, dissociated, and cultured on tissue culture plastic so that all cells differentiate into melanocytes, or on other substrates able to promote the development of adrenergic cells (13) (see Section 4.2). Optimal conditions for the formation of NC clusters have been described in detail (14). One simple procedure is to explant isolated neural tubes on a non-adhesive substratum, such as a plastic bacteriological Petri dish, and thus cellular clusters will form on the freely floating neural tubes, provided that the medium contains both fetal calf serum (FCS) and chick embryo extract (CEE). After 24–48 h, clusters are isolated from the tubes using sharpened needles and transferred onto a plastic or collagen-coated tissue culture dish (in approximately 100 μl of medium) where the cells can disperse (day 0 of secondary culture).

Protocol 2. Secondary cultures of NC cells

1. Isolate the entire neural primordium as described in *Protocol 1*, and initially plate it in tissue culture Petri dishes in DMEM, 10% FCS, 2% CEE.
2. After 24 h in culture, carefully scrape away the neural tubes with a thin Pasteur pipette, pulled over a Bunsen flame.
3. Rinse the cultures twice with CMF-PBS.
4. Replace the medium with 0.25% trypsin/0.01% EDTA solution in CMF-PBS for 1 min at 37 °C. The cells will detach.
5. Collect the cells and centrifuge in 10 ml medium for 10 min at 300 *g*.
6. Resuspend the pellet in 300 μl medium and count the cells in a haemocytometer (there should be from 5000 to 8000 cells per tube).
7. Adjust the volume of the suspension to 20 000 cells per 50 μl.

Protocol 2. *Continued*

8. Plate each 50 µl as a drop in the centre of 35 mm culture Petri dishes.

9. After 3 h, cells are attached and fed with 2 ml medium.

3.3 Clonal cultures of NC cells

Clonal techniques have been first applied to trunk NC cells by Cohen and Königsberg (11). In this method, further improved by work from M. Sieber-Blum's laboratory, crest cells are obtained from neural tube outgrowths (see Section 3.2.2) and subcultured at clonal density on collagen. A second method has been devised recently, involving the seeding of individual cells, under the microscope, on a feeder-layer of 3T3 mouse fibroblasts. Initially applied to the study of cephalic migrating crest cells (15), this method has been extended recently to the analysis of sclerotomal crest cells, which can be isolated from the surrounding mesenchymal cells by labelling the total suspension with monoclonal antibody HNK-1. Then, only crest-derived HNK-1-positive cells are chosen under epifluorescence illumination for subsequent clonal seeding (16). Both techniques, detailed in *Protocols 3* and *4*, have demonstrated the large heterogeneity of the progenitors contained in the NC, as far as their proliferative and differentiation abilities are concerned.

Protocol 3. Colony assay of NC cells

1. Start by culturing the whole neural primordium of E2 quail embryos as described in *Protocols 1* and *2*.

2. After 18–24 h, remove the tubes and detach the remaining crest cells as indicated in *Protocol 2.*

3. Count the cells in the resulting suspension in a haemocytometer and dilute appropriately for cloning by limiting dilution (17).

 (a) For cloning, adjust the suspension to about 16 cells/ml. One drop from a Pasteur pipette is placed in each well of multiwell plates (previously coated with poly-D-lysine or collagen and air-dried). Repeated inspection under the microscope (just after plating, then 2 and 18 h after attachment) are necessary to establish unequivocally the clonal nature of each culture.

 (b) For simultaneous processing and analysis of relatively large numbers of cells, the colony assay is a more convenient method. The NC single cell suspension is diluted to about 50 cells/ml, and 1 ml aliquots dispensed into collagen-coated 35 mm culture dishes. Such low seeding density makes the chance confluence of two individual colonies unlikely.

4. The culture medium consists of 75% MEM, 15% horse serum, and 10% 11-day CEE, and is supplemented with 50 µg/ml gentamycin sulphate. Human soluble fibronectin (50 µg/ml) can be added so as further to improve differentiation of adrenergic cells (18).

Protocol 4. Clonal cultures of NC cells on 3T3 cells

The clonal procedure (15) has been applied initially to migrating cephalic crest cells.

1. Dissect the mesencephalic crests of a 9- to 13-somite quail embryo bilaterally and dissociate them into single cells as described in Section 3.2.1. The suspension should contain approximately 1000 crest cells/ml.

2. From this suspension, aspirate individual cells with a micropipette or a pulled Pasteur pipette whilst observing under the microscope (e.g. Olympus CK2; 100× magnification). This ensures that only a single cell is seeded per culture.

3. Place each cell on a feeder layer[a] of growth-inhibited mouse 3T3 fibroblasts in eight-well Lab Tek chamber slides (Miles)[b].

[a] For preparation of feeder layers, mouse Swiss 3T3 cells growing in DMEM supplemented with 10% calf serum are treated for 3 h by addition to the medium of 4 µg/ml mitomycin-C (Sigma). Rinse the cells twice with DMEM and trypsinize, for subsequent replating on glass or plastic Lab-Tek chamber slides, at the seeding density of 5×10^4 cells/well. Treated 3T3 cells can be used as feeder-layers for NC clonal cultures as soon as they have attached.

[b] The medium for clonal cultures consists of: Ham's F-12 nutrient mixture/Dulbecco's modified Eagle's medium/BGjb medium (6:3:1, v/v) supplemented with 10% (v/v) fetal calf serum, 25% (v/v) embryo extract (from 11-day chick embryos), adenine (24.3 ng/ml), hydrocortisone (0.4 µg/ml), insulin (5 µg/ml), triiodothyronine (13 ng/ml), transferrin (10 µg/ml), epidermal growth factor (10 ng/ml), isoproterenol (0.25 µg/ml), and choleragen (8.4 ng/ml).

---

4. Culture substrates and media

The *in vitro* culture method is finding increasing use in the study of the differentiation of crest-derived, particularly PNS, phenotypes. The data acquired revealed that many properties associated with differentiation can indeed develop *in vitro*, although their emergence is highly dependent on culture conditions, among which the composition of culture medium and the presence of particular substrate molecules are very important. In addition, certain NC or NC-derived cell populations were shown to have specific requirements for growth and differentiation *in vitro*, according to their development stage and origin.

4.1 Culture media

NC and NC-derived cells can be cultured in various media, containing serum and supplemented with chick embryo extract (CEE), or in defined media without serum. Their compositions are detailed in *Table 2*. A simple medium supplemented with serum (medium 1, *Table 2*) ensures cell attachment and growth of NC cells. Addition of 2% CEE (medium 2, *Table 2*) enhances cellular proliferation. It should be noted that in both media 1 and 2, crest cells (from neural folds, whole neural primordium, and migratory cephalic crest) proliferate actively but acquire few differentiated traits, as judged by the appearance of various phenotypic markers. However, enrichment of culture media with 10% CEE (medium 3, *Table 2*) strongly promotes NC differentiation, mainly along the melanocytic and adrenergic pathways (19). In contrast, it should be noted that the presence of high levels of extract is not required for expression of the adrenergic phenotype by more mature crest cells from somites or sclerotomes. Indeed, cultures of somites and sclerotomes give rise to neurones and catecholaminergic cells even in the absence of serum, that is, during culture in a defined medium (BBM) (8). The advantage of using BBM for growing early crest cells (i.e. neural folds and migrating cephalic crest) is that it triggers the differentiation of a sensory-like population as soon as 24 h after culture (20). In addition, cultures in BBM

Table 2. Composition of culture media

- medium 1: Dulbecco modified Eagle's medium (DMEM), 10% fetal calf serum (FCS)
- medium 2: DMEM, 10% FCS, 2% chick embryo extract (CEE)
- medium 3: DMEM, 10% FCS, 10% CEE
- basic Brazeau medium (BBM) (20)

DMEM/Ham's F12/BGjb medium, Fitton-Jackson modification (3:6:1)	
Bovine serum albumin	2 g/l
Hepes	2.38 g/l
Gentamycin	50 mg/l
Hydrocortisone	100 µg/l
Insulin	1 µg/l
3,3',5-triiodothyronine	0.4 µg/l
Parathyroid hormone	0.2 µg/l
Glucagon	10 ng/l
Transferrin	10 mg/l
Epithelial growth factor (EGF)	0.1 µg/l
Basic fibroblast growth factor (bFGF)	0.2 µg/l

There are modified compositions for media 1, 2, and 3, where FCS is replaced by horse serum (HS), and DMEM by other standard media such as Eagle's minimum essential medium (MEM), or Ham's F12. All sera are used after heat-inactivation. Media are sterilized by 0.45 µm filtration before use, and all cultures incubated at 37 °C in a humified atmosphere with 5% CO_2.

---

provide defined conditions (see *Table 2*) for evaluating the effect of neurotrophic factors on NC cell development (10).

Another defined medium, MCDB 202, to some extent allowing the *in vitro* differentiation of adrenergic cells from trunk NC cells, has also been described (21).

4.2 Substrates

The migratory cephalic NC cells and the migrating trunk crest cells contained in, and cultured with, the sclerotome do not have very strict substratum requirements for adhesion and growth since, in most cases, they can be cultured simply on tissue culture plastic. The trunk crest cells migrating *in vitro* from the explanted neural tube appear more sensitive to the substrate. For example, poly-D-lysine or rat tail collagen-coating of culture dishes improve adhesion and spreading of dissociated crest cells cultured in serum-containing media. In serum-deprived medium (BBM), attachment and growth of NC cells require previous treatment of the culture surfaces by 10% serum-containing medium (for 1 h at 37 °C) prior to cell seeding. Pre-migratory crest cells (from neural folds) do not adhere to plastic simply coated with collagen or serum. To overcome this poor ability to adhere, it is necessary to use a surface on which a preculture of two-day-somitic cells has deposited an extracellular matrix-rich substrate (20). Serum, as well as preculture with somitic cells, must cause the adsorption of fibronectin onto the substrate. This extracellular matrix component has been shown to be particularly suitable for NC cell adhesion and migration both in culture (22) and *in vivo* (23, 24). It has to be noted that the cephalic NC yields fibroblasts that are able to produce fibronectin (25). Similarly, trunk NC cells cocultured with sclerotome can also benefit from the fibronectin-rich substrate produced by somitic mesenchymal cells. This is very likely to account for the capacity of these NC cells to grow on plastic.

Numerous investigations have attempted to identify discrete matrix components involved in NC cell migration and to define their role in crest cell differentiation (see ref. 26). Coating of surface cultures with purified extracellular matrix components, (e.g. glycosaminoglycans, collagens, laminin, and fibronectin) modulates not only the adhesive properties and migratory behaviour of crest cells, but can also influence the expression of differentiated phenotypes. In cultures of NC clusters, matrix molecules affect patterns of cell dispersal which are correlated with the extent of pigmentation. Melanogenesis is reduced in the presence of collagen and fibronectin, and enhanced by laminin (27). In addition to suppressing melanogenesis, fibronectin promotes development of adrenergic cells (13), as do other more complex substrates:

(a) Substrate-attached material from fibroblast and somitic cell cultures decreases pigmentation and stimulates adrenergic differentiation in cultures of NC cells from clusters or neural tube outgrowth (13, 17). Extracellular matrix is obtained in such studies by detaching cells prepared from half-confluency cultures with 0.02% EDTA and subsequent irradiation.

(b) Feeder layers of 3T3 cells have a similar effect. In addition, enhanced adrenergic cell differentiation can be obtained even in the absence of high concentrations of CEE (15).

(c) Reconstituted basement membrane (RBM) components from Engelbreth–Holm–Swarm tumour (Matrigel; Collaborative Research) specifically stimulates the adrenergic development of trunk NC cells (28). The number of catecholamine-positive cells that differentiate in 10% CEE-containing medium is increased about 50-fold in the presence of RBM gel overlay. For applying RBM overlay, secondary cultures of trunk NC cells are allowed to attach for 1 h. The medium is then removed and replaced by addition of the RBM gel which will solidify following 30 min incubation at 37 °C. Cultures are then fed with growth medium.

5. Concluding remarks

This chapter gives an account of the multiple methods for performing *in vitro* cultures of avian NC. Analysis of cultures have concerned mainly the characterization of the migratory properties of crest cells, and the study of the onset of phenotypic traits during NC cell differentiation *in vitro*. Most of these properties were shown largely to differ according first to the developmental stage and axial level of the excised NC cell populations, and second to the choice of the experimental conditions used. *In vitro* studies have confirmed results of *in vivo* experiments, in showing the large developmental plasticity of the NC. In appropriate conditions, virtually all cell types generated by the crest can be obtained *in vitro* from all axial levels of the neural primordium, except the mesectodermal derivatives, which are limited, as in the embryo, to the head NC. *In vitro* clonal cultures of NC cells have been instrumental in providing important data on cell lineage segregation during NC differentiation. This technique, together with complementary *in vivo* approaches, has demonstrated unambiguously the presence of highly pluripotent as well as more restricted cells in the NC. In addition, *in vitro* clonal analysis of NC populations taken from progressively older embryos may help elucidate when the restrictions of developmental potentialities occur during development of NC derivatives. This raises the crucial question of how the microenvironment influences the fate of NC cells. Since the same crest cell population can give rise *in vitro* to different cell types depending on the choice of medium and substrate, avian NC cultures can be used as a model system to analyse the mechanisms through which phenotypic diversity is generated in NC-derived cells. Characterization of extrinsic factors as well as the corresponding surface receptors in responsive NC cells will be necessary to help resolve these issues.

Acknowledgements

We thank M. Scaglia for secretarial assistance. This work was supported by Centre National de la Recherche Scientifique, Institut National de la Santé et de la Recherche Médicale, Fondation pour la Recherche Médicale Française, Ligue Nationale contre le Cancer, Association pour la Recherche contre Le Cancer, Association pour la Recherche contre la Sclérose en Plaques.

References

1. Dorris, F. (1936). *Proc. Soc. Exp. Biol. Med.*, **34**, 448.
2. Cohen, A. M. (1972). *J. Exp. Zool.*, **179**, 167.
3. Zacchei, A. M. (1961). *Arch. Ital. Anat. Embriol.*, **66**, 36.
4. Hamburger, V. and Hamilton, H. L. (1951). *J. Morphol.*, **88**, 49.
5. Smith, J., Fauquet, M., Ziller, C., and Le Douarin, N. M. (1979). *Nature*, **282**, 853.
6. Cheney, C. M. and Lash, J. W. (1981). *Dev. Biol.*, **81**, 288.
7. Fauquet, M., Smith, J., Ziller, C., and Le Douarin, N. M. (1981). *J. Neurosci.*, **1**, 478.
8. Nataf, V. and Monier, S. (1992). *Dev. Brain Res.*, **69**, 59.
9. Kalcheim, C. and Gendreau, M. (1988). *Dev. Brain Res.*, **41**, 79.
10. Kalcheim, C., Carmeli, C., and Rosenthal, A (1992). *Proc. Natl. Acad. Sci. USA.*, **89**, 1661.
11. Cohen, A. M. and Königsberg, I. R. (1975). *Dev. Biol.*, **46**, 262.
12. Loring, J. F., Glimelius, B., Erickson, C. A., and Weston, J. A. (1981). *Dev. Biol.*, **82**, 86.
13. Loring, J., Glimelius, B., and Weston, J. A. (1982). *Dev. Biol.*, **90**, 165.
14. Glimelius, B. and Weston, J. A. (1981). *Cell Differ.*, **10**, 57.
15. Baroffio, A., Dupin, E., and Le Douarin, N. M. (1988). *Proc. Natl. Acad. Sci. USA.*, **85**, 5325.
16. Sextier-Sainte-Claire Deville, F., Ziller, C., and Le Douarin, N. M. (1992). *Dev. Brain Res.*, **66**, 1.
17. Sieber-Blum, M. and Cohen, A. M. (1980). *Dev. Biol.*, **80**, 96.
18. Sieber-Blum, M., Sieber, F. and Yamada, K. M. (1981). *Exp. Cell Res.*, **133**, 285.
19. Howard, M. J. and Bronner-Fraser, M. (1985). *J. Neurosci.*, **5**, 3302.
20. Ziller, C., Dupin, E., Brazeau, P., Paulin, D., and Le Douarin, N. M. (1983). *Cell*, **32**, 627.
21. Sieber-Blum, M., and Chokski, H. R. (1985). *Exp. Cell Res.*, **158**, 267.
22. Rovasio, R. A., Delouvée, A., Yamada, K. M., Timpl, R., and Thiéry, J. P. (1983). *J. Cell Biol.*, **96**, 462.
23. Boucaut, J. C. Darribère, T., Poole, T. J., Aoyama, H., Yamada, K. M., and Thiéry, J. P. (1984). *J. Cell Biol.*, **99**, 1822.

24. Bronner-Fraser, M. (1985). *J. Cell Biol.*, **101**, 610.
25. Newgreen, D. F. and Thiéry, J. P. (1980). *Cell Tissue Res.*, **211**, 269.
26. Perris, R. and Bronner-Fraser, M. (1989). *Comments Dev. Neurobiol.*, **1**, 61.
27. Rogers, S. L., Bernard, L. and Weston, J.A. (1990). *Dev. Biol.*, **141**, 173.
28. Maxwell, G. D. and Forbes, M. E. (1987). *Development*, **101**, 767.

19

Embryonic stem cells and gene targeting

ELIZABETH J. ROBERTSON and GAIL R. MARTIN

1. Introduction

Embryonic stem cells (ES cells), in conjunction with protocols for targeted gene disruption, have made it possible to introduce mutations into specific genetic loci in mammalian embryos. ES cells are permanent tissue culture lines of pluripotential stem cells derived from the inner cell mass of the normal mouse embryo (1, 2). They can be readily manipulated in culture and genetically altered cell clones then used to generate chimaeric animals following their microinjection into preimplantation stage carrier embryos. As ES cells reproducibly populate both the somatic cell lineages and the germ line, they afford a means of introducing mutations into the mouse genome. The discovery that mammalian cells can undergo a process of homologous recombination (site-specific recombination between exogenously introduced cloned DNA and the chromosomal cognate gene), has facilitated the generation of mice carrying a range of targeted genetic alterations, from large deletions to single nucleotide changes (3).

This chapter presents protocols for the culture and stable transfection of ES cells, plus methods for drug selection and expansion of ES cell clones carrying integrated DNA sequences. Protocols for the derivation of ES cells, the production of chimaeric mice, and the construction of gene targeting constructs, can be found elsewhere (4, 5).

2. Culturing embryonic stem cells

2.1 General considerations

ES cell lines are generally freely available, and can normally be obtained on request from any laboratory routinely performing gene targeting experiments. It is advisable to ask for details of the specific cell culture conditions used in the donating laboratory. Cultures are usually shipped as viable cells at room temperature or as frozen stocks; in either case, the culture should be expanded immediately and frozen stocks made (these remain viable for years).

To minimize the rate at which ES cells acquire chromosomal abnormalities, cells should be grown only when required for experiments. Cultures should be karyotyped routinely or checked for their ability to generate extensively chimaeric animals, and both ES and feeder cell cultures must be tested routinely for mycoplasmal contamination.

ES cells can be cultured in either circular dishes (with loose lids) or flasks (with screw caps). Although flasks are often preferred for routine passage of cell stocks, dishes must be used for any experiments that involve colony isolation. Flasks and dishes must be made of tissue culture plastic; do not use Petri dishes made for bacteriological work, as cells do not adhere to them. To avoid confusion in the laboratory, tissue-culture dishes and bacteriological dishes could be purchased from different suppliers. Density of the cell monolayer has a pronounced effect on the rate of division, so surface area must be taken into account when deciding the number of cells to be plated.

To maintain ES cells in an undifferentiated state, they must be grown either:

(a) in the presence of feeder cells (living fibroblasts which have been inactivated by irradiation or mitomycin C)

(b) in medium containing LIF (Leukaemia Inhibitory Factor)

(c) in medium pre-conditioned by cells of the BRL (buffalo rat liver) cell line

Here we describe culture of ES cells in the presence of feeder cells.

2.2 Growing ES cells on fibroblast cell feeder layers

Although many cell lines of fibroblastic cells can be used as feeders, either STO cells or primary mouse embryo fibroblasts (MEFs) are preferred. For propagation, they should be seeded at $2–5 \times 10^5$ cells per 10 cm dish and passaged by trypsinization (*Protocol 1*) when they reach confluence (after 3–5 days). They do not require feeding between passaging. Continued incubation after confluence has been reached causes selection for cells that do not remain flat at confluence; these do not make particularly good feeders. The actual feeder layers to be used for ES cell culture should be just confluent (*Protocol 2*).

Protocol 1. Trypsinizing fibroblast cells or ES cells

1. Warm the CMF-PBS[a], trypsin solution[b], and fresh medium[c] to 37 °C.

2. Using an inverted microscope, examine the cell cultures for contamination by yeast or bacteria.

3. Aspirate off the culture medium.

4. Rinse the cultures with CMF-PBS (approximately 3 ml per 10 cm dish).

5. Add trypsin solution (approximately 3 ml per 10 cm dish).

6. Incubate for 3–10 min at 37 °C, until cells are floating or separated from dish. Do not leave in trypsin for longer than required to achieve this.

7. Add fresh medium (approximately 3 ml per 10 cm plate) to stop the action of the trypsin. Pipette vigorously to disaggregate the cells from each other and the dish.

8. Transfer this mixture into a 15 ml centrifuge tube and pellet cells by centrifugation at 300 g for 3–5 min.

9. Aspirate supernatant, add fresh medium (twice the previous volume) and resuspend the cells by pipetting vigorously.

10. Count the cells using a haemocytometer[d].

---

[a] CMF-PBS. 10.0 g NaCl, 0.25 g KCl, 1.44 g $Na_2HPO_4.12H_2O$, 0.25 g KH_2PO_4 per litre H_2O; adjust pH to 7.2.

[b] Trypsin. 0.25% trypsin, 0.02% EDTA in CMF-PBS.

[c] Dulbecco's Modified Eagle's Medium (DMEM), plus 4.5 g/l glucose, 10–15% calf or fetal calf serum (depending upon the cell type), 2 mM glutamine, 0.5 units/ml penicillin, 5 µg/ml streptomycin (optional: 10 µM β-mercaptoethanol). Fungicides are generally avoided when culturing feeder cells or ES cells.

[d] For example, a confluent 10 cm dish of fibroblasts usually contains 0.5–2 $\times$ 10^7 cells.

Protocol 2. Using feeder cell layers

To prepare the feeder cell layers (the day before seeding the ES cells), either follow steps **1–3**, or **4-7**:

1. *Method 1.* Trypsinize confluent fibroblast cells following *Protocol 1*, resuspend in fresh medium (equal to the original volume of medium).

2. Irradiate the cells with approximately 5000 rads (50 Gy) with a cobalt or caesium irradiation source.

3. Pellet the cells again by centrifugation, and resuspend in fresh tissue culture medium, count the cells and calculate the cell concentration. Go to step **8**.

4. *Method 2.* To a confluent monolayer of fibroblasts, add mitomycin C to the cultures to a final concentration of 10 µg/ml.

5. Incubate for 2–3 h.

6. Trypsinize as described in *Protocol 1*.

7. Wash the cells free of mitomycin C by centrifuging and resuspending in medium, three times. Count the cells and calculate concentration. Go to step **8**.

Protocol 2. *Continued*

8. Adjust the concentration of the treated fibroblasts to 3×10^5 cells/ml in fresh medium. Seed these onto gelatin-treated tissue culture plates[a] at 5×10^4 cells/cm^2. Culture overnight.

9. Seed the ES cells at high density ($3–10 \times 10^6$ per 10 cm plate for cells that will be cultured for 2–3 days, $2–5 \times 10^6$ for cells that will be cultured for 4–5 days).

10. Maintain the ES cell cultures as described in Section 2.3.

[a] To prepare: make a 0.1% gelatin (swine skin Type 1, Sigma G-2500) solution in H_2O and sterilize by autoclaving. Apply enough gelatin solution to the tissue culture plates to cover the surface, and leave to stand at room temperature until ready for use (at least 5 min). Just before use, aspirate the excess gelatin solution from the plates.

2.3 Maintenance of ES cell cultures

Beginning two days after seeding, the ES cells need to be fed daily, by replacement of the medium. A typical schedule would be to change the medium on days 2 and 3, and passage on day 4 (by trypsinization as in *Protocol 1* and reseeding). To prevent differentiation, ES cells should be passaged at 3–4 day intervals. The ES cell number will be $2–4 \times 10^7$ cells per 10 cm dish after 3 days of culture; $5–6 \times 10^7$ cells after 4 days.

3. Differentiation of ES cells: embryoid body formation

Although the most widespread use of ES cells is for gene targeting (Section 4), they are a useful model system for *in vitro* studies of cell differentiation. ES cells can differentiate in culture into a variety of cell types in the absence of chemical inducers; the trigger for differentiation is the formation of three-dimensional aggregates. Such aggregates may mimic the inner cell mass of the normal mouse blastocyst (6–8).

Protocol 3. Formation of embryoid bodies

1. Harvest ES cells and seed at high density ($5–10 \times 10^7$ cells per 10 cm dish). Incubate at 37 °C for 20 min, then collect the medium from the dish (this will contain the ES cells; any feeder cells carried over during the passage will have adhered).

2. Seed the clean ES cells at 3×10^6 cells per 10 cm bacteriological (not tissue culture) dish. Culture at 37 °C. The cells will form loosely attached aggregates.

3. To replace the medium (daily), transfer the culture to a centrifuge tube, allow the cell aggregates to settle, aspirate off the old medium, add fresh medium, and return the cells to a fresh bacteriological dish.

---

4. Genetic manipulation and gene targeting using ES cells

The generation of mice carrying targeted gene mutations involves:

(a) design and construction of a targeting vector for the gene of interest

(b) transfection of the construct into ES cells, recovery of clonal derivatives, and screening for homologous recombination events

(c) generation of chimaeric animals and test-breeding to identify germ line transmission of the desired mutation

The targeting vector will be designed to allow either a 'replacement' event in the genome or a more simple 'insertion' event at the target site (discussed in detail in ref. 5). The parameters that govern the frequency of homologous recombination are poorly understood, but this frequency varies between different loci, targeting constructs, and sources of cloned DNA.

The most commonly used procedure for generating chimaeras from ES cells involves microinjection of small numbers of cells into the blastocoele cavity of a preimplantation mouse embryo, using a microinjection apparatus (4). Details concerning maintenance of mouse colonies, embryo manipulation and microinjection are described in ref. 9.

Methods for introduction of constructs into ES cells, selection for lines with integration and screening for homologous recombination events, are given below.

4.1 Electroporation of ES cells

Protocol 4. Electroporation of ES cells

Requirements:

- exponentially growing cultures of ES cells[a]
- 4 × 10 cm pre-made plates of feeder cells[b]
- an electroporator (eg: BioRad gene pulsor plus capacitance extender, with 0.4 mm electrode gap cuvettes)
- linearized and purified DNA for electroporation[c]
- sterile CMF-PBS, Pasteur pipettes, and culture medium

Protocol 4. *Continued*

1. Each electroporation (including control) will require 2×10^7 exponentially growing ES cells; prepare sufficient plates for the experiment.

2. Wash the growing ES cell cultures once with CMF-PBS, and trypsinize (*Protocol 1*).

3. Pool the cells into a 50 ml centrifuge tube and pellet by low speed centrifugation (300 *g*, 5 min).

4. Aspirate the supernatant and resuspend the cell pellet in 20–30 ml CMF-PBS. Perform a cell count using a haemocytometer and calculate the total cell number.

5. Spin down the cells as in step 3, aspirate and resuspend in appropriate volume of CMF-PBS to give a final density of 4×10^7 cells/ml.

6. For each electroporation, transfer 0.5 ml of the ES cell suspension into electroporation cuvette.

7. Add 15–20 μg linearized DNA to the cuvette and mix well using a Pasteur pipette. Also prepare one cuvette of cells as a 'no DNA' control.

8. Place each cuvette into the electroporation chamber and electroporate once at 200 V, 960 μF capacitance[d].

9. Using a Pasteur pipette, transfer the suspension into 40 ml culture medium (pre-warmed to 37 °C) and mix thoroughly (many cells are killed and will aggregate to form sticky clumps). Wash out the cuvette with culture medium to retrieve all the cells.

10. Disperse the suspension into 4×10 cm pre-made feeder plates, and place in 37 °C incubator. For the 'no DNA' control, simply plate one-quarter onto one plate of feeder cells.

---

[a] As with other cell types, the best transfection efficiencies are obtained using exponentially growing cultures. For example, seed the cells at high density the day preceding the electroporation (8×10^6 ES cells per 10 cm feeder plate), replace the medium on the following morning, and perform the electroporation about 4 to 6 h later.

[b] Calculate the magnitude of the experiment in advance, to ensure that sufficient numbers of healthy STOs or MEFs are available to prepare the required feeder plates.

[c] The DNA is prepared by linearizing aliquots of caesium-banded plasmid by restriction enzyme digestion, followed by phenol extraction and ethanol precipitation. Resuspend in TE and determine the concentration. A final concentration of between 0.5 to 1 μg/μl is ideal.

[d] These conditions are for the BioRad electroporator. For other instruments, determine the best conditions by trial and error (using a plasmid such as *pgk-neo*).

4.2 Drug selection of ES cells

The most commonly used method for selection of clones with integrated DNA sequences involves incorporation of a neomycin-resistance gene (*neo^r*) into the construct, coupled with G418 treatment to select for expressing cells.

Many vectors also include a 'counter-selection' cassette, typically provided by *pMCItk* (10), flanking the genomic sequences. Addition of GANC or FIAU will then kill those G418-resistant cells that randomly incorporated the vector.

4.2.1 G418 selection

Protocol 5. G418 selection

1. Refeed the ES cell culture plates 24 h after electroporation, to remove the majority of dead, unattached, cells[a].

2. Approximately 36 to 48 h after electroporation, add G418 to a final concentration of 0.2 mg/ml *active* G418 (typically G418 is sold at about 50% active, so take into account the batch specifications).

3. Refeed the plates as necessary, according to the density of the ES cells. As it generally takes 3 days for the selection to start to take effect, the cultures are normally fed daily for the first 3 days, then every 2 days after the first wave of cell death eliminates the majority of cells.

---

[a] It is important not to stress the ES cells; for example, do not allow cells to become too dense and or highly acidic.

[b] To prepare G418 stocks, dissolve the powder at 100 mg/ml in H_2O, and filter sterilize (0.2 μm pore size). Stocks are stable at $-20\ °C$. ES cell medium plus G418 can be stored at 4 °C and used for at least 10 days.

4.2.2 GANC counter-selection

If the targeting construct incorporates a *tk* counter-selection cassette, ES cells that carry random integrations of the vector can be eliminated the addition of GANC or FIAU to the selection medium. It is not necessary to add the selection until 72 h after the G418 selection is initiated, as only a small minority of cells will be killed. Typically, add GANC to a final concentration of 2×10^{-6} M in ES cell medium plus G418. The stock is a 2×10^{-3} M solution of GANC (obtained from Syntex) in H_2O, stored frozen at $-20\ °C$.

To determine the 'enrichment' obtained by the inclusion of the *tk* cassette in the targeting construct, it is important to leave at least one plate in G418 alone. At the end of the experiment stain representative plates with a dilute solution of Giemsa or Coomassie blue and count the number of colonies obtained under single and double selection.

4.3 Isolation and expansion of drug-resistant ES cell colonies

The progress of the drug selection is monitored by following cell viability on the 'no DNA' control plate. When the feeder layer of this plate is devoid of

ES cells, the selection is considered complete. Drug-resistant colonies on experimental plates will be macroscopically visible after about 5 days in selection, but should be allowed to grow until they consist of several thousand cells. Typically, colonies are ready for picking and expansion 7–8 days after the addition of G418. At this point the selection should be removed, to avoid further stress on the cells. Each resistant colony is expanded to give enough cells for freezing, and for preparation of DNA or cell lysates for PCR, following *Protocol 6*.

Protocol 6. Expansion of drug-resistant ES cell colonies

Requirements

- 6 × 10 cm tissue culture dishes, Gilson P200 Pipetman
- CMF-PBS, ES cell medium and trypsin/EDTA (*Protocol 1*).
- stereo dissecting microscope
- for day 1 of the protocol: six multiwell trays (24 × 1 cm wells), each well pre-seeded with 5×10^4 STO or MEF feeder cells in 1 ml medium
- for day 7 of experiment: up to 144 plates (6 cm) pre-seeded with feeder cells

1. 7–8 days after the addition of G418, wash each plate containing drug resistant ES cell colonies once with CMF-PBS.
2. Flood each ES cell plate with 10 ml CMF-PBS.
3. In each of approximately six 10 cm tissue culture plates, spot out a 6 × 4 array of 50 μl drops of trypsin/EDTA.
4. Transfer the flooded plate containing ES cell colonies to the stage of a stereo dissecting microscope. Using low power magnification (10× objective), locate, and gently dislodge 24 individual colonies from the feeder layer with a pulled Pasteur pipette[a].
5. Pick up single floating colonies in a few microlitres of CMF-PBS using either a pulled Pasteur and mouth tube or a wide-bore yellow tip and Gilson Pipetman. Transfer single colonies to separate drops of trypsin from step 3. Picking 24 colonies should take about 5 min.
6. Transfer the plate of 24 drops to a 37 °C incubator and leave for 5 min.
7. Remove the plate from the incubator. Disperse each colony into a suspension by taking up 50 μl ES cell medium in a narrow-bore yellow tip, adding it to the cells in a trypsin drop, and vigorously pipetting with the tip vertical and touching the bottom of the plate. Monitor the disaggregation under the dissecting microscope.
8. Transfer the contents of each drop into a well of a pre-seeded multiwell tray and culture at 37 °C.

9. Repeat steps 4 to 8 until all preseeded multiwell trays are used; generally six trays of 24 colonies are picked for each experiment.

10. Refeed the trays with fresh culture medium every day for the following 6–7 days (no further drug selection is required), by which time the cells should have reached confluence.

11. Expand each well into a 6 cm pre-formed feeder plate using standard trypsinization procedure (there is no need to spin out the cells, just transfer the suspension directly).

12. Refeed daily until confluent.

13. Wash, trypsinize and collect the cells from each plate in 2 ml ES cell medium. Aliquot the suspension into three sterile microfuge tubes, pellet the cells in microfuge, aspirate the supernatant, and place the tubes on ice.

14. Resuspend one pellet from each clonal culture in 50 to 75 µl of ice-cold freezing medium (90% serum, 10% DMSO) and immediately place at −70 °C as a frozen stock of cells[b].

15. Store the two remaining pellets from each clonal culture −20 °C for DNA preparation and analysis.

---

[a] It is important to time the picking of colonies to ensure that optimal expansion. If the colonies are allowed to get too large, many will start to exhibit overt differentiation. If they are picked when too small, the wells will not become confluent and the growth rate of the ES cells will slow considerably. There is substantial variation in the size of colonies; pick a range of sizes.

[b] Unnecessary exposure to DMSO before freezing will reduce cell viability. The frozen cells will be viable for several months at −70 °C. This should be sufficient time to screen DNAs and identify putative homologous recombinants. If a longer time is anticipated, then it is safer to freeze the cells in conventional freezing vials and store under liquid nitrogen.

4.4 Identifying homologous recombination events

A number of strategies can be used to identify the homologous recombination events among the drug-resistant clones. The most labour intensive is to purify DNA from each clone, and to use Southern analysis to identify which clones have the predicted rearrangements at the target locus. PCR-based strategies have also been described (11); typically these use one primer specific to the target gene and one specific to the drug selection cassette. These techniques require few cells for the analysis and samples can be assembled into pools of 5–10; these are therefore particularly useful when the targeting frequencies are low (less than 1 per 1000). The disadvantages of PCR-based strategies are that they can give false positives, and they limit the extent of genomic homology which can be incorporated into the vector (if greater than 2–3 kb either side of the drug selection cassette, PCR amplification will be inefficient). Often, PCR is used for an initial screen; candidate positive clones are then tested by Southern analysis.

4.5 Generation of chimaeras using targeted cell clones

After identification of targeted ES cell clones, they must be expanded prior to generation of chimaeric mice.

Protocol 7. Expanding targeted clones for generation of chimaeras

1. Identify 6–12 candidate targeted cell clones by DNA analysis (Section 4.4).

2. Thaw the frozen stock (prepared in *Protocol 6*, step **14**) of these clones rapidly by holding the microcentrifuge tube in a 37 °C water bath for a few seconds. Immediately[a] fill the tube with fresh ES cell medium and pellet the cells by low speed centrifugation. Aspirate, resuspend the pellet in ES cell medium, transfer into a 1 cm multiwell tray preseeded with feeder cells and culture at 37 °C.

3. Feed daily until confluent.

4. Expand using routine protocols[b], and make at least six frozen stocks from each clone for storage under liquid nitrogen.

---

[a] The DMSO must be removed from the cells quickly; prolonged exposure to DMSO at 37 °C reduces viability of the cells.
[b] Keep records of the relative passage number of the clonal isolates.

As the production of chimaeric mice is a laborious and lengthy process, it is imperative to initiate the gene targeting experiment using cell lines with demonstrated germ line potential. For example, the majority of starting cells must be euploid if germ line transmission of a mutation is to be achieved form a small panel of targeted ES cells. The karyotype of any targeted clones should also be determined after selection and expansion, prior to injection, since aneuploid cells can differentiate to give very extensive somatic tissue derivatives, but will not form germ cells. Protocols for the generation of ES cell chimaeras are given in ref. 4.

References

1. Evans, M. J. and Kaufman, M. H. (1981). *Nature*, **292**, 154.
2. Martin, G. R. (1981). *Proc. Natl Acad. Sci. USA*, **78**, 7634.
3. Hooper, M. L. (1992). *Embryonal stem cells: introducing planned changes into the animal germline.* Harwood Academic Publishers, Switzerland.
4. Robertson, E. J., ed. (1987). *Teratocarcinomas and embryonic stem cells: a practical approach.* IRL Press at Oxford University Press, Oxford.
5. Joyner, A., ed. (1993). *Gene targeting: a practical approach.* IRL Press at Oxford University Press, Oxford.

6. Martin, G. R. and Evans, M. J. (1975). *Proc. Natl Acad. Sci. USA*, **72**, 1441.
7. Martin, G. R. and Evans, M. J. (1975). *Cell*, **6**, 467.
8. Martin, G. R., Wiley, L. M., and Damjanov, I. (1977). *Dev. Biol.*, **61**, 230.
9. Hogan, B., Costantini, F., and Lacy, E., eds. (1986). *Manipulating the mouse embryo: a laboratory manual*. Cold Spring Harbor Press, Cold Spring Harbor, NY.
10. Mansour, S. L., Thomas, K. R., and Capecchi, M. (1988). *Nature*, **336**, 348.
11. McMahon, A. and Bradley, A. (1990). *Cell*, **62**, 1073.

20

Introduction of genes using retroviral vectors

JACK PRICE

1. Introduction

Retroviruses are naturally evolved agents of gene transfer. Because they work under normal physiological conditions, they can be used to introduce genes into cells *in vivo*. Potentially, therefore, they can introduce genes into embryos over a wide range of developmental stages. The principal developmental study to which they have been applied is cell lineage, where the transferred gene is non-functional; a more recent application is the use of retroviruses as a localized agent of transgenesis.

1.1. Retroviruses and transgenesis

A retrovirus is an encapsidated, membrane-bound particle that gains entry to a cell via an interaction between the viral *env* coat protein and a cell surface receptor. The specificity of this interaction defines the viral host range. Once inside the cell, a DNA copy of the RNA viral genome becomes integrated into the host cell DNA (1), a process requiring the cell to go through S phase of the cell cycle within a few hours of infection. Once integrated, the viral genome will be inherited by all the progeny of the infected cell. Retroviruses also spread horizontally and will eventually infect a group of adjacent cells. This means that viruses can be used to disperse a gene through the cells of the tissue. A useful vector for this application is the RCAS type-A replication-competent avian retrovirus which has been used for targeted misexpression of genes in chick embryos (2).

1.2. Retroviruses as lineage labels

The use of retroviral vectors as a genetic cell lineage marker (3, 4) has the advantage that, unlike injected dyes, the marker is not diluted as the cell divides. All the progeny of a labelled cell carry a replicated copy of the viral genome. Unlike applications for transgenesis, the ability of viruses to spread horizontally is a disadvantage for lineage studies. This has been overcome by the development of replication-incompetent retroviral vectors (see ref. 5). Several retroviral vectors of this type have been adapted to carry lineage markers such as β-galactosidase and alkaline phosphatase.

One limitation of retroviruses as lineage markers is that neither the precise number of cells that become infected, nor their exact location, can be known at the time of injection. When a virus is injected, the viral titre is adjusted to give roughly the desired number of infections, but the precise number has to be inferred from the final clonal analysis. This chapter describes methods for lineage analysis with retroviruses, but they are also applicable to other retroviral applications.

2. Using retroviruses as lineage markers

There are three major steps in a retroviral lineage project:

(a) engineering the virus and introducing it into a packaging cell line

(b) production and titration of the virus and checking for contamination with helper virus

(c) introducing the virus into the embryo and analysing the labelled clones

Methods for the first step are covered elsewhere (6) and several engineered viruses suitable for lineage analysis are already available. In the following sections I will concentrate on the methods involved in the second and third steps; they involve no procedures beyond the capabilities or facilities of most biological laboratories.

2.1 Selection of experimental strategy

A number of points need to be considered when planning a cell lineage project using retroviruses.

(a) What is the appropriate viral type for the species to be used?

    i.   'Ecotropic' murine viruses will infect mouse and rat cells, but not other species[a]. The most popular producer lines of this type are Ψ_2 and Ψ_{cre} (7, 8). Both produce Moloney murine leukaemia virus derivatives.

    ii. Rous sarcoma virus (RSV) derivatives. These are chick viruses which will infect many, but not all, chick strains (9, 10)[a].

    iii. 'Amphotropic' murine viruses infect a wider range of species (e.g. rabbit, chick, some primates)[a]. These include human, which raises some safety issues, particularly if the viruses carry potential oncogenes. The popular producer lines are Ψ_{am}, Ψ_{crip} (7, 8), and PA317 (11); these lines tend to produce lower titres than the ecotropic equivalents.

Each of these categories contains viruses carrying *lacZ* and have been used for lineage studies (3, 4, 12). More recently, a virus encoding alkaline phosphatase has become available (13). All the cell lines (both parent cells and cells expressing viruses) described in this section are available from the American Type Culture Collection (ATCC).

(b) Are the cells to be labelled dividing? Virus will only integrate into actively cycling cells; if the cell cycle is longer than about 20 h, obtaining a significant infection ratio can be a problem (e.g. some stem cells, which divide very slowly). To study such cells in culture, the infection rate must be increased, for example by increasing the viral titre.

(c) Can the virus reach the cells? Viruses do not penetrate tissue well, so it is important to avoid any barriers between the site of injection and the cells to be infected. For example, a virus will not penetrate basement membranes.

(d) The retroviral labelling technique described here requires that the retroviral marker gene be expressed. This is a function of the promoter elements used in the viral construct. There are two options with different advantages and disadvantages:

- the endogenous retroviral promoter of the viral long terminal repeat (LTR)
- internal promoters

The LTR is expressed in most cell types, but all cells of preimplantation mouse embryos turn off expression from this promoter (14). Once turned off, it remains inactive in most cells, even after the cells have developed beyond the point at which they will express newly introduced virus. The LTR is also turned off in undifferentiated ES cells. Thus, viruses with endogenous promoters cannot be used for generating transgenic mouse embryos, and they are of no use for cell lineage studies in the preimplantation mouse embryo.

Internal promoters do not have this problem, but they are often weaker than the viral LTR, and they are also often regulated in different cell types, sometimes in quite unexpected patterns (12). Clearly, if some cells turn off expression of the retrovirus, the lineage relationships could be distorted since clones might appear to be composed of a single cell type because all the cells of a second type had turned off expression. There are two plausible mechanisms by which the LTR could be specifically turned off:

ι. A particular cell type could specifically repress the LTR promoter.

ii. The virus could by chance insert in the genome at a locus that was repressed in that cell type.

If the first of these mechanisms were to occur, the particular cell type would not appear anywhere in the clonal analysis, but no examples of this have been described. The second mechanism will operate in only a tiny minority of insertion sites, causing a negligible overall distortion.

Other approaches have used detection of the integrated provirus as a lineage marker (15), but they have lower resolution than is required for most lineage analysis.

(e) Will there be much dispersion of the progeny of the infected cell? If the progeny spread widely, interpretation is difficult. At present there is no easy way around this problem, although a method using 100 different viral variants in conjunction with the PCR technique has recently been described (16). This technique is labour-intensive and technically demanding, and if the different viruses are not present in equal titre, interpretation of the results is complicated (17).

[a] According to the British Advisory Committee on Genetic Modification, work with mouse and chick ecotropic viruses carrying marker genes must be under ACGM level 1 containment. Amphotropic viruses require ACGM level 2 containment.

3. Preparation of retrovirus

The preparation of virus has three steps, plus an optional fourth:

(a) virus is collected from producer cells

(b) its concentration is adjusted by titration

(c) it must be checked to ensure that it is not producing helper virus.

The optional step is viral concentration. Before embarking on these procedures the packaging cells must be established in culture, and their quality ascertained. Many problems of viral titre arise from sub-optimal producer cells.

(a) Grow the cells in culture using the culture medium appropriate for the cell line. For NIH-3T3 based lines (Ψ_2, Ψ_{am}, Ψ_{crip}, Ψ_{cre}), this will be Dulbecco's modified Eagle's medium (DMEM), supplemented with 10% either newborn (NCS) or fetal calf serum (FCS). For other cells (e.g. PA317 cells), a more complex medium is required (11).

(b) Ascertain that the cells are dividing at an appropriate rate. A confluent dish should regain confluency in 3–4 days after being split 1:10.

(c) Check that the cells are free of mycoplasma.

(d) Freeze down stocks of healthy cells. Always collect virus from freshly thawed stocks.

(e) Ascertain that all the cells are resistant to the selectable marker encoded by the virus (generally *neo*). Most producer cells, once selected, are relatively stable in this regard. Others, however, have to be kept under constant selection.

(f) Stain the cells to be sure that they are all positive for the histochemical marker encoded in the virus (generally *lacZ*: see *Protocol 7*).

3.1 Production of retroviral stocks

The collection of virus from producer cells is simple: it amounts to the collection of tissue culture supernatant from a fibroblast cell line. The details will vary for different cell lines, but the following protocol is appropriate for the most popular lines, those derived from NIH 3T3 cells.

Protocol 1. Collection of virus

1. Set up producer cells on 10cm dishes such that they are just sub-confluent on the day required. Do this by splitting a confluent plate 1:10, and letting them grow for 2–3 days.

2. Change to fresh growth medium. Use half the normal volume of medium, i.e. 5 ml for a 10 cm dish.[a]

3. 20–24 h later collect medium on ice[b] and filter through 0.45 μm Nalgene filter.

4. Aliquot[c] and freeze[d] this stock and store at −20 °C. If virus is to be concentrated, proceed to *Protocol 4*.

[a] If virus must be collected in medium other than that in which the cells normally grow, it can be switched at this point. Producer cells are relatively hardy and usually survive for 24 h in any isotonic medium supplemented with serum without the viral titre being affected, but this should be checked for any given medium. There are two particular circumstances in which switching medium is recommended:

(a) Some producer cells must be kept under drug selection, but viral stocks should not contain these drugs.

(b) If virus is to be concentrated (see *Protocol 3*), virus should be collected in medium containing FCS rather than NCS. The latter gives a larger protein precipitate.

[b] The viral half-life is temperature dependent. Collected virus should always be kept on ice.
[c] Repeated freezing and thawing will destroy virus. Choose an aliquot size that will not require aliquots to be reused.
[d] Do not freeze virus on solid CO_2. Carbon dioxide causes the pH of the viral suspension to fall, which destroys the virus.

3.2 Titration of viral stocks

Viral stocks prepared as above will contain a suspension of viral particles, and some measure of the effective concentration is required before use. The virus can be assessed for its ability to transmit either its selectable marker, or its histochemical marker, to a control cell type (usually NIH-3T3 cells for mouse viruses). Titration by selection is more widely applicable, but titration for the expression of the selectable marker is simpler. Both give quantitatively similar results, and both are given here.

Protocol 2a. Titration of virus by selection

The most commonly used selectable marker is the *neo* gene. This selects for mammalian cells resistant to the drug G418.

1. Maintain NIH-3T3 cells (or another appropriate cell line) in routine culture.

2. Plate NIH-3T3 cells at moderately low density, so that after 24 h they are just beginning exponential growth. Generally this involves splitting a confluent plate 1:10, and use the following day. Use 3 or 6 cm dishes.

3. One day after plating the cells, remove the medium and add virus at several concentrations. For viral supernatant (see *Protocol 1*) use 1 and 10 µl; for concentrated virus (see *Protocol 4*) use 0.1 and 1 µl per plate. Titrate each concentration in duplicate, and count whichever concentration gives an appropriate number of clones. Use a previously assayed virus as a positive control. Also add polybrene[a] (Sigma; final concentration 10 µg/ml). Return the cells to 37 °C. Gently agitate the plates several times for 2–3 h.

4. After about 3 h, bring volume to 1 ml (2 ml for a 6 cm dish) and incubate at 37 °C overnight.

5. Add medium to final volume of 2 ml (5 ml for 6 cm dish).

6. After 2–3 days, split cells 1:10 or 1:20 into medium containing G418 (1 mg/ml)[b]. Change medium every 2–3 days.

7. After 5–7 days, most of the cells will begin to die. A few days later, clones of resistant cells will become apparent. Count the number of clones per plate, or make an estimate by counting a representative fraction of a plate. This represents the number of colony-forming-units (c.f.u.) in the volume of virus added to the plate. Multiply by the appropriate factor to express the results as c.f.u./ml. Since two different viral concentrations were used, two different (but similar) values for the titre will be obtained. Accept the higher of the two.

---

[a] Polybrene is a positively charged molecule which aids the binding of virus to cells. It is essential for good titres in this type of assay. Note, however, that it is toxic to some cells.
[b] The dose of G418 required to kill normal but not resistant cells will vary depending on the batch of drug and on the cell line used. This should be determined in advance.

For the viruses generally used for lineage marking, titration by expression of the histochemical marker is quicker and more appropriate.

Protocol 2b. Titration of virus by histochemical staining

1. Follow steps **1–6** in *Protocol 2a*.

7. 2–3 days after adding virus, when the plate has become confluent, stain for the histochemical marker. For *lacZ*, fix the cells with 0.5% glutaraldehyde in phosphate buffered saline (PBS) and stain with X-Gal (*Protocol 7*).

8. Estimate the number of clones per plate. Multiply by the appropriate factor to give the titre in cfu/ml.

As a rough guide, the titre of ecotropic mouse virus (e.g. Ψ_2) producers should be about 10^5 per ml. From amphotropic lines (e.g Ψ_{crip}) expect 2–10-fold less.

3.3 Helper virus

Once it has been established that the producer line is producing adequate titre, it is important to check that it is not producing helper virus. 'Helper' virus in this context is virus that can spread the viral vector horizontally; wild type virus has the ability to infect a cell, then spread to neighbouring, unrelated cells. Replication-incompetent retroviruses cannot spread in this way. However, some early producer cell lines were made in such a way that, at low frequency, transmissible virus could arise through genetic recombination. This applies to Ψ_2 and Ψ_{am} and explains their gradual replacement with Ψ_{cre} and Ψ_{crip} respectively (see ref. 9 for details). Ψ_2 is still very popular, because it is a good producer line and because production of helper occurs relatively infrequently. Consequently, these lines are useful if checked regularly. For viruses carrying a histochemical marker, their ability to spread the marker can be assayed simply.

Protocol 3. Checking for helper

1. Split NIH-3T3 cells 1:10 and infect them as in *Protocol 1*. Use a large amount of virus (e.g. 0.5–1 ml of viral supernatant).

2. Allow these cells to grow to confluence as for a titration experiment.

3. Change medium, washing the cells in culture medium once.

4. Leave the cells in fresh medium for 24 h then collect the medium and filter through a 0.45 μm Nalgene filter.

5. Titrate this medium on NIH-3T3 cells as in *Protocol 2b*, but use 1 ml rather than 1 μl.

The titre from this experiment should be zero. If labelled clones appear, then NIH-3T3 cells infected with the virus have transmitted virus to the NIH-3T3 cells in the assay. Any producer cells found to be positive in this assay should be discarded.

3.4 Concentration of virus

For many embryological experiments, the concentration of virus obtained from the above protocols will be insufficient, as the volume of virus that can be introduced into an embryo may be small. The cells can be made to produce more virus (18), but it is simpler to concentrate the viral stock by centrifugation.

Protocol 4. Concentrating viral stocks

1. Prepare viral stock as in *Protocol 1*, steps **1–4**.

2. Keeping the stock cold at all times, centrifuge the stock[a] overnight ($\geqslant 6$ h) at 14 000 r.p.m. on a Beckman SW28 rotor and L8 ultracentrifuge or equivalent.

3. Keeping the centrifuge tubes on ice, drain the pellet and discard the supernatant.

4. Resuspend the pellet in 1/100th. of the volume of the original stock[b] using isotonic buffer containing some protein, such as DMEM + 10% FCS.

5. Aliquot and freeze the concentrated viral stock (see footnotes *a* and *d* in *Protocol 1*).

6. Titrate as in *Protocol 2*.

[a] We usually begin with 100 ml of stock, from twenty 10 cm tissue-culture dishes. Keep some unconcentrated stock to titrate alongside the concentrated virus.
[b] With virus from the amphotropic producer lines, PA317 and Ψ_{crip}, which tend to have a lower titre than ecotropic virus, the viral titre can be increased 100 times by concentrating 400 times with regard to volume (e.g. 40 ml to 100 μl).

Although concentration typically gives a 10-fold increase in titre, the cost is a 100-fold reduction in volume, i.e. only 10% of the virus is recovered. This procedure is inefficient, therefore, and only worthwhile when a small volume of high titre virus is required.

4. Introducing virus into embryos

Potentially, any dividing population of cells at any stage of development can be labelled with a virus. Most of the retroviral cell lineage studies have so far fallen into one of three categories:

- labelling embryonic precursor cells in the living embryo
- labelling cells in culture
- labelling cells in suspension and subsequently implanting them into a host animal

Methods for the latter two are relatively unsophisticated; they involve incubating cells with an appropriate dilution of viral stock plus polybrene for up to several hours. Following at least 2–3 days for expression of the virus, the cells are processed for the histochemical marker. The main problem likely to be encountered is that the cells might not divide fast enough for a large enough number of cells to become labelled. This can be overcome either by inducing faster division using growth factors, or by co-culturing the cells with the viral producer cells.

More technically demanding are experiments in which virus is introduced into the neighbourhood of dividing cells *in vivo*. Such studies have included studies on avian embryos *in ovo*, rodent embryos *in utero*, and neonatal rodents. The following protocol gives a method for rodent embryos *in utero* with some notes on alternatives.

Protocol 5. Injection of virus into rodent embryos

1. Timed pregnant rat embryos are anaesthetised with 0.68 ml per 250 g body weight of a 1:1:2 mixture of Hypnorm (Janssen Pharmaceuticals Ltd): Hypnovel (Roche): water, delivered intraperitoneally[a].

2. Once anaesthetized, open first the skin, then the abdominal cavity by midline incisions. This reveals the uterus, which can be gently pulled from the abdominal cavity, one horn at a time (see Chapter 7).

3. The embryos can be visualized with a fibre-optic lamp fitted with a fine, flexible light-guide by shining this light through each conceptus from behind. Any part of the embryo can now be seen clearly enough to be injected through the uterine wall while causing minimal damage. The injection can be made with hand-held Hamilton syringe, fitted with a 30-gauge needle. Alternatively, a micropipette can be used with a micromanipulator. Either way, volumes of less than 1 μl of virus/polybrene[b] can be introduced into specific regions of the embryo.

4. Following injection, the uterus can be tucked gently back into the abdominal cavity, and the wound sutured closed. 5/0 Prolene filament (Ethicon Ltd) is suitable.

5. The animal recovers in about 45 min. Generally, the injected embryos are born on schedule with no apparent ill effects[c]. They can be raised to the required age before they are killed and the tissue of interest is prepared for histochemistry. If the histochemical marker is *lacZ*, they should be killed by perfusion through the heart with a fixative consisting of 0.2%

Protocol 5. *Continued*

glutaraldehyde, 2% paraformaldehyde, 0.1 M PIPES/NaOH pH 6.9, mM MgCl$_2$, 1.25 mM EGTA.

---

[a] This is a controlled procedure in the UK and most other countries, and should be cleared with the appropriate authorities before any study is embarked upon.
[b] A dye (e.g. 1% Indian ink) can be added to the inoculum to help visualization.
[c] Recovery rates are generally 50–100%. Less than 10% show any signs of damage.

Similar injection procedures can be used for newborn animals (e.g. rats, mice, rabbits) without having to inject through the uterus. The best anaesthesia for neonates is probably cryoanaesthesia. Put the animal at −20 °C; its body temperature quickly drops and its heart stops. This takes between 10 and 25 min depending on the species. The animal can now be injected into regions where cell division is still taking place, e.g. cerebellum, retina. If the animal is warmed to 37 °C in a water bath, it recovers quickly without any apparent ill effects.

5. Detection of the histochemical marker

Almost all the studies done to date have used *lacZ* as a marker. Tissue can be stained to detect *lacZ* using the X-Gal reaction. Because this versatile stain works on tissue prepared in a number of different ways, there are a number of options to be considered.

(a) Generally tissue is better if perfused with fixative, but pieces of tissue can be fixed by immersion. If the tissue is to be sectioned on a cryostat, it can be frozen fresh, then fixed on the slide.

(b) The best fixative is the glutaraldehyde/paraformaldehyde combination described in *Protocol 5*. Formalin is not recommended because it includes some methanol. All fixatives containing organic solvents inactivate the β-galactosidase enzyme. For this reason neither plastic nor wax sections can be X-Gal stained.

(c) Tissue can be stained as whole mounts, as thick vibratome sections, or as cryostat sections. Cells can be stained in monolayer cultures or in suspension. The following protocol is for vibratome or cryostat sections, and small variations are given for other forms of tissue.

Protocol 7. X-Gal histochemistry

1. Fix sections for 10 min in ice-cold 0.5% glutaraldehyde (unnecessary if the tissue was well perfused with fixative).

2. Wash twice for 10 min on ice in PBS containing 2 mM MgCl$_2$.

3. Make up an X-Gal buffer comprising:

- 20 mM potassium ferricyanide[a]
- 20 mM potassium ferrocyanide
- 0.01% sodium deoxycholate
- 0.02% Nonidet P-40 (NP-40)
- 2 mM $MgCl_2$

4. Make up X-Gal stock at 40 mg/ml in dimethylformamide[b].

5. Make X-Gal solution by diluting X-Gal stock to 1 mg/ml in X-Gal buffer[c].

6. Incubate sections in X-Gal solution[d].

7. Wash the sections several times in PBS. Sections may be counterstained at this time.

8. Mount the sections. Cryostat sections can be dehydrated through alcohol, cleared, then mounted in DPX (BDH). Free-floating vibratome sections can be floated onto slides in 1% gelatin in 40% ethanol, dried, then dehydrated and mounted. X-Gal stained cells will appear indigo blue.

[a] The optimum concentration for both ferricyanide and ferrocyanide ions is 5 mM, and this is recommended for staining monolayer cultures. At this concentration, however, the blue reaction product tends to spread in tissue sections. 20–30 mM gives better overall results.
[b] This solution should be made up entirely in glass using glass pipettes.
[c] This solution keeps at 4 °C and can be reused.
[d] The time and temperature of incubation can be varied as required. For monolayer cultures try 1–3 h at 37 °C; for sections, overnight at 4 °C seems better. Note also that some tissues and cells (e.g. blood vessel endothelial cells, choroid plexus) have high endogenous activity and will turn blue if the staining continues too long. The inclusion of EGTA in the fixative (see *Protocol 5*) reduces this problem by chelating the calcium required for the endogenous activity. Include some uninfected tissue to control for possible endogenous activity.

References

1. Varmus, H. E. (1982). *Science*, **216**, 812.
2. Morgan, B. A., Izpisúa-Belmonte, J.-C., Duboule, D., and Tabin, C.J. (1992). *Nature*, **358**, 236.
3. Sanes, J. R., Rubenstein, J. L. R., and Nicolas, J. F. (1986). *EMBO J.*, **5**, 3133.
4. Price, J., Turner, D., and Cepko, C. (1987). *Proc. Natl. Acad. Sci. USA*, **84**, 156.
5. Danos, O. (1991). In *Practical molecular virology* (ed. M. K. L. Collins), pp. 17–28. Humana, Clifton, NJ.
6. Vile, R. (1991). In *Practical molecular virology* (ed. M. K. L. Collins), pp. 1–16. Humana, Clifton, NJ.
7. Mann, R., Mulligan, R. C., and Baltimore, D. (1983). *Cell*, **33**, 153.
8. Cone, R. D. and Mulligan, R. C. (1984). *Proc. Natl. Acad. Sci. USA*, **81**, 6349.
9. Stoker, A. W. and Bissell, M. J. (1988). *J. Virol.*, **62**, 1008.

10. Savatier, P., Bagnis, C., Thoraval, P., Poncet, D., Belakebi, M., Mallet, F., Legras, C., Cosset, F.-L., Thomas, J. L., Chebloune, Y., Faure, C., Verdier, G., Samarut, J., and Nigon, V. (1989). *J. Virol.*, **63**, 513.
11. Miller, A. D. and Buttimore, C. (1986). *Mol. Cell. Biol.*, **6**, 2895.
12. Beddington, R. S. P., Morgenstern, J., Land, H., and Hogan, A. (1989). *Development*, **106**, 37.
13. Fields-Berry, S. C., Halliday, A.L., and Cepko, C. L. (1992). *Proc. Natl. Acad. Sci. USA*, **89**, 693.
14. Jaenisch, R., Fan, H., and Croker, B. (1975). *Proc. Natl. Acad. Sci. USA*, **72**, 4008.
15. Soriano, P. and Jaenisch, R. (1986). *Cell*, **46**, 19.
16. Walsh, C. and Cepko, C. L. (1992). *Science*, **255**, 434.
17. Kirkwood, T. B. L., Price, J., and Grove, E. A. (1992). *Science*, **258**, 317.
18. Russell, S. J. (1991). In *Practical molecular virology* (ed. M. K. L. Collins), pp. 29–44. Humana, Clifton, NJ.

IV
Molecular techniques

21

Immunocytochemistry of embryonic material

CLAUDIO D. STERN

1. Introduction

The localization and characterization of protein and carbohydrate molecules in embryos is an essential part of any study on the molecular bases of developmental processes. Methods for immunocytochemistry and for the characterization of antigens are in wide use and well established, and have been reviewed extensively elsewhere (1, 2). Embryonic material, particularly at early stages of development, presents special problems. First of all, it is sometimes difficult to find the correct compromise between antigen preservation, sensitivity of the method of localization, and good structural preservation. Second, when characterizing the antigen(s), it is sometimes difficult to obtain sufficiently large amounts of antigen for detection by conventional immunoblots (Western blots). The following small set of protocols work particularly well for the localization of embryonic antigens in tissue sections at the light microscope level and in whole mounts. A simple, reliable, and sensitive method for the characterization of antigens from small amounts of embryonic material is also included.

2. Histological localization of antigens

There are two main ways to localize specific antigens in embryonic material at the light microscope level: in tissue sections and in whole mounts. They are complementary and both usually informative. In sections it is relatively easy to determine the specific cell layers or tissue types expressing the antigen in question, but more difficult to obtain a general view of the distribution of the antigen in the embryo. Whole mounts are better to study regional distribution, but it is more difficult to ascribe this expression to particular cell populations. It is sometimes possible to stain whole embryos with an antibody, detected by immunoperoxidase or alkaline phosphatase, and then to section the same embryos to obtain more detailed information about the patterns of expression in different tissues.

Both mehods have specific technical problems associated with them. In tissue sections it is important to balance sensitivity and antibody preservation with structural preservation. Plastic and wax sections, which give the best morphological preservation, often interfere with immunological detection. Some antigens do survive wax or plastic embedding and conventional histology, as well as harsher fixation techniques (e.g. Karnovsky, Zenker's, Saint-Marie, Carnoy, Bouin's). In whole mounts it is important to ensure even penetration of the antibody throughout the embryo, often achieved with the help of detergents, whilst preventing the antigen from being extracted from the tissue by the detergents. Whole mounts often present additional problems in terms of background staining: even a light background staining is made more prominent by the thickness of the tissue.

The methods given below, using 'light' fixation and either whole-mount staining or frozen (cryostat) sections of gelatin-embedded material have given us the most consistent results with many different antibodies and the best compromise between resolution of antigen localization and good structural preservation of delicate, very early embryos. If a cryostat is not available, then the method of choice, at least for young embryos, is probably to stain the embryos as whole mounts with either immunoperoxidase or alkaline phosphatase, and then to section them. It is important to remember always to include a control in the absence of primary antibody, particularly when trying out a method for the first time.

2.1 Immunocytochemistry of tissue sections

2.1.1 Fixation

(a) Fix embryo in appropriate fixative for antigen, as determined by trial and error. Good general procedures for protein and carbohydrate antigens are: absolute methanol at 4 °C for 3 h (or longer), absolute ethanol at 4 °C for 24–48 h, buffered formol saline (4% paraformaldehyde[a] in PBS, pH 7.0) for 30 min to 1 h.

(b) If fixed in methanol or ethanol, rehydrate slowly down a series of the alcohol (70%, 50%, 25%) to PBS until specimen sinks. If fixed in aldehyde, wash many times in PBS to remove traces of aldehyde.

(c) If the fixed embryos are to be stored for longer than about 12 h before further use, then keep in PBS containing 0.02% thimerosal as a preservative against microbial growth, at 4 °C. However, some antigens can suffer on prolonged storage.

(d) Embryos, particularly older ones, which contain some blood cells, may have endogenous peroxidase activity. This can be blocked by including 0.1% H_2O_2 during fixation.

---

[a] 4% paraformaldehyde (w/v) is made by dissolving paraformaldehyde powder in PBS with constant stirring at about 60 °C; it will dissolve slowly and only at alkaline pH. To protect against the *toxic vapours*, do this in a fume hood. Check the pH again after dissolving it. If too acid,

neutralize with 1 M NaOH. Use within 24 h or store as frozen aliquots. For many applications, however, it is sufficient to use 1:10 formaldehyde solution (commercially available as 37–40% stock, giving final concentration about 4%), but this should be mixed with PBS just prior to use and the pH checked.

2.1.2 Cutting frozen sections of embryos

The method given here (*Protocol 1*) requires a cryostat. In our experience, the best cryostats for cutting young embryonic material are made by Bright and by Leitz. Young embryos are best cut by first embedding in gelatine. This provides a transparent yet firm support, allowing embryos to be orientated as required. Provided that the block has a regular shape, gelatin-embedded sections should produce good ribbons (like wax sections) in the cryostat. Older embryos (more than about 8 mm long) can be embedded in OCT compound or can be sectioned in a vibratome.

Protocol 1. Gelatin embedding and cutting frozen sections

1. Transfer embryo to 5% sucrose/PBS at 4 °C until the specimen sinks. Transfer to 20% sucrose/PBS at 4 °C and leave overnight.

2. Embed: 7.5% (final, w/v) gelatin (Sigma G2500, 300-Bloom) in 15% (final, w/v) sucrose/PBS. Melt mixture with gentle stirring at about 45 °C, then leave at 38–39 °C until any bubbles disappear. Preheat embryos in 20% sucrose to 38 °C. Place them into the molten gelatin/sucrose mixture and leave at 38 °C to infiltrate until they sink (1–5 h). Then allow mixture to gel in a mould at room temperature. It is useful to have a flame nearby and some instruments (watchmakers' forceps, mounted needle, Pasteur pipette cut to a broad end) to help manipulate the embryos to the desired orientation in the gelatin mould.

3. Store blocks at 4 °C until required, wrapped in cling-film (not longer than 5 days).

4. Section in cryostat: trim gelatin block with embryo with a razor-blade so that its edges are parallel. Mount on cryostat chuck with OCT compound and freeze for 20–30 sec (longer time may lead to cracking) in a small plastic beaker of isopentane standing on solid CO_2. Place inside the cryostat chamber to equilibrate to this temperature. The optimum temperature for specimen and chamber seems to be around −22 °C, with the specimen holder and knife slightly colder if possible (about −25 °C). Section at 5–25 μm.

5. Pick up each group of sections, as they emerge over the knife, by briefly bringing up to it a gelatinized (subbed) slide (see *Protocol 2*) at room temperature. The sections will adhere to the slide. Allow them to dry at room temperature for a few minutes, and store them at 4 °C with some silica gel for no longer than 6 days.

Protocol 2. Making gelatin-subbed slides

1. Soak slides overnight in chromic acid (made by dissolving 100 g potassium dichromate in 850 ml H_2O and 100 ml concentrated H_2SO_4).
2. Wash with running tap water and rinse twice with distilled H_2O.
3. Make gelatin solution: 5 g gelatin (less strong than for embedding; e.g. Sigma G2625 or G6144), H_2O to 1 litre. Dissolve by gentle heating: when dissolved, add 0.5 g chromic potassium sulphate ('chrome alum'). Filter through Whatman No. 1 paper.
4. Place slides vertically in a rack and dip in the gelatin solution. Dry slides in a dust-free atmosphere. Can be stored indefinitely at room temperature.

2.1.3 Antibody incubations and detection

Protocol 3. Immunocytochemistry of tissue sections

1. Mark a ring around a group of sections on the glass slide, using a diamond pen.
2. Remove gelatin by placing slides with sections into PBS, prewarmed to 38 °C, for about 30 min (less for freshly cut sections; longer if the sections have been stored for a few days). Wash three times in PBS at room temperature. Sections should stay on the slides.
3. Using paper tissue, quickly dry around diamond mark on slide, without letting the sections dry out.
4. Block non-specific binding. Place about 100 µl of blocking solution in each ring to cover the sections and place slides horizontally in humid chamber for 30 min at room temperature. *Important*: do not let sections dry out at this stage or at any other subsequent stage. A general purpose blocking solution is 1% bovine serum albumin (BSA), 0.02% Tween-20, 1% heat-inactivated goat serum[a] or fetal calf serum in PBS. If non-specific binding is observed the composition of this should be altered.
5. Remove blocking solution, but do not wash. Cover sections with antibody (supernatants containing monoclonal antibodies are usually used neat, but add 100 µl 1–2% BSA for every millilitre of supernatant, to buffer. Purified antibodies or antisera can be diluted up to 1:5000 depending on immunoglobulin concentration; dilute antibodies in blocking solution). Incubate slides horizontally in a humid chamber at room temperature for 1 h.
6. Wash[b] well with PBS or PBST (3–5 times).

7. Incubate in 1:50–1:1000 of the appropriate secondary antibody[c], labelled as required with fluorescein (FITC), rhodamine (TRITC), horseradish peroxidase, alkaline phosphatase, etc. for 30 min to 1 h.

8. Wash[b] well in PBS or PBST (3–5 times).

9. The remaining part of the procedure varies depending on the detection method chosen. For immunofluorescence, follow *Protocol 4*, for immuno-peroxidase, *Protocol 5*, and for alkaline phosphatase, *Protocol 6*.

[a] To make heat-inactivated goat serum, warm goat serum to 55 °C for 30–45 min in a water bath. The purpose of this is to denature any remaining immunoglobulins in the serum.
[b] Washing procedure. A quick method for washing sections requires a combination of a plastic wash bottle filled with PBS, or PBS with 0.02% Tween-20 (PBST), and three to five Coplin jars filled with the same solution. First pour out some solution from the wash bottle over the slide (taking care not to squirt it directly onto the sections). Shake off the excess and dip slide into the first Coplin jar for a few seconds. Repeat these two procedures 3–5 times.
[c] It is usually advantageous to reduce the non-specific binding of the secondary antibody by preabsorbing it against the embryonic material being stained. Two simple procedures can be used. The first requires incubation of the working solution of secondary antibody for about 1 h at 4 °C with about 1/10 volume of embryonic tissues fixed in the same way as those being stained, and washed thoroughly to remove the fixative. This is followed by a brief centrifugation to remove the embryos from the solution. Alternatively, the secondary antibody may be absorbed against acetone powder made from embryos of the species being stained. A procedure for this is given in Chapter 26.

Protocol 4. Mounting for immunofluorescence (FITC, TRITC, Cascade Blue, Texas Red, etc.)

1. Mount under a coverslip in Gelvatol[a] (to reduce fluorescence quenching and photobleaching).

2. Observe under epifluorescence with appropriate filters for dye used.

3. Slides stained by immunofluorescence can be stored for at least a few days in the dark at 4 °C.

[a] Gelvatol (1, 3, 4) is prepared by mixing 36 ml 0.02 M NaH_2PO_4, 14 ml 0.02 M KH_2PO_4 and adjusting to pH 7.2. To this, add: 0.327 g NaCl, 0.024 g NaN_3, 0.6 g DABCO (diazobicyclo-octane; Aldrich); 10 g polyvinyl alcohol 20/30 (Fisons). Stir overnight at room temperature, wrapped in foil as DABCO is light-sensitive. When dissolved, add 20 ml glycerol. Stir overnight. Centrifuge at 7000 g for 15 min. Store at room temperature, protected from light.

Protocol 5. Detection for peroxidase-coupled antibodies

1. After the three to five washes in PBS following incubation in the secondary antibody, wash slides three further times (10 min each) with 0.1 M Tris-HCl pH 7.4.

Protocol 5. *Continued*

2. Make substrate just before use: 5 mg 3,3′-diaminobenzidine tetrahydro-chloride[a] (DAB; Aldrich) in 10 ml Tris-HCl pH 7.4. Keep in the dark. (*Note*: the pH of this solution is important. If the solution is more than slightly tinted the pH is likely to have ventured too far from 7.4 and it may form a precipitate over the sections. The pH can be checked with pH test strips without risking DAB contamination of a pH electrode).

3. Place slides in DAB solution in the dark for about 5 min.

4. Add H_2O_2 to a final concentration of 0.003% (make a 1:100 stock of commercial, 100 vol. H_2O_2 and add 100 μl of this, 1:10 000 from the stock, to each 10 ml DAB solution). Incubate about 2–10 min, occasionally watching under a dissecting microscope. Higher concentrations of peroxide cause very rapid reactions but much higher background levels, and may inactivate the enzyme.

5. Stop reaction by washing under running tap water, which will also intensify (blacken) the reaction product.

6. Dehydrate with a series of alcohols (50%, 70%, twice in absolute ethanol), clear in xylene or Histoclear and mount under a coverslip in D.P.X. (BDH). It is possible to counterstain lightly with a dilute solution of light green in absolute ethanol or with haematoxylin before clearing and mounting.

[a] *DAB is a suspected carcinogen.* Place all utensils and solutions that have come into contact with DAB into a solution of household *bleach* (sodium hypochlorite) to inactivate. Wear gloves and mask to weigh out DAB.

Protocol 6. Detection for alkaline-phosphatase-coupled antibodies

1. Make alkaline phosphatase buffer (APB): 0.1 M Tris-HCl pH 9.5, 50 mM $MgCl_2$, 100 mM NaCl, 0.1% Tween-20, 1 mM levamisole. Make on day of use.

2. Wash slides two further times (10 min each) in APB.

3. Make reaction mixture immediately before use: to each 10 ml APB add: 45 μl Nitro Blue Tetrazolium (NBT; 75 mg/ml in 70% dimethylformamide) and 35 μl 5-bromo-4-chloro-3-indolyl-phosphate (BCIP; 50 mg/ml in 100% dimethylformamide). Gibco-BRL sell a kit (Immunoselect) with these solutions ready to use.

4. Add reaction mixture to sections, and incubate at room temperature (30 min to 6 h; further time is of no advantage as exposure to the high pH of the buffer will inactivate the enzyme) in the dark, until colour reaction develops.

5. Wash twice with PBS.

Alkaline phosphatase stained specimens cannot be dehydrated with ethanols or cleared, but the following allows them to be cleared and mounted permanently: fix 30 min in 4% paraformaldehyde. Then *either*: mount in aqueous mountant (e.g. Hydromount, Aquamount, Gelvatol—see *Protocol 4*) *or*: dehydrate 5 min in methanol, 10 min in propan-2-ol, clear twice for 30 min in tetrahydronaphthalene and mount in D.P.X.

2.1.4 Biotin/streptavidin, PAP, and APAAP for increasing sensitivity

In some cases, biotinylated secondary antibodies can be used, followed by streptavidin or avidin coupled to an enzyme (peroxidase or alkaline phosphatase). Many workers report further increases in sensitivity using this method. However, many embryos (such as chick) contain large amounts of endogenous biotin and avidin, and we have found these methods to offer no advantage. The best way to find out for each particular system is to purchase the kit from Vector Laboratories (ABC kit) which is self-explanatory and includes appropriate blocking reagents to avoid problems with endogenous biotin and avidin. Always set up a control in the absence of primary antibody to check for possible artifacts.

Some workers also find that the use of peroxidase–anti-peroxidase (PAP) or alkaline phosphatase–anti-alkaline phosphatase (APAAP) systems also increase sensitivity. In our experience in embryos, they increase the background just as much or more than the signal. For details on how to use them, consult an appropriate book on immunological methods such as the excellent book by Harlow and Lane (1).

2.1.5 Double immunostaining of sections

The easiest way to visualize two antigens in the same section is to use double-immunofluorescence, using two different fluorochromes. Even three fluorochromes can be combined, if one of them is Cascade Blue (Molecular Probes, Inc.) whose emission is sufficiently separate from those of FITC and TRITC. The primary antibodies can be mixed together, as can the secondaries. Alternatively, immunoperoxidase and alkaline phosphatase can be combined, and in this case the DAB peroxidase reaction has to be carried out first. However, both methods require that the antibodies used be different immunoglobulin classes or raised in different species (e.g. a polyclonal rabbit serum and a mouse IgG). If this is not possible, double staining has to be performed using antibodies coupled directly to fluorochromes, enzymes or biotin (for subsequent detection with streptavidin coupled to an enzyme or to a fluorochrome).

199

2.2 Whole mount immunocytochemistry

The general procedure for staining whole embryos or organs with antibodies is similar to those described above for tissue sections. However, incubation and washing times are much longer, to help penetration of the antibodies, and it is more important to include some detergent; during all incubation and washing periods, gentle agitation is important. Large embryos, as a rule, will require more detergent and longer times; those given work well on chick embryos of about 1–2.5 days' incubation and on mouse embryos up to about 13 days. Amphibian and insect embryos require some additional modifications to allow the antibodies to penetrate through the rigid outer covering and to make them sufficiently transparent for observation of the stained specimen; these modifications are mentioned in the chapters on these species. It is also important to remember that some surface and soluble antigens may be extracted by the presence of any detergent; in these cases, the fixation and permeabilization procedures may have to be altered.

2.2.1 Incubations and detection for whole mount immunocytochemistry

---

Protocol 7. Incubations for whole mount immunocytochemistry

1. Fix as for tissue sections (Section 2.1.1). If using immunoperoxidase, include 0.1% H_2O_2 in the fixative.

2. Wash: three times for 10 min and three times for 30 min to 1 h in PBST (PBS containing between 0.01–0.1% Tween-20 or 0.5–1% Triton-X100 and 0.02% thimerosal as preservative), with gentle rocking, at room temperature.

3. Block overnight with gentle rocking at 4 °C in blocking buffer (PBST containing 1% BSA), 1–10% fetal calf or heat-inactivated goat serum). To make heat-inactivated goat serum, warm goat serum to 55 °C for ʽ30–45 min in a water bath. The purpose of this is to denature any immunoglobulins in the serum.

4. Incubate in primary antibody made up to the appropriate concentration (monoclonal supernatants usually diluted 1:1, or dilute up to 1:10 000 for ascites fluid, purified immunoglobulins or some antisera) in blocking buffer for 1–4 days with gentle agitation at 4 °C. If using undiluted supernatants, add about 100 μl of 1% BSA in PBST to buffer and the appropriate concentration of Tween-20 or Triton-X100 to help penetration. Larger embryos and IgMs will require longer incubations than IgGs or smaller embryos.

5. Wash: three times for 10 min and three times for 30 min to 1 h in PBST, with gentle rocking, at room temperature.

6. Incubate in appropriate secondary antibody (see footnote *c*, *Protocol 3*) at an appropriate dilution in blocking buffer, overnight at 4 °C, with gentle agitation. The following peroxidase-coupled secondary antibodies have given us the best results:

- mouse IgG: Jackson anti-mouse IgG-peroxidase
- mouse IgM: Sigma goat anti-mouse IgM peroxidase
- rabbit IgG: Amersham donkey anti-rabbit IgG-peroxidase
- rat IgM: Calbiochem anti-rat IgM peroxidase

7. Wash: three times for 10 min and three times for 30 min to 1 h in PBST, with gentle rocking, at room temperature.

8. The remaining part of the procedure depends on the detection method chosen, peroxidase (*Protocol 8*) or alkaline phosphatase (*Protocol 9*).

Protocol 8. Immunoperoxidase detection for whole mounts

1. Wash two further times for 20 min each in 0.1 M Tris pH 7.4.

2. Follow the procedure described for sections (*Protocol 5*). Make substrate just before use: 5 mg DAB in 10 ml 0.1 M Tris pH 7.4. Keep in the dark. Place embryos in 10 ml DAB solution in the dark at 4 °C with gentle agitation for 30 min to 1 h.

3. Add H_2O_2 to a final concentration of 0.003% (make a 1:100 stock of commercial, 100 vol. H_2O_2 and add 100 μl of this, 1:10 000 from the stock, to the DAB solution). Incubate 5–20 min, occasionally watching under a dissecting microscope. Higher concentrations of peroxide cause very rapid reactions but much higher background levels, and may inactivate the enzyme.

4. Discard the spent DAB into a bucket of dilute bleach (see footnote *a*, *Protocol 5*). Stop the reaction by washing many times in tap water, which will also intensify (blacken) the reaction product.

5. If desired, counterstain and/or dehydrate with a series of alcohols (50%, 70%, twice with absolute ethanol), clear in xylene or Histoclear.

6. If desired, stained embryos can be embedded in paraffin wax and sectioned conventionally. The oxidized DAB will even stand acids, all histological stains, heat, all organic solvents tried to date, and proteases. It is therefore possible to follow the immunoperoxidase procedure given here with whole mount *in situ* hybridization methods using digoxigenin-labelled probes.

Protocol 9. Whole mount immunodetection for alkaline phosphatase

1. After the washes following incubation in secondary antibody, wash embryos three further times (10 min each) with APB (see *Protocol 6*).

2. Add reaction mixture (10 ml APB containing 45 μl Nitro Blue Tetrazolium [NBT; 75 mg/ml in 70% dimethylformamide] and 35 μl 5-bromo-4-chloro-3-indolylphosphate [BCIP; 50 mg/ml in 100% dimethyl-formamide]) to embryos and incubate in the dark at room temperature (30 min to 2 h; further time is rarely of advantage as exposure to the high pH of the buffer will inactivate the enzyme, but, provided no precipitation is visible in the reaction mixture, can lead to further sensitivity) until colour reaction develops.

3. Wash twice with PBS.

4. Alkaline-phosphatase-stained specimens cannot be dehydrated with ethanol or cleared, but the following allows them to be subjected to conventional wax histology. Fix 30 min in 4% paraformaldehyde, dehydrate 5 min in methanol, 10 min in propan-2-ol. Clear twice for 30 min in tetrahydronaphthalene. Infiltrate in 50:50 tetrahydronaphthalene/fibrowax (or Paraplast) at 57 °C for 30 min. Immerse twice in fibrowax or Paraplast at 57 °C for 30 min. Embed in moulds and section conventionally.

2.2.2 Double immunostaining of whole mounts

If two antibodies of interest were raised in different species of hosts (e.g. mouse, rat, rabbit) and/or if they are different immunoglobulin classes (e.g. IgM, IgG), then it may be possible to co-localize them in the same whole-mounted embryo. To do this, use a mixture of the primary antibodies for the first incubation and a mixture of secondary antibodies for the second incubation. Then perform the DAB reaction, and wash well in PBS. Finish by processing the alkaline phosphatase reaction. It is usually advantageous to keep the intensity of the reactions low (particularly the peroxidase detection) because otherwise the embryos may be very dark.

3. Biochemical characterization of antigens

Space does not permit an in-depth review of all the possible procedures for characterizing antigens or checking the specificity of the antibodies used in histological localization. However, because of the importance of characterizing the specificity of antibodies used in the procedures outlined above, two methods will be given here. Both work well and consistently from small amounts of material and are rapid (results within a day) and easy to perform in any laboratory, even if not set up for biochemistry. The absolute requirement, however, is for a tank capable of running small gels (a simple one is manufactured by Atto Corporation and can be purchased from Genetic Research Instrumentation), and a simple power supply capable of delivering about 150 V DC (constant voltage). Together these should cost less than £500. Immunoblotting requires, in addition, a horizontal blotting apparatus (e.g. Horizblot, or LKB Transphor). These can be obtained from about £500. The power supply required should be able to operate in constant current mode and deliver up to about 60 mA.

The procedures given here are based on the Atto minigel tank and the LKB Transphor system but are easily adaptable to any other minigel and horizontal blotting systems.

3.1 Characterization of antigens by immunoblotting (Western blots)

The process of antigen characterization by immunoblotting consists of three steps: separation of proteins from embryos by SDS-polyacrylamide electrophoresis (SDS/PAGE), transfer of the separated proteins onto a nitrocellulose membrane (blotting), and probing the blot with the antibody to be tested. To prepare a sample for SDS/PAGE, first collect some embryonic material. As a guide, you will need about 5 µl of packed embryo volume to run about three to five tracks on a minigel set-up. It should be easy to collect this amount of material for any of the species of embryos in common use. Simply collect the embryos in the normal saline for the species, and transfer them in as small a volume of the saline as possible (such as in a Gilson tip, in 1–5 µl per embryo) to an Eppendorf tube kept on solid CO_2, to freeze them quickly. The saline used for collecting the embryos should contain a cocktail of as many as possible of the protease inhibitors shown in *Table 1*.

Table 1. Protease inhibitors[a]

Inhibitor	Stock solution	Working concentration
Phenylmethylsulphonylfluoride (PMSF)	10 mg/ml in isopropanol	50 µg/ml (1:2000)
N-Tosyl-L-phenylalanine chloromethyl ketone (TPCK)	3 mg/ml in ethanol	50 µg/ml (1:60)
N-Ethyl maleimide (NEM)	100 mM in ethanol	1 mM (1:100)
Pepstatin	1 mg/ml in methanol	1 µg/ml (1:1000)
Leupeptin	1 mg/ml in saline or water	1 µg/ml (1:1000)
Aprotinin	100 µg/ml in saline or water	0.2 µg/ml (1:500)
N-α-p-tosyl-L-lysine chloromethyl ketone (TLCK)	1 mg/ml in acetate buffer, pH 5.0	50 µg/ml (1:20)
$α_2$-Macroglobulin	100 units/ml in saline	1 unit/ml (1:100)

[a] The stock solutions can be kept frozen for weeks if necessary.

3.1.1 Solutions required to cast protein minigels

On the day when the experiment is to be run, cast the minigels. The following stock solutions are needed:

- solution 1: 40 g acrylamide[a] and 1.07 g *N,N'*-methylene-bis-acrylamide[a], made up to 100 ml with distilled water
- solution 2: 0.8% sodium dodecyl sulphate (SDS)
- solution 3: 3 M Tris-(hydroxymethyl)-aminomethane (Trizma base, Sigma) adjusted to pH 8.5 with concentrated HCl
- solution 4: 0.28% ammonium persulphate
- solution 5: distilled water
- solution 6: 1 M Tris pH 6.8
- solution 7: *N,N,N',N'*-tetramethylethylenediamine (Temed)

[a] Unpolymerized acrylamide solutions and acrylamide powder are powerful neurotoxins, with cumulative effects. Wear gloves whenever using them, and a mask whenever the powder is being weighed.

3.1.2 Pouring the gel

The Atto minigel system consists of the tank itself, a pair of glass plates on which the glass spacers are already fixed, a plastic comb for casting the channels, a rubber sealing gasket, and a pair of strong plastic clips to hold the plates together during pouring of the gel. First make sure that the plates are very clean. Wash them with water and detergent, and finally rinse with 70% alcohol. Wear gloves to avoid leaving proteins on the plates. Now assemble the glass plates together, with the spacers between them, and the gasket around the edge just outside the spacers, and secure with the plastic clips. Stand the assembly upright on the bench, making sure that it is level. Now insert the plastic comb between the plates, as far as it will go. With a marker pen, make a mark about 2 cm *below* the end of the comb. This will mark the extent of the separating gel. Remove the plastic comb.

Table 2. Mixing proportions for pouring polyacrylamide gels

To make 20 ml volume:

Solution No.	3%	4%	5%	7.5%	10%	12%	15%
1	1.5 ml	2.0 ml	2.5 ml	3.75 ml	5.0 ml	6.0 ml	7.5 ml
2	2.5	2.5	2.5	2.5	2.5	2.5	2.5
3	2.5	2.5	2.5	2.5	2.5	2.5	2.5
4	1.0	1.0	1.0	1.0	1.0	1.0	1.0
5	12.5	12.0	11.5	10.25	9.0	8.0	6.5
7	25 μl	25 μl	25 μl	25 μl	25 μl	25 μl	25 μl

Now make 20 ml of the gel mixture as indicated in *Table 2*. For a first attempt, use a 10% gel (if separating high molecular weight components, make a more dilute gel, or a stronger gel for low molecular weight components; 10% is a good compromise). Add solution 7 (Temed) at the very last minute, as this will initiate polymerization. As soon as this is done, carefully (to avoid air bubbles) pour the mixture between the plates up to the level of the mark. With a Pasteur pipette, add about 0.5 ml isobutanol to the top; this will ensure a straight edge to the separating gel. Wait for the gel to polymerize completely (may take as long as 2 h; light will accelerate the process). The solution left in the beaker where it was mixed will serve as an indicator.

Once polymerized, pour the isobutanol off the gel, and rinse many times with distilled water with a Pasteur pipette to eliminate any traces of isobutanol. Now mix the stacking gel. 10 ml is sufficient. Make a 4% gel, using solution 6 instead of solution 3. Finally add 100 µl of Temed (solution 7) for 10 ml stacking gel mixture. Immediately pour between the plates, up to the brim, and insert the comb into the unpolymerized stacking gel. It will polymerize quickly (a few minutes or even seconds).

3.1.3 Electrode and sample buffers

(a) The electrode buffer to run the system consists of: 1 g SDS, 14.4 g glycine, 3.025 g Tris. Make up to 1 litre with distilled water. The pH will be around 8.3, but do not adjust it.

(b) The buffer in which the samples are processed (sample buffer) consists of the following: for 11 ml (pH 6.8): 1 ml 10% SDS, 5 ml 50% glycerol, 5 ml 0.3 M Tris pH 6.8, 500 mg dithiothreitol (DTT) (omitted for non-reducing conditions), 5 mg Bromophenol Blue. This buffer may be stored frozen in aliquots for up to about 2 months.

3.1.4 Sample preparation and running the gel

Protocol 10. Sample preparation and running the gel

1. To the Eppendorf tube with frozen tissue sample, add ¼ of the volume of sample buffer, vortex briefly, and place the Eppendorf upright in a beaker of boiling water for 3 min.

2. Vortex again, and then centrifuge at high speed in microcentrifuge (e.g. MicroCentaur) for 2 min to pellet any undissolved materials. The samples are now ready to run on the gel. It is also important to have suitable molecular weight markers, which can be purchased commercially. It may be an advantage to use prestained molecular weight markers (obtainable from Sigma, Amersham and other companies), so that separation of different molecular weights can be followed directly.

Protocol 10. *Continued*

3. Remove the plastic clips and gasket from the gel and insert it into the tank so that the notched plate faces inwards and to the top (into the upper buffer container).

4. Fill upper and lower containers with electrode buffer, making sure that no bubbles are trapped at the bottom.

5. Remove the channel forming comb, carefully so as not to break the thin acrylamide separators between channels. Gently rinse each channel with a syringe filled with electrode buffer to remove any unpolymerized acrylamide.

6. Now place each sample in a channel using either a Gilson pipette fitted with a duckbill (very fine) tip, or with a fine glass or plastic pipette or with a fine syringe (e.g. Hamilton syringe). One of the channels should contain the molecular weight markers (concentration as indicated by the manufacturer), and the other(s) should contain the sample(s). Run everything in duplicate in two halves of the gel; one half will be used to stain for protein, whilst the other half will be blotted and probed with the antibody.

7. Put the lid on the tank and connect it to the power supply. Run for about 1.5–3 h at 100–120 V (constant voltage) until the blue line (dye front) nearly reaches the bottom of the gel.

3.1.5 Staining for total protein

It is important, in addition to characterization of the antigen(s) on blots, to be able to confirm that the initial protein sample is representative across the whole molecular mass range. For this, one can stain a gel similar to the one that will be blotted. Two are available. The first (*Protocol 11*) is simple but insensitive, based on Coomassie Blue. The second (silver staining, *Protocol 12*) is slightly more involved but very sensitive; despite some workers believing that silver staining is temperamental, the protocol given here works very consistently provided that very clean deionized water is used to make all the solutions.

Alternatively, total proteins can be stained in the blot itself using either Aurodye (colloidal gold, sold by Janssen; very sensitive but fairly expensive, and does not allow probing of the blot with antibodies afterwards) or Ponceau-S (a reversible red dye, but very insensitive). Methods for these can be found in standard immunological textbooks (e.g. ref. 1). However, it is generally better to silver stain the duplicate half of the gel.

To process the freshly run gels for protein staining and transfer to nitrocellulose, start by removing the gel plate from the tank. Carefully prise the two glass plates away from each other. With a clean razor-blade, cut off the stacking gel and discard it, and split the gel into the half to be stained for proteins and the half to be blotted. Pick up each half carefully off the plate with gloved fingers, and place the half for protein staining into a clean plastic box with fixer (50% methanol, 10% acetic acid) for 15 min. Meanwhile, keep the half destined for blotting in a clean plastic box with distilled water.

Protocol 11. Coomassie Blue staining

1. After fixation, place the gel in a plastic box containing 0.1% solution of Coomassie Blue R250 made up in water/methanol/acetic acid (5:5:2) for 1–2 h, until it has become evenly blue.

2. Transfer the gel (keep the dye for reuse) into destain: 5% acetic acid, 7.5% methanol. Put a piece (about 10 cm square by 3 cm deep) of plastic foam into the solution to help absorb the dye; alternatively, an old teabag (wash it well if used!) filled with activated charcoal and stapled securely to seal will do just as well or better.

3. Leave overnight, rocking gently to clear the background and reveal the protein bands.

Protocol 12. Silver staining

1. After fixation, place the gel to be stained into a clean plastic box with destaining solution (5% acetic acid, 7.5% methanol) for 15 min.

2. Then fix again for 10 min in 10% glutaraldehyde.

3. Place overnight in a large volume of clean deionized water with gentle rocking to wash out the glutaraldehyde.

4. The next day, place for 30 min into 5 μg/ml dithiothreitol (DTT).

5. Immerse for 20 min, on a rocker, in 100 ml 0.1% $AgNO_3$, made with the best quality deionized water.

6. After a very quick rinse in deionized water, rinse quickly in a small volume of 3% Na_2CO_3 containing 100 μl formalin for every 200 ml.

7. Place into 100 ml of fresh 3% Na_2CO_3/formalin and shake gently until good staining appears, watching over a light box. Stop the reaction (before the background begins to turn yellow) using 5 ml 2.3 M citric acid to neutralize, and leave 10–15 min. Place it in distilled water containing a little glycerol if desired for storage for up to 2 days.

3.1.6 Blotting

The half of the gel destined for immunoblotting will have been placed in distilled water. The procedure described here for blotting is called the semi-dry, discontinuous buffer transfer system and is particularly good for even transfer of components of different molecular weights.

Protocol 13. Semi-dry discontinuous buffer transfer of proteins

1. The following materials are needed:
   - Buffer I: 0.3 M Tris (36.3 g/litre), 20% methanol
   - Buffer II: 0.025 M Tris (3.025 g/litre), 20% methanol
   - Buffer III: 0.04 M ε-aminohexanoic acid, 0.025 M Tris, 20% methanol (pH 9.3)
   - 18 pieces of clean filter paper (e.g. Whatman No. 1), cut slightly larger (about 1–2 mm larger all around) than the gel to be transferred
   - 1 piece of nitrocellulose (handle with gloves and with *great care*, *very fragile*) cut to the same size as the filter papers

2. Wet the graphite electrodes of the horizontal blotting apparatus with distilled water and assemble the system as follows:
   - cathode (positive electrode) (usually at the bottom)
   - six filter papers soaked briefly in buffer I
   - three filter papers soaked briefly in buffer II
   - nitrocellulose, rinsed quickly in distilled water
   - gel, dipped quickly in buffer II
   - nine filter papers soaked briefly in buffer II
   - anode (negative electrode)

3. Transfer at constant current, at a rate of 0.8 mA per cm^2 of gel area, for 1 h.

4. After running, disconnect the power supply, disassemble the transfer unit and remove top filter papers.

5. The gel will adhere to the nitrocellulose; immerse briefly in distilled water to help separate them. If any remaining pieces of stacking gel still adhere to nitrocellulose, remove them carefully.

6. Wash nitrocellulose in 20 mM Tris-saline (20 mM Tris, 0.9% NaCl, azide if wanted, pH 7.4–8.2). Do not let it dry out.

3.1.7 Probing the blot with antibodies

Nitrocellulose blots can be stained using the procedures described above for sections and whole mounts above, using peroxidase- or alkaline-phosphatase-coupled secondary antibodies. For sensitivity and speed, however, a newly available method using chemiluminescence (ECL system) will also be described.

Protocol 14. Detection of antigens in blots

1. The following solutions are needed:

   - TBS: 20 mM Trizma base, 0.9% NaCl. Adjust pH to about 7.4–8.0 with HCl
   - TFT: TBS containing: 5% fetal calf serum and 0.1% Tween-20
   - blocking buffer: TFT containing: 2% Marvel (dried skimmed milk) (warm up slightly to dissolve) *or* 5% BSA

2. Wash the blot two or three times in TBS and two or three times in TFT (10–15 min per wash).

3. Block for 1 h or overnight in blocking buffer, on rocker.

4. Replace with 1–10 ml (as required to just cover blot) of fresh blocking buffer, and add appropriate volume of primary antibody directly to the blocking solution to give the recommended dilution (1/10 to 1/5 of the dilution necessary for staining sections or whole mounts is usually sufficient). Incubate at room temperature with gentle rocking for 3 h or overnight.

5. Pour off primary antibody and wash three times with TBS and two to three times with TFT (10–15 min each, in a large volume). Incubate in solution of appropriate peroxidase- or alkaline-phosphatase-labelled secondary antibody (see footnote *c*, *Protocol 3*) made up in blocking buffer for 1–3 h at room temperature, on a rocker. Pour off secondary antibody and wash three times in TBS and two to three times with TFT (10–15 min each, large volume).

6. There are three simple methods for detection:

   (a) for alkaline phosphatase-labelled secondary antibodies, proceed as described in *Protocol 6*
   (b) for peroxidase-labelled secondary antibodies, proceed as described in *Protocol 5*
   (c) for peroxidase-labelled secondary antibodies, you can also use the ECL system (Amersham) following the manufacturer's instructions. The ECL system requires exposure of a photographic plate (X-ray film type, also supplied by Amersham specifically for the ECL detection system, as is the developer required) to the chemo-luminescence emmitted by the peroxidase substrate used, but the

Protocol 14. *Continued*

>   whole process is extremely fast and far more sensitive than DAB-
>   based methods of detection. Exposures can be done under darkroom
>   safelight and range from 5 sec to 1 min; making multiple exposure for
>   antigens of different abundance is therefore very easy.

3.2 Characterization of antigens by immunoprecipitation (method courtesy of Dr Andrea Streit)

This method should be used if a blotting apparatus is not available. However,
it is important to realize that it tends to reveal more proteins than
immunoblotting, and that not all of these proteins are specifically recognized
by the antibodies being tested. In part, this may be because some of the
antigens will themselves bind other proteins, which will also be detected by
this method. The protocol included here, although a little involved, includes a
set of stringent washing steps to remove much of this non-specific binding. It
is also important in each experiment to include a control in which the primary
antibody is omitted, to control for the possibiltiy that the Sepharose beads
themselves may be precipitating some specific proteins.

Protocol 15. Immunoprecipitation

1. The materials needed are as follows:
    (a) solubilization buffer: 0.15 M NaCl, 20 mM Tris pH 7.4, 1 mM
        EDTA, 1 mM EGTA, 0.5% Nonidet P-40 or Triton X-100, 1 mg/ml
        bovine serum albumin (BSA) and a cocktail of protease inhibitors
        (see above and *Table 1*)
    (b) sucrose cushion: 1 M sucrose in solubilization buffer
    (c) washing buffer I: 0.4 M NaCl, 1 mM EDTA, 1 mM EGTA, 0.5%
        Nonidet P-40, 50 mM Tris pH 7.4, 0.05% NaN_3
    (d) washing buffer II: 0.1% SDS, 50 mM Tris pH 7.4, 1 mM EDTA, 1
        mM EGTA, 0.05% NaN_3
    (e) washing buffer III: 0.1% SDS, 0.5% Nonidet P-40, 1% sodium
        deoxycholate (DOC), 0.15 M NaCl, 10 mM Tris pH 8.1, 0.05%
        NaN_3
    (f) sample buffer (see Section 3.1.3b)
    (g) Sepharose beads coupled to an appropriate 'secondary' antibody
        (i.e. one appropriate for the antibody being tested), equilibrated in
        solubilization buffer
    (h) antibody to be tested and sample to be tested

210

2. Collect embryos (the amount needed can be ten times that described for electrophoresis above, but will depend on the abundance of the antigen being detected) in about 1 ml solubilization buffer (at least 10–20 times the volume of tissue collected).

3. Incubate on ice with gentle shaking for 30 min.

4. Centrifuge at 100 000 g at 4 °C for 30 min.

5. Add the antibody to be tested (10 µg of a pure monoclonal antibody or 100–200 µg of polyclonal antiserum). Incubate for at least 1 h or overnight at 4 °C with gentle agitation.

6. Add about 120 µl Sepharose beads coupled to 'secondary' antibody. Incubate 1 h (not longer) at 4 °C with gentle agitation. Add 20 µl Phenol Red solution.

7. Layer carefully over 600 µl sucrose cushion and centrifuge in a Biofuge-A for 8 min at 4000 g (7000 r.p.m.). Remove upper phase carefully and discard.

8. Coat cushion carefully with washing buffer I and centrifuge for 5 min at 800 g (3000 r.p.m.). Remove upper phase and discard.

9. Coat cushion with washing buffer II and centrifuge for 5 min at 800 g. Discard upper phase.

10. Remove cushion carefully and wash the pellet: suspend the pellet in 1 ml washing buffer II, shake for 5 min and centrifuge it for 3 min at 800 g. Discard the supernatant.

11. For monoclonal antibodies only: suspend the pellet in washing buffer III, shake for 5 min and centrifuge 3 min at 800 g. Discard the supernatant.

12. Suspend pellet in 1ml washing buffer II, shake 5 min, coat on a fresh sucrose cushion (600 µl in solubilization buffer) in a fresh Eppendorf tube. Centrifuge 8 min at 4000 g, remove upper phase and sucrose cushion carefully and discard them.

13. Suspend pellet in 50 µl sample buffer (see methods for SDS/PAGE above) and boil for 5 min in a water bath.

14. This can now be run on a gel as described in Sections 3.1.1–3.1.4. The resulting gel should be stained for total proteins by Coomassie Blue or silver staining (see Section 3.1.4). Remember to place molecular weight markers in a parallel track on the gel, as well as to process a control sample as above but with the primary antibody incubation omitted. In the final gel, two bands other than those due to the sample will be seen, corresponding to the heavy and light chains of the antibodies used.

References and further reading

1. Harlow, E. and Lane, D. (1988). *Antibodies: a laboratory manual*. Cold Spring Harbor, New York.
2. Catty, D. (ed.) (1988). *Antibodies: a practical approach*. IRL Press at Oxford University Press, Oxford.
3. Heimer, G. V. and Taylor, C. E. (1974). *J. Clin. Pathol.*, **27**, 254.
4. Osborn, M. and Weber, K. (1982). *Methods Cell Biol.*, **24**, 97.
5. Roitt, I. M. (1988). *Essential immunology*. Blackwells Scientific Publications, Oxford.
6. Hames, B. D. and Rickwood, D. (ed). (1981). *Gel electrophoresis of proteins: a practical approach*. IRL Press at Oxford University Press, Oxford.

22

RNAase protection assays

ARIEL RUIZ i ALTABA

1. Introduction

The development of a sensitive assay with which reliably to measure and map messenger RNAs (1, 2) has proved very useful in the study of the molecular mechanisms underlying inductive and differentiation events. The transcription of specific genes is often taken as a reliable marker of the commitment of the expressing cell towards one fate or another. For example, embryonic cells expressing α-actin are thought of as cells differentiating along a muscle pathway and those expressing N-CAM along a neural pathway. Early regulatory genes, the expression of which may be crucial in the commitment of a cell to a given fate, are also used as markers, although the expression of these genes should always be interpreted cautiously. Here I describe a method, based on that of Krieg and Melton (3), to measure the levels of mRNAs in embryonic cells by RNAase digestion of RNA:RNA hybrids. For general molecular techniques the reader is advised to consult ref. 4.

To test for the presence or abundance of a specific mRNA it is necessary first to have a cDNA or coding genomic piece cloned downstream of a phage promoter to perform an *in vitro* transcription reaction. A small portion of the cDNA is transcribed in the antisense orientation yielding a short antisense RNA. A radioactive nucleotide is included in the polymerization reaction to ensure that the antisense RNA will be uniformly labelled to a high specific activity so that small amounts can be detected by autoradiography. Excess labelled antisense RNA is then mixed with total RNA extracted from the embryos or cells to be tested under conditions that drive RNA:RNA hybrid formation. The hybridization mixture is reacted with RNAases that cleave single standed RNAs, leaving stable RNA:RNA hybrids intact. To minimize hybrid degradation, the conditions of RNAase digestion should be optimized by testing hybrids made with synthetic unlabelled sense and labelled antisense RNAs. Since the only labelled macromolecules remaining after digestion will be those that were protected from the RNAases by virtue of having formed a stable hybrid, these can be separated in denaturing polyacrylamide gels to measure their length. The intensity of full-length protected fragments will reflect the abundance of the mRNA that is tested in the RNA sample. It is always necessary to test for the presence of RNAs the abundance of which is known. These controls allow for the calibration and standardization of the results with other RNAs.

2. Extraction of small amounts of RNA from embryonic cells

Protocol 1. Isolation of RNA from embryonic cells

1. Transfer early frog, chick, rat, mouse, or fish embryos to be processed to a small glass homogenizer with 1 ml PK[a] buffer containing 10 μl of a stock of proteinase K (Boehringer) at 25 mg/ml. This stock must be kept frozen. 1 ml of buffer is usually sufficient for 10–50 embryos, depending on their size. The homogenizer and pestle should be rinsed before and after every use with 1% SDS.

2. Rapidly homogenize embryos by stroking the pestle five to ten times. The solution should appear homogeneous.

3. Transfer the homogenate to a sterile 15 ml plastic tube and incubate for 1 h at 37 °C. Vortex occasionally.

4. Add an equal volume of equilibrated phenol/chloroform (1:1) and vortex vigorously for about 1 min.

5. Spin tubes in a clinical centrifuge for 10 min at top speed (3000 *g*).

6. Transfer the upper (aqueous) phase to a new 15 ml tube without disturbing the denatured proteins at the interphase. Extract once with an equal volume of chloroform. Vortex vigorously for about 30 sec.

7. Spin as in step **5**.

8. Transfer the aqueous phase to a new 15 ml tube and add 2.5 vol. of 100% ethanol. Incubate at −70 °C for more than 1 h or overnight at −20 °C.

9. Spin tubes as in step **5**.

10. Decant the supernatant and dry the nucleic acid pellet. The pellet should be white. If small amounts of nucleic acids are expected, a pellet may not be obvious.

11. Resuspend the pellet in 12.5 μl of DEPC-treated[b] and autoclaved water per embryo or embryo equivalent. Ensure that all the nucleic acid is dissolved; the liquid should reach the sides of the tube. Heating the sample to 68 °C for 1–3 min speeds up solubilization.

12. Add an equal volume of 8 M LiCl (DEPC-treated and autoclaved[b]). Vortex and incubate at −20 °C overnight. A quick precipitation at −70 °C will not give the best yield. This step selectively precipitates

RNA species of more than about 200 bases. Store RNA in this form at 25 μl per embryo or embryo equivalent.

13. Spin the desired volume as in step **5**. If the volume is small, vortex and transfer the solution to an Eppendorf tube and spin in a microcentrifuge at maximum speed for 5 min.

14. Aspirate the supernatant and remove the remaining liquid with a capillary pipette. This is important since the supernatant contains genomic DNA.

15. Rinse the pellet with 1 ml 100% ethanol at −20 °C.

16. Dry the pellet and resuspend it in the desired amount of DEPC-treated water (for poly(A) selection or measuring RNA concentration) or in 100% formamide (for RNAase protection).

[a] PK buffer: 50 mM NaCl, 50 mM Tris-HCl pH 7.5, 5 mM EDTA, 0.5% SDS.
[b] Diethylpyrocarbonate (DEPC)-treated water: add DEPC (Sigma) to a glass bottle of deionized water to a concentration of 0.1%. Shake vigorously for a few seconds, and incubate at room temperature overnight. Then autoclave for 20 min. *Caution*: DEPC is highly toxic. Other solutions, except those containing primary amines (e.g. Tris) may be DEPC-treated.

Some embryonic tissues and old embryos may not homogenize easily and may contain high levels of ribonucleases. In this case it is advised to proceed with a protocol involving quick protein denaturation by guanidinium compounds (5) (see Chapter 24).

3. RNA synthesis

Protocol 2. *In vitro* synthesis and purification of high specific activity labelled RNA

1. Digest DNA at a site 3′ to the fragment to transcribe. Transcription will initiate upstream of the polylinker sequences of the vector, and proceed through the insert to the end of the linear DNA molecule. There is no need to purify the DNA template after digestion.

2. Thaw and bring to room temperature all components to be added to the transcription reaction. This is important as the spermidine in the transcription buffer will precipitate the DNA if added cold. Nucleotides should be kept on ice for extended periods.

3. Assemble the transcription reaction in a microcentrifuge tube in the following order:

   ● 1–2 μl (~1 μg) cut DNA template

Protocol 2. *Continued*

- 0.5 μl 200 mM dithiothreitol (DTT)
- 0.5 μl 2 mg/ml BSA
- 0.5 μl 10 mM rATP
- 0.5 μl 10 mM rCTP
- 0.5 μl 10 mM rGTP
- 1 μl 10 × transcription buffer[a]
- 2 μl [^{32}P]UTP (400 Ci/mmol) (*fresh*)
- 1–2 μl DEPC-treated water (see *Protocol 1*)
- 0.5 μl RNAsin (20–40 units)
- 1 μl T7, T3, or SP6 (as appropriate for vector and strand being transcribed) RNA polymerase (40–80 units)

Final volume: 20 μl

4. Incubate the reaction at 37–40 °C for 1 h.
5. Add 1.5 μl RNAase-free DNAase (1 unit/μl). Incubate for 15 min at 37 °C.
6. After the incubation, either add 20 μl 100% formamide and then proceed to step **16**, or first clean up the sample as instructed in steps **7–15**.
7. Bring volume to 100 μl with DEPC-treated water.
8. Extract once with an equal volume of phenol/CHCl$_3$ (1:1).
9. Extract once with an equal volume of CHCl$_3$ alone.
10. Precipitate with 0.1 vol. of 7 M ammoniun acetate or 3 M sodium acetate pH 4.5 and 2.5 vol. 100% ethanol.
11. Incubate in solid CO$_2$ at −70 °C for 1 h.
12. Spin at maximum speed in a microcentrifuge for 5 min.
13. Aspirate supernatant with a drawn-out pipete. Discard radioactive liquid appropriately.
14. Check pellet with a Geiger counter for incorporation (It should be *very* 'hot'). Dry pellet, for example, in a dry block set at 45 °C.
15. Resuspend the pellet in sequencing gel sample buffer (containing 80% formamide and dyes). Vortex vigorously to get all the material on the sides of the tube.
16. Denature the RNA probe by heating for 1–2 min at 68 °C.
17. Load all the sample in a well of a thin (1 mm) 6% polyacrylamide/8.3 M urea sequencing gel. A small gel (e.g. 20 × 20 cm) is highly recommended. If necessary, run freshly prepared end-labelled DNA markers in the same gel. Make sure these are well denatured. It is also

important to clean the well scrupulously using a jet of electrode buffer before application of the sample to remove unpolymerized acrylamide.

18. Run the gel. Separate glass plates and tightly wrap the gel on top of one of the plates with plastic wrap. Check gel with Geiger counter. There should be a spot that is very 'hot'.

19. Place wrapped gel in a cassette and expose an X-ray sensitive film for 0.5–2 min. Mark the position of the film in relation to the gel by making small holes in the periphery of by aligning one corner of both gel/plate and film.

20. Develop the film. There should be a major band and several other minor ones in the sample lane (*Figure 1*). The shorter fragments represent premature terminations. Any major band is suitable for use as a probe. Note the apparent size of the dyes in a 6% gel for reference: Bromophenol Blue, 60 bases; xylene cyanol, 220 bases.

21. Align the developed film and the gel and make small incisions with a needle around the band to be extracted. Remove the film and cut a narrow gel slice with a razor or scapel. It is more important to obtain pure probe than to recover all the radioactivity, so it is best to cut as close as possible to the 'hottest' part of the band. Remove the plastic wrap from the slice and place this in a microcentrifuge tube.

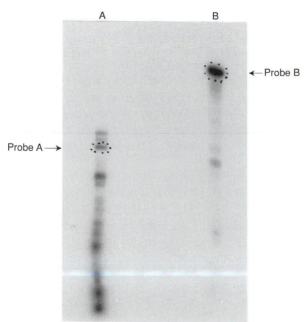

Figure 1. Probe purification on urea/polyacrylamide gel, 2 min exposure. A, Probe A; B, probe B.

217

Protocol 2. *Continued*

22. Briefly spin tube to bring gel slice to the bottom and add 10–50 μl (depending on how 'hot' the band is) of 50% formamide (all formamide should be deionized; it should not freeze at −20 °C). Incubate at 45 °C for 1–12 h to allow the RNA molecules to exit the gel.

a 10 × transcription buffer: 400 mM Tris-HCl pH 7.5, 60 mM MgCl$_2$, 20 mM spermidine-HCl.

Any vector, such as those of the pSP and pGem (Promega) or pBluescript (Stratagene) series, containing T3, T7, or SP6 promoter sequences and a polylinker for cloning can be used. A pure population of RNA molecules of the same length is required for RNAase protection. Antisense probes should be small (~80–300 nucleotides) to allow the visualization of minor differences in length after digestion with RNAases and gel electrophoresis. In the case of cDNAs, polylinker sequences present in the 5′ end of the antisense transcript (10–40 bases in length) will be cleaved during digestion as these do not form hybrids and thus are not protected. The presence or absence of these sequences will distinguish the undigested from the digested, protected fragments. T7 and T3 RNA polymerases are usually much more efficient than SP6 RNA polymerase. For digestion of the template, avoid using restriction enzymes that generate a 3′ overhang, such as *PstI*, since polymerases may use this to initiate transcription of the wrong strand.

The choice of antisense probe to be used will depend on several factors including the absence of sequences that provoke premature termination of transcription, absence of a long mononucleotide stretch, such as a long poly(A) tail in cDNAs, and the ability to form stable hybrids due, in part, to the absence of large A+T-rich regions. Although premature terminations may be avoided by lowering the temperature of the transcription reaction, it is recommended to try two to four different fragments when choosing an antisense probe.

Differences in the stability of the RNA:RNA hybrids (which are more stable than contaminating RNA:DNA hybrids) may be reflected in the sensitivity of the assay. For this reason it is also recommended that the stability to RNAase digestion on hybrids made with synthetic sense and antisense RNAs should be tested. Normally, a good antisense RNA probe is capable of detecting less than one picogram of complemetary sense RNA in an RNAase protection assay. A transcription reaction routinely yields RNA probe for more than 100 protection assays. The purified probe can be kept in 100% formamide for two days, allowing a new assay to be set up the following morning.

4. RNAase protection

Protocol 3. RNAase protection assay

1. Transfer the desired amount of RNA, measured either in micrograms or in embryo equivalents, to a microcentrifuge tube. RNA precipitated in LiCl can be used at this point (from step 12, *Protocol 1*). Before taking aliquots of the RNA, vortex the precipitate repeatedly. RNA precipitates do not aggregate and it is possible to use volume as a reliable measure of RNA quantity. For abundant RNA species, 0.5–1 embryo equivalents is sufficient. For rare RNA species, 10–15 embryo equivalents are often necessary. If more RNA is needed, use poly(A)-selected RNA.

2. Spin in a microcentrifuge at maximum speed for 5 min.

3. Aspirate *all* the liquid.

4. Wash with 500 µl 100% ethanol at −20 °C.

5. Aspirate all the liquid and dry the pellet. A portion of the RNA will be found on the sides of the tube.

6. Add 24 µl 100% deionized formamide.

7. Vortex vigorously, making sure liquid reaches all sides.

8. Add 3 µl 10 × salts[a].

9. Add 3 µl DEPC-treated water or tRNA for the negative control.

10. Vortex vigorously. All the RNA should be in solution. Heat to 68 °C for 5–10 min if necessary.

11. Add about 200 caps in less than 1 µl of eluted probe in 100% formamide. This amount represents excess probe to drive the hybridization of all target sequences to completion in an average RNA sample. Therefore, the assay is quantitative.

12. Heat to 68 °C for 5 min to denature all RNAs.

13. Immediately incubate microcentrifuge tubes at 45 °C overnight (minimum 8 h).

14. Bring hybridizations to room temperature and thaw RNAase stocks. RNAase A (Sigma) at 2 mg/ml (50 × stock). RNAase T1 is usually not required

15. Prepare RNAase digestion buffer containing RNAase A[b].

16. Add 350 µl RNAase digestion buffer with RNAase A to each hybridization sample. Vortex vigorously and centrifuge briefly.

Protocol 3. *Continued*

17. Incubate for 1 h at room temperature. The time and temperature of incubation will vary for each probe, but this is a good starting point.

18. Add 20 μl of 20% SDS.

19. Add 2 μl of 25 mg/ml proteinase K.

20. *Vortex vigorously and centrifuge briefly.* Incubate at 37 °C for 15 min.

21. Add an equal volume of phenol/CHCl₃ (1:1) and vortex.

22. Spin in a microcentrifuge at maximum speed for 5 min.

23. Prepare a new set of labelled microcentrifuge tubes with 10 μl (10 μg) carrier tRNA. To prepare tRNA, phenol extract extensively and precipitate at least twice with ethanol.

24. Add top aqueous phase of step 22 to the new labelled microcentrifuge tubes containing tRNA. Mix.

25. Fill the tube with 100% ethanol and incubate at −70 °C for 1 h or at −20 °C overnight.

26. Spin at maximum speed in a microcentrifuge.

27. Aspirate the supernatant and draw all remaining liquid. Pellet should be visible and should be slightly radioactive.

28. Wash the pellet with 500 μl 100% ethanol. Repeat steps **26** and **27**.

29. Dry the pellet. Open tubes in a dry heating block dry within 10 min.

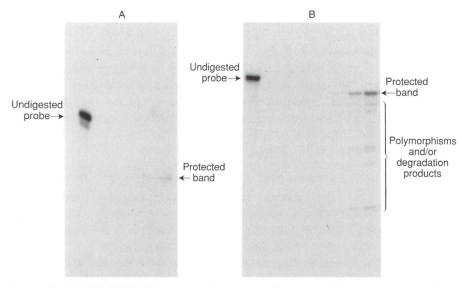

Figure 2. Results of RNAase protection assay. Urea/polyacrylamide gel, overnight exposure. A, Probe A; B, probe B.

30. Make sure pellets are completely dry. Add 5 μl sequencing loading buffer containing 80% formamide and dyes.

31. Vortex hard, up and down. Make sure all pellets are fully resuspended. Spin briefly and heat at 100 °C for 2 min, or longer if necessary.

32. Load the samples in a small polyacrylamide/urea sequencing gel. Make sure wells are clean. This is very important as diffuse or wavy bands can reduce sensitivity. Include a lane containing a small amount (e.g. 5–20 cps) of undigested purified probe to distinguish between digested and undigested molecules (*Figure 2*).

33. Run the gel. Wrap gel on the glass plate that it sticks to with plastic wrap. Put wrapped gel in cassette and expose to X-ray sensitive film. Exposure times will vary from 30 min to 2 weeks. For long exposures make sure cassette is not contaminated.

---

a 10 × salts: 4 M NaCl, 400 mM Pipes pH 6.4, 10 mM EDTA. Treat with DEPC (see *Protocol 1*) and autoclave.

b RNAase digestion buffer: 300 mM NaCl, 5 mM EDTA, 10 mM Tris-HCl pH 7.5, 40 μg/ml RNAase A. Make NaCl/EDTA solution, DEPC-treat (see *Protocol 1*) and autoclave. Make Tris buffer in DEPC-treated water and autoclave.

Controls should be performed in all assays. These include a negative control with an RNA sample that will not form hybrids, such as tRNA. Positive controls include testing for the presence and abundance of known or standard RNA molecules. For example, *β*-tubulin or actin. In *Xenopus* embryos the levels of the elongation factor EF-1α are routinely tested as a positive control (6). These controls also provide a means of standardization. If possible, it is recommended that two or more different probes be mixed in the sample. This enables testing of different RNA species from the same sample. Moreover, this method provides the best way to have internally consistent controls for RNA recovery and abundance. In this case, the probe used as control or the probe detecting the most abundant species should be smaller. This is important as degradation products of abundant species could mask the protected fragments of less abundant ones. Since RNAase A cleaves single mismatched nucleotides, allelic variations in the RNAs also result in partially protected fragments.

When working with cDNAs, the difference between digested and undigested fragments will be small (derived from the digested polylinker sequences, see above) and it is imperative to run a bit of undigested probe every time. If the results are not satisfactory, the probe and digestion conditions should be checked. Some probes require digestion at 37 °C for 30 min, while others give best results after digestion at 4 °C for 2 h!! RNAase protections with two or more probes require the optimization of the digestion conditions.

Acknowledgements

I am grateful to Doug Melton, Paul Krieg, Richard Harvey, Chris Kintner, and Jim Smith for teaching me about frogs and/or what to do with them. I thank Sam Pfaff for comments on the protocols.

References

1. Melton, D. A., Krieg, P. A., Rebagliati, M. R., Maniatis, T., Zinn, K., and Green, M. R. (1984). *Nucleic. Acids Res.*, **12**, 7035.
2. Krieg, P. A. and Melton, D. A. (1984). *Nucleic Acids Res.*, **12**, 7057.
3. Krieg, P. A. and Melton, D. A. (1987). *Methods enzymol. Part F.*, **155**, 397.
4. Sambrook, J., Fritsch, E. F.. and Maniatis, T. (1989). *Molecular cloning: a laboratory manual*, 2nd. ed. Cold Spring Harbor Laboratory Press, Cold Spring Harbor, NY.
5. Chomczynski, P. and Sacchi, N. (1987). *Anal. Biochem.*, **162**, 156.
6. Krieg, P. A., Varnum, S. M., Wormington, M., and Melton, D. A. (1989). *Dev. Biol.*, **133**, 93.

23

cDNA library construction

AKIRA KAKIZUKA, RUTH T. YU, RONALD M. EVANS,
and KAZUHIKO UMESONO

1. Introduction

DNA libraries allow the identification, isolation, and characterization of genes expressed in particular cell types or at specific developmental stages. They provide information on primary structures of mRNAs for comparative studies, and can yield molecular markers for use in descriptive or experimental developmental studies.

A technical difficulty often encountered in developmental biology is the limited amount of material. Whilst 1–10 mg of total RNA can be obtained from 1 g of tissue, this amount is not feasible for most embryonic tissues. In this chapter, we describe a simple and successful method of mRNA purification which enables a cDNA library to be constructed from a much smaller amount of starting material. At each stage we describe the results we obtained when constructing a cDNA library from the chick Hensen's node: a minute region located at the tip of the primitive streak.

2. RNA purification

2.1 Isolation of total RNA from embryonic tissues

We recommend starting with a minimum of 0.1 to 0.5 mg of embryonic tissue, depending on the abundance of RNA in the tissue of interest. An appropriate number of embryos should be dissected, the tissue of interest isolated, rinsed with Ringer's solution or PBS (phosphate-buffered saline) and then used immediately for RNA extraction, or quick-frozen in liquid nitrogen for storage at −80 °C. To isolate total RNA, several methods can be used, such as the lithium chloride/urea method (1) or the acid guanidinium/phenol/chloroform method (2; Chapter 24). Both methods work well with early embryonic tissues; we generally use the latter (e.g. this allowed isolation of 300 μg total RNA from 300 chick Hensen's nodes; unpublished results). Materials for this method are readily obtained; Promega supply them in kit form.

2.2 Purification of poly(A)⁺ RNA

Several methods have been developed for purifying poly(A)⁺ RNA from total RNA. Classically, oligo(dT) cellulose chromatography has been used (3), but this method gives poor recovery of poly(A)⁺ RNA from small amounts of total RNA (less than 1 mg). We prefer oligo(dT)-Latex beads (4), which are simpler to use and more reliable with small amounts of total RNA (5). From 1–3% of the total RNA can be recovered as poly(A)⁺ RNA. Utilizing the methodology described in *Protocol 1* we recovered 6 μg of polyA⁺ RNA from 300 μg of Hensen's node total RNA.

Protocol 1. Purification of poly(A)⁺ RNA

1. Suspend the total RNA in DEPC-water[a] (typically 1–2 μg/μl) in an autoclaved microcentrifuge tube[b].

2. Add oligo(dT)-Latex beads[c] (0.5–1 μl/μg total RNA).

3. Heat at 65 °C for 10 min (to denature the secondary structure of the RNA) and immediately cool on ice.

4. Add 1/10 the volume of 5 M NaCl[b] and incubate at 37 °C for 20 min (allows attachment of the poly(A)⁺ RNA to the beads).

5. Recover the poly(A)⁺ RNA-Latex bead complex by centrifuging the tube at 15 000 g (high speed in a microcentrifuge) for 10 min. Discard the supernatant.

6. To remove non-specifically bound ribosomal RNA and transfer RNA, resuspend the pellet in 500 μl 0.1 M NTE solution[b] (a mixture of 10 μl 5 M NaCl and 500 μl TE[d] solution). Mix by pipetting up and down, using a 1 ml disposable tip[b].

7. Centrifuge at 15 000 g for 10 min and carefully discard the supernatant.

8. To detach the poly(A)⁺ RNA from the beads, suspend the pellet in 400 μl DEPC-water by pipetting as described above, heat the tube at 65 °C for 10 min, then cool rapidly on ice.

9. Centrifuge at 15 000 g, at 4 °C, for 10 min, and transfer the supernatant (which contains the poly(A)⁺ RNA) to a sterile microcentrifuge tube.

10. Recover poly(A)⁺ RNA by ethanol precipitation (add 24 μl 5 M NaCl and 1 ml ethanol to the supernatant, mix well, centrifuge tube at 15 000 g for 10 min, and discard the supernatant).

11. Rinse the pellet and inside of tube with 1 ml 70% ethanol, and centrifuge again at 15 000 g for 10 min.

12. Remove supernatant and suspend pellet in 10–50 μl DEPC-water.

13. Estimate the final yield by measuring the O.D. at 260 nm with a spectrophotometer (e.g. dilute 1.2 μl poly(A)⁺ RNA with 58.8 μl water

and measure the 60 μl sample in a microcuvette; an absorbance reading of 1 at 260 nm = 40 μg/ml).

[a] DEPC-water: water purified using a cartridge filtration system (e.g. Milli-Q, Millipore Inc.), treated for > 12 hours with 0.1% (v/v) diethyl pyrocarbonate (DEPC), and autoclaved.
[b] In *Protocols 1–7*, all tips, tubes, and aqueous solutions must be sterile (e.g. autoclaved).
[c] Oligo(dT)-Latex beads: 2% (w/v) suspension (Qiagen).
[d] TE: 10 mM Tris-HCl pH 8.0, 1 mM EDTA.

3. Construction of the cDNA library

Construction of good quality cDNA libraries requires full-length cDNA, which necessitates the use of high quality reverse transcriptase, and cDNA which has been efficiently blunt-ended (avoiding hairpin structure formation). The method described here uses the Pharmacia-LKB cDNA synthesis kit (code 27-9260-01), with modifications, and has been used to construct high quality cDNA libraries from a variety of sources (5–8). The rationale behind cDNA library construction has been documented elsewhere (9–12); here we focus on the detailed procedures.

3.1 First strand synthesis

We recommend starting with 1–5 μg of poly(A)$^+$ RNA, of which 10–50% is expected to be converted to cDNA. For the Hensen's node cDNA library, we used 1.2 μg poly(A)$^+$ RNA.

Protocol 2. First strand synthesis

1. Prepare 1–5 μg poly(A)$^+$ RNA in 20 μl DEPC-water in an autoclaved microcentrifuge tube.

2. Heat at 65 °C for 10 min and cool immediately on ice.

3. Transfer the heat-denatured RNA to a tube containing 11.5 μl first strand synthesis mixture[a]; add 1 μl DTT solution[a] and 0.5 μl [α^{32}-P]dCTP (3000 Ci/mmol, 10 mCi/ml).

4. Incubate in a 37 °C water bath for one hour, then cool on ice. Label as 'tube A'. Measure efficiency of cDNA synthesis using *Protocol 3*.

[a] Pharmacia-LKB cDNA synthesis kit.

Protocol 3. Monitoring cDNA synthesis by TCA precipitation

This protocol is applicable to first strand or second strand synthesis.

1. In a beaker, soak several glass filters[a] with ice-cold TCA-P solution[b].

Protocol 3. *Continued*

   This treatment reduces non-specific binding of the radioactive nucleotide.

2. Mix 92 µl TE with 5 µl 10 µg/µl salmon sperm DNA in a microcentrifuge tube. Add 3 µl from tube A (from *Protocol 2*; if monitoring first strand synthesis) or tube B (from *Protocol 4*; if monitoring second strand synthesis) and mix well.

3. Spot 2 µl of this mixture onto a non-treated glass filter[a]. Call this filter T1 for first strand synthesis or T2 for second strand synthesis. This filter will monitor total radioactivity.

4. Add 1 ml ice-cold TCA-P solution[b] to the rest of the mixture and mix well.

5. Keep the tube on ice for 15 min.

6. Place one pre-treated filter from step 1 in position on a filtration apparatus (e.g. Millipore).

7. Transfer the TCA-P precipitates from step 5 onto the filter. Rinse the inside of the microcentrifuge tube with 1 ml ice-cold TCA-P solution[b] and transfer onto the filter.

8. Wash the filter with 50 ml ice-cold TCA-P solution[b], then with 20 ml 100% ethanol. This filter will monitor precipitated counts (call this filter P1 for first strand synthesis or P2 for second strand synthesis).

9. Dry both filters (T and P) and measure their radioactivity in a liquid scintillation counter.

10. Calculate the yield of first strand cDNA as: (P1/T1) × 1.75 µg. Calculate the conversion efficiency as: (µg first strand cDNA)/(µg mRNA starting material) × 100[c].

---

[a] Whatman glass filters 2.4 cm GF/C.
[b] TCA-P solution: 5% trichloroacetic acid in 25 mM sodium pyrophosphate solution.
[c] In our example, T1 = 23 826 c.p.m.; P1 = 4139 c.p.m.; first strand cDNA yield = (4139/23 826) × 1.75 = 0.3 µg; conversion efficiency is: (0.3 µg cDNA/1.2 µg mRNA) × 100 = 25%. Since 3 µl from 33 µl were removed to monitor efficiency, the final yield is: 30/33 × 0.3 µg = 0.27 µg.

3.2 Second strand cDNA synthesis and blunt-ending of the cDNA

Protocol 4. Second strand cDNA synthesis

1. Transfer the remaining 30 µl in tube A (from *Protocol 2*) to the tube containing second strand synthesis mixture[a] (label as tube B).

2. Add 5 µl [α-^{32}P]dCTP (3000 Ci/mmol, 10 mCi/ml) to tube B.

3. Incubate at 12 °C for one hour, transfer to 22 °C for another hour, then cool on ice.

4. Monitor the T2 and P2 counts as described in *Protocol 3*. The yield and efficiency of second strand sythesis are calculated as follows. Second strand cDNA yield = [P2 − (P1 × 0.31)]/[T2 − (P1 × 0.31)] × 5.57 µg. Conversion efficiency = (µg second strand cDNA)/(µg first strand cDNA) × 100. Between 70 and 100% of first strand cDNA is expected to be converted to double-stranded cDNA[b].

---

[a] From Pharmacia-LKB cDNA synthesis kit.
[b] In our example, T2 = 81 200 c.p.m.; P2 = 4583 c.p.m.; second strand cDNA yield = [4583 − (4139 × 0.31)]/[81 200 − (4139 × 0.31)] × 5.57 = 0.23 µg; second strand conversion efficiency is: 0.23 µg/0.27 µg × 100 = 83% Because 3 µl from 100 µl was used for monitoring second strand synthesis, the final yield of double-stranded cDNA is: (0.27 µg first strand + 0.23 µg second strand) × 97 µl/100 µl = 0.485 µg. Since 4583 c.p.m. was measured for 98% of the 3 µl sample, the total count we now have is: [4583 c.p.m./(3 µl × 0.98)] × 97 µl = 151 208 c.p.m. This corresponds to 0.485 µg cDNA, so each ng cDNA is (151 208 c.p.m./485) = 312 c.p.m. In *Protocol 7* Cerenkov counting is used (which is only 50% efficient); hence, ~150 Cerenkov c.p.m. corresponds to 1 ng cDNA.

Protocol 5. Blunt-ending of the cDNA

1. Add 1 µl Klenow enzyme[a] to the remaining contents of tube B from *Protocol 4* .

2. Incubate product from at 37 °C for 30 min (this step generates blunt ends).

3. To stop the reaction, add 100 µl phenol/chloroform/isoamyl alcohol (50:49:1).

4. Vortex the tube and centrifuge for 5 min.

5. Transfer the aqueous phase to a new microcentrifuge tube (label as tube C).

6. Add 100 µl 4 M ammonium acetate[b] to the organic phase of tube B, vortex, and recentrifuge.

7. Add this aqueous phase to tube C and mix well.

8. Add 400 µl ethanol, mix and centrifuge at 15 000 *g* for 10 min[c].

9. Remove the supernatant and suspend the pellet in 200 µl TE solution.

10. Add 12 µl 5 M NaCl and 0.5 ml ethanol. Mix well and centrifuge at 15 000 *g* for 10 min.

11. Discard the supernatant, rinse the pellet and inside of the tube with 70% ethanol, and centrifuge again at 15 000 *g* for 10 min[d].

Protocol 5. *Continued*

12. Discard supernatant and suspend the pellet in 40 μl TE solution. This tube (C) now contains blunt-ended cDNA ready for adaptor ligation.

[a] From Pharmacia-LKB cDNA synthesis kit.
[b] Sterilize 4 M ammonium acetate by passing through a 0.2 μm pore size filter.
[c] The first precipitation removes free nucleotides.
[d] The second precipitation removes ammonium ions which may interfere with the subsequent kinase reaction.

3.3 Adaptor ligation and phosphorylation

During the adaptor ligation reaction, only one adaptor becomes ligated to each end of the blunt-ended cDNA, because only one strand is phosphorylated. After phosphorylation of the other end of the ligated adaptor and separation of the cDNA from unligated or self-ligated adaptor, the cDNA is ready to be ligated to a phage or plasmid vector. The resulting cDNA clones can be easily separated from the vector by either *Eco*RI or *Not*I restriction enzyme digestion.

Protocol 6. Ligation and phosphorylation of the adaptor

1. Add to the 40 μl blunt-ended cDNA in tube C:
   - 10 × ligation buffer[a]   5 μl
   - *Eco*RI-*Not*I adaptor[b,c]   1–5 μl[d]
   - T4 DNA ligase[c]   1 μl

2. Incubate at 15 °C overnight.

3. Heat at 65 °C for 15 min to stop the ligase activity and cool the tube to room temperature.

4. Add 5 μl ATP solution[c] and 1 μl T4 polynucleotide kinase[c]. Incubate at 37 °C for 30 min.

5. Add 50 μl phenol/chloroform/isoamyl alcohol (50:49:1), vortex, centrifuge (5 min), and transfer the aqueous phase into a sterile microcentrifuge tube (labelled D).

6. Add 50 μl 4 M ammonium acetate to the organic phase of tube C, vortex, centrifuge, and add the aqueous phase to tube D.

7. Optional: to facilitate loading of the sample and monitoring of the quality of chromatography, a trace of xylene cyanol can be added to tube D.

[a] 10 × ligation buffer: 500 mM Tris-HCl pH 8.0, 70 mM $MgCl_2$, 10 mM DTT, 10 mM ATP. Store at −20 °C.
[b] *Eco*RI-*Not*I adaptor AATTCGCGGCCGCT
                          GCGCCGGCGA-P
[c] From Pharmacia-LKB cDNA synthesis kit.
[d] 1 μl adaptor per 1 μg of original poly(A)$^+$ RNA.

3.4 Purification of cDNAs by Sepharose CL-4B column chromatography

In order to separate the cDNA from excess adaptor, prior to ligation to the vector, we recommend chromatography in small polypropylene columns.

Protocol 7 Purification of cDNAs by Sepharose CL-4B column chromatography

1. Transfer 20 to 30 ml of Sepharose CL-4B (Pharmacia-LKB) to a sterile 50 ml centrifuge tube (e.g. Falcon), and wash with CL-4B buffer[a] (11) several times.

2. Set up a polypropylene column (e.g. Poly-Prep, Bio-Rad) as shown in *Figure 1* and fill the column with washed CL-4B to the edge of the narrow portion (2 ml bed volume).

3. Equilibrate the column with 10–20 ml of CL-4B buffer.

4. Load the cDNA solution from tube D and then add several millilitres of CL-4B buffer.

5. Remove the tri-way stop cock to reduce the dead volume.

6. Collect about 24 fractions (two drops each) of eluate into sterile microcentrifuge tubes.

7. Estimate the radioactivity by scintillation counting in the Cerenkov mode (i.e. no scintillant, tritium range) (13).

8. Plot the counts on a graph. Typically one sharp peak should be observed (see *Figure 2*).

9. Combine the fractions from the beginning of the peak, to one fraction after the peak, into one microcentrifuge tube (tube E). About 50–70% of the total count should be recovered in these fractions.

10. Add 2 vol. ethanol to tube E, mix well, and centrifuge at 15 000 g for 10 min.

11. Remove the supernatant, rinse pellet with 70% ethanol, recentrifuge for 10 min and remove the supernatant.

12. Measure the total radioactivity recovered in Cerenkov mode[b].

13. Suspend the cDNA in TE solution to a final concentration of 10 ng/µl. The cDNA can be stored at −20 °C for several months or −80 °C for several years.

[a] CL-4B buffer: 10 mM Tris-HCl pH 8.0, 600 mM NaCl, 1 mM EDTA, 0.1% sarkosyl. Filter through 0.2 µm pore size and store at 4 °C for up to a few months.
[b] In our example, about 33 000 Cerenkov c.p.m. were recovered. In our experiment, 150 Cerenkov c.p.m. is 1 ng cDNA (see *Protocol 4*); thus, 33 000 c.p.m. = 220 ng cDNA recovered from 485 ng starting material.

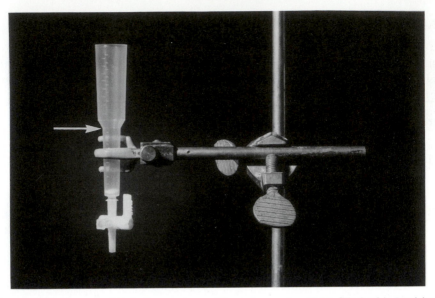

Figure 1. Setting up of Poly-Prep column. The column should be filled with 2 ml bed volume of Sepharose CL-4B (indicated by the arrow).

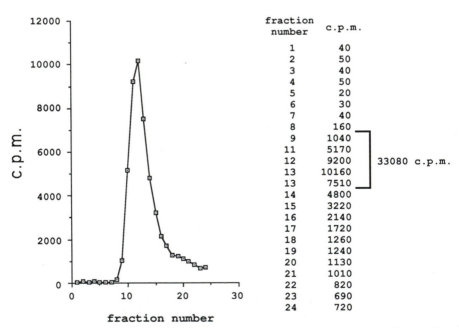

fraction number	c.p.m.
1	40
2	50
3	40
4	50
5	20
6	30
7	40
8	160
9	1040
11	5170
12	9200
13	10160
13	7510
14	4800
15	3220
16	2140
17	1720
18	1260
19	1240
20	1130
21	1010
22	820
23	690
24	720

33080 c.p.m.

Figure 2. An example of Sepharose CL-4B column chromatography. Twenty four fractions were collected and their Cerenkov counts plotted. Samples 9–13 were recovered for the library (see text).

230

3.5 Vector ligation and *in vitro* packaging

The vector should be chosen according to screening requirements. λgt10 is good for hybridization screening, especially by oligonucleotide probes; λgt11 is suitable for antibody screening or screening for specific DNA binding proteins; λZAPII is convenient since it permits easy excision of the phage insert directly into a plasmid vector without subcloning. One disadvantage of λZAPII is that it gives weaker hybridization signals than λgt10.

Protocol 8. Vector ligation and *in vitro* packaging

1. In a microcentrifuge tube mix:
   - 1 μg/μl *Eco*RI-cut, CIP-treated phage vector — 1 μl
   - 10 ng/μl cDNA (from *Protocol 7*) — 1 μl
   - 10 × ligation buffer (see *Protocol 6*) — 0.5 μl
   - TE — 2 μl
   - T4 DNA ligase[a] — 0.5 μl
2. Incubate at 4 °C overnight.
3. Use 2.5 μl of the mixture for one packaging reaction, following the procedure described in the *in vitro* packaging kit (e.g. Gigapack II Gold, Stratagene). The remaining half of the ligated library may be stored at 4 °C (short term) or −20 °C (longer term). From 5–20 × 10^6 independent plaques can be expected from each packaging reaction.

[a] From Pharmacia-LKB cDNA synthesis kit.

Acknowledgements

We thank Dr David Mangelsdorf for critical reading of the manuscript. K. U. is a research associate and R. M. E. is an investigator of the Howard Hughes Medical Institute at the Salk Institute for Biological Studies.

References

1. Auffray, C. and Rougeon, F. (1979). *Eur. J. Biochem.*, **107**, 303.
2. Chomczynski, P. and Sacchi, N. (1987). *Anal. Biochem.*, **162**, 156.
3. Aviv, H. and Leder, P. (1972). *J. Mol. Biol.*, **134**, 743.
4. Kuribayashi, K., Hirata, M., Hiraoka, O., Miyamoto, C., and Furuichi, Y. (1988). *Nucleic Acids Symp. Ser.*, **19**, 61.

5. Kakizuka, A., Miller, W. H. Jr., Umesono, K., Warrell, R. P. Jr., Frankel, S. R., Murty, V. V. V. S., Dmitrovsky, E., and Evans, R. M. (1991). *Cell*, **66**, 663.
6. Kakizuka, A., Sebastian, B., Borgmeyer, U., Hermans-Borgmeyer, I., Bolado, J., Hunter, T., Hoekstra, M. F., and Evans, R. M. (1992). *Genes Dev.*, **6**, 578.
7. Inoue, J., Kerr, L. D., Kakizuka, A., and Verma, I. M. (1992). *Cell*, **68**, 1109.
8. Mangelsdorf, D. J., Borgmeyer, U., Heyman, R. A., Zhou, J. Y., Ong, E. S., Oro, A. E., Kakizuka, A., and Evans R. M. (1992). *Genes Dev.*, **6**, 329.
9. Okayama, H. and Berg, P. (1982). *Mol. Cell. Biol.*, **2**, 161.
10. Gubler, U. and Hoffman, B. J. (1983). *Gene*, **25**, 263.
11. Ausubel, F. M., Brent, R., Kingston, R. E., Moore, D. D., Seidman, J. G., Smith, J. A., and Struhl, K., ed. (1987). *Current protocols in molecular biology.* John Wiley and Sons, New York.
12. Sambrook, J., Fritsch, E. F., and Maniatis, T., ed. (1989). *Molecular cloning: a laboratory manual.* 2nd edn. Cold Spring Harbor Press, Cold Spring Harbor, New York.
13. Slater, R. J., ed. (1990). *Radioisotopes in biology.* IRL Press at Oxford University Press, Oxford.

24

Analysis of gene expression by reverse transcriptase–polymerase chain reaction

PETER KOOPMAN

1. Introduction

A molecular genetic approach to development often involves isolating genes controlling cellular developmental processes, analysing their structure and regulation, and gaining clues about their function by examining their expression. Two constraints commonly encountered in studying gene expression in embryos are scarcity of tissue and low levels of expression. When the gene of interest is expressed in small groups of cells, or when transcripts are present in vanishingly low amounts, it is difficult to perform Northern blotting, RNAase protection assays, or *in situ* hybridization. In these cases the exquisite sensitivity of the polymerase chain reaction (PCR) can provide a solution. PCR has been used to detect transcripts present at down to one copy per thousand cells (1), and in samples as small as a mouse blastocyst (2–4), or even a single cell (5, 6).

As well as sensitivity, the specificity of PCR can often be an important consideration. When analysing members of multigene families, or transgenes that differ only slightly from endogenous genes, it is difficult to obtain a specific probe for many methods. PCR is capable of distinguishing transcripts that differ by as little as one nucleotide. PCR is quick (from tissue to result in one day), inexpensive, and can be used to assay several types of transcript simultaneously. This chapter describes methods for the application of PCR to the study of gene expression, emphasizing animal embryo studies, but the techniques described can be used in any situation where the advantages of PCR are called for.

2. Methods

2.1 RNA preparation

An extremely versatile method of RNA extraction is the acid guanidinium/ phenol/chloroform (AGPC) method (7). Minimum handling leads to high

RNA yields, although DNA contamination seems to be inevitable with this method. The following adaptation (*Protocol 1*) is suitable for one cell up to 10 mg of tissue, but can be scaled to accommodate larger amounts if necessary.

Protocol 1. RNA extraction[a]

1. Add 100 µl solution D[b] to < 10 mg tissue in a microcentrifuge tube. Wait until the tissue disintegrates before proceeding[d].

2. Add 10 µl 2 M sodium acetate pH 4 and mix.

3. Add 100 µl *unbuffered* water-saturated phenol and mix.

4. Add 20 µl chloroform/isoamyl alcohol (49:1), and mix very well.

5. Chill on ice for 15 min, centrifuge at 15 000 *g* for 15 min at 4 °C.

6. Transfer aqueous (top) phase to a new tube. If dealing with less than 10^5 cells or 1 mg tissue, add 1 µl 20 mg/ml RNAase-free glycogen (Boehringer Mannheim).

7. Add 240 µl ethanol, mix, and freeze on solid CO_2.

8. Thaw and centrifuge at 15 000 *g* for 15 min at 4 °C.

9. Wash the pellet with 80% ethanol, drain, and dry.

10. Redissolve in 5–100 µl DEPC-water[c]. Heat to 65 °C for 10 min to ensure dissolution and inhibit secondary structure, then snap chill on ice. Store at −70 °C.

[a] Although the guanidinium thiocyanate present in solution D should inactivate contaminating ribonucleases, standard precautions should be taken for RNA work (8) including the use of sterile dissection instruments, gloves, DEPC-treated aqueous solutions[c] (except those containing Tris), reserved chemical stocks, and baked glassware. Precautions should also be taken to avoid nucleic acid contaminants that will invalidate the PCR (see Section 4).
[b] Solution D: 25 g guanidinium thiocyanate, 29.3 ml DEPC-water[c] 1.76 ml 0.75 M sodium citrate pH 7, 2.64 ml 10% sarkosyl (sodium *N*-laurylsarcosinate), 38 µl 2-mercaptoethanol. Store aliquots at −70 °C; thawed aliquots can be kept in the dark at room temperature for up to 1 month.
[c] DEPC-water: distilled water with 0.01% (v/v) diethyl pyrocarbonate (DEPC), left for at least 12 h and autoclaved.
[d] Small pieces of soft embryonic tissue should disintegrate within 10 min in solution D. Tougher tissue may need agitation: vortex or flick tissue repeatedly with a needle (wear safety glasses). Very tough tissue may call for the use of a Dounce homogenizer.

Ideally, primers to be used for RT-PCR should correspond to two different exons, so that any genomic DNA will give a larger product than will cDNA from the transcript being assayed (or no product at all). Where both primers are within one exon, the RNA preparation must be treated with DNAase I, so that PCR bands will only reflect mRNA expression. If DNAase I treatment is necessary, follow *Protocol 1* steps **1–9**, then follow *Protocol 2*.

Protocol 2. DNAase I treatment of RNA preparations

1. Resuspend the pellet from *Protocol 1* step **9** in 50 µl DNAase buffer[a] containing 2 units RNAase-free DNAase I (e.g. Boehringer Mannheim 776 785), and 2 units RNAase inhibitor (e.g. Promega Biotec N2511). Incubate 2 h at 37 °C.

2. Extract once with phenol[b]/chloroform/isoamyl alcohol (25:24:1), centrifuge at 15 000 *g* for 5 min, and transfer the aqueous phase to a new microcentrifuge tube.

3. Add 6 µl 2 M sodium acetate and 150 µl ethanol, freeze on solid CO_2, then centrifuge at 15 000 *g* for 15 min.

4. Wash the pellet twice with 80% ethanol, drain, and dry.

5. Redissolve in 5–100 µl DEPC-water. Heat to 65 °C for 10 min to ensure dissolution and inhibit secondary structure, then snap-chill on ice. Store at −70 °C.

[a] 5 × DNAase buffer: 200 mM Tris-HCl pH 8.0, 50 mM NaCl, 30 mM $MgCl_2$.
[b] Use unbuffered water-saturated phenol.

2.2 Reverse transcription (RT)

The RNA yield from 10 mg tissue should be at least 1 µg; sufficient for a 30 µl 'large scale' RT reaction. If less tissue is used, or if the gene of interest is expressed at very low levels, a 7.5 µl 'small scale' reaction is recommended. If PCR is to be performed using primers from within one exon, duplicate reactions should be set up with and without reverse transcriptase. Common priming strategies for RT reactions include using gene-specific primers, random hexamers or oligo-dT (12-mers to 18-mers). Oligo-dT is strongly recommended, and has the advantage of allowing the RT product to be assayed for several transcripts, either simultaneously or sequentially.

Protocol 3. Reverse transcription (RT)

For 'large-scale RT' start at step **1**; for 'small scale' start at step **3**.

1. For large-scale RT[a]: in a sterile microcentrifuge tube on ice, mix in order:
   - DEPC water   *x* µl (to make 30 µl final volume)
   - 5 × RT buffer[b]   6 µl
   - 3.75 mM dNTPs[c]   1 µl
   - 500 ng/µl primer[d]   1 µl

Protocol 3. *Continued*

- 100 mM DTT   1 µl
- 1 µg RNA   *y* µl (to make 30 µl final volume)
- 200 U/µl reverse transcriptase[c]   1 µl

2. Incubate at 42 °C, 30 min (ample time to reverse-transcribe 2.5 kb). The sample is now ready for PCR (*Protocol 4*).

3. Small scale RT: resuspend the RNA in 5 µl DEPC-water. Transfer to a 0.5 ml microcentrifuge tube.

4. Add 2.5 µl 'small-scale RT mix'. This mix is made as in step 1, but omitting the water and RNA.

5. Overlay with 30 µl light mineral oil (Sigma M5904/M3516).

6. Incubate at 42 °C, 30 min. In *Protocol 4*, PCR reagents can be added directly to this tube.

---

[a] Quantities are given for single reaction. If multiple reactions are to be run, make a stock mix and aliquot to pre-chilled tubes containing the template RNA. Make 10% more stock mix than actually required, to allow for pipetting losses.
[b] 0.25 M Tris-HCl pH 8.3, 375 mM KCl, 15 mM MgCl$_2$; supplied by Gibco BRL.
[c] Mix equal volumes of 15 mM dATP, dGTP, dCTP, and dTTP.
[d] Gene-specific primer, random hexamers or oligo-dT (e.g. Pharmacia LKB 27-7858-01).
[e] MoMuLV reverse transcriptase (e.g. Gibco BRL 8025SA). Reverse transcriptase is a finicky enzyme, so add it to the mix at the last minute. The use of RNAase H$^-$ reverse transcriptase (e.g. Gibco BRL SuperScript) does not generally improve PCR signal (tested for transcripts where the distance between the poly(A) tail and the region of amplification is up to 3 kb).

2.3 PCR

The products of RT reactions can be used for PCR amplification without further purification. Usually 5 µl from a large-scale RT reaction is used, or the whole of a small-scale RT reaction. However, residual reverse transcriptase can inhibit PCR involving some primer–template combinations. This can be overcome by phenol/chloroform extraction of the RT reaction product or the use of less RT reaction in the PCR.

Protocol 4. PCR

1. On ice[a], prepare a stock mixture containing the following (multiply the amounts by the number of reactions to be performed, and add 10% to allow for pipette error):

- H$_2$O   *x* µl (to make a 45 µl final volume)
- 3.75 mM dNTPs[b]   20 µl
- 5 × PCR buffer[c]   10 µl
- PCR primers   500 ng each

- *Taq* DNA polymerase   0.5–2.5 units[d]

  Aliquot 45 μl per 0.5 ml microtube containing 5 μl of an RT reaction.

2. Cover each reaction with mineral oil.

3. Denature for 2 min at 94 °C, then perform 30 cycles as follows. For plate temperature-driven machines (e.g. Cetus): 95 °C, 30 sec[e]; T °C[f], 1 min; 72 °C, 1 min[g]. For tube thermocouple-driven machines: 94 °C, 5 sec[e]; T °C[f], 30 sec; 72 °C, 30 sec[g].

4. Cool slowly to room temperature. Analyse 5–10 μl on a 2% agarose/TBE gel[h].

[a] PCR reactions should be set up on ice to reduce the variability between experiments and minimize false priming.

[b] Mix equal volumes of 15 mM dATP, dGTP, dCTP, and dTTP.

[c] 5 × PCR buffer: 250 mM Tris-HCl pH 9.0, 75 mM ammonium sulphate, 35 mM MgCl$_2$, 0.85 mg/ml molecular biology grade BSA (e.g. Boehringer Mannheim 711 454), 0.25% Nonidet P-40. Store in aliquots at −20 °C and test each new batch. This buffer has given good results with *Taq* DNA polymerase from many suppliers, whilst some commercial buffers are incompatible with this protocol. This buffer requires a high nucleotide concentration (1.5 mM final); the use of commercial dNTP solutions can therefore prove expensive. Cost can be reduced by making dNTP solutions from powder stocks (e.g. Boehringer Mannheim 103 985, 104 043, 104 108, 104 272), adjusting each to pH 7 using 1 M unbuffered Tris.

[d] The amount of *Taq* DNA polymerase should be titrated for the primer-template combination used. Even slightly too much *Taq* polymerase can introduce background smearing and ghost bands. Use the minimum without sacrificing signal intensity.

[e] Denaturation times should be kept to a minimum to avoid excessive degradation of polymerase activity.

[f] T (annealing temperature) should be determined empirically for each primer pair (use the highest temperature possible without reducing the amount of PCR product), but is usually between $T_m - 10$ and $T_m + 10$, where: $T_m = [2 \times {}^\#(A + T)] + [4 \times {}^\#(G + C)]$ °C.

[g] These extension times at 72 °C are suitable for products up to 1 kb. They should be increased 30 sec for each additional kilobase. Some workers favour a long extension step (5–10 min) in the final cycle to ensure complete synthesis of all PCR products; I have not found this practice necessary. All of the step times given can be adjusted; the important thing is to be consistent between experiments.

[h] 5 × sample loading dye: 10 mM Tris-HCl pH 7.0–8.5, 1 mM EDTA, 20% Ficoll, orange G to taste (e.g. 1 mg/ml).

---

3. Refinements

3.1 Improving sensitivity

Several strategies can be used to manipulate the sensitivity of detection:

(a) Increasing the number of cycles. Up to 60 cycles has been suggested, but it is doubtful that much *Taq* polymerase activity survives this punishment. Also, the amount of specific product may plateau beyond 30 to 35 cycles, favouring the amplification of undesired products.

(b) Additional enzyme may be added after about 20 cycles. This is suggested since the amount of DNA synthesis to be done doubles with each cycle, yet enzyme activity decreases with each denaturation step.

(c) Primers can be designed to generate a shorter fragment, as these amplify more efficiently than longer ones.

(d) The agarose gel can be blotted onto a nylon filter that then hybridized to a radioactive probe for the gene of interest. The probe may include the primer sequences, as the short length of probe/primer homology will not permit probe binding to undesired 'ghost' PCR bands at high stringency.

(e) A second round of PCR can be undertaken using a second pair of primers nested within the first pair. After 20 cycles of PCR, 1 µl is transferred into a second reaction containing the nested primers, and further cycles executed. This is a very effective method of improving both sensitivity and specificity.

The paradox of all these strategies is that the greater the sensitivity, the greater the risk of revealing a false positive caused by contamination, thus invalidating the experiment. Great care needs to be taken to avoid this problem (see Section 4).

3.2 Quantitation

Applied to studies of gene expression, quantitative PCR is an attempt to measure the abundance of transcripts in a tissue; quantification is, however, full of pitfalls. For example, because PCR proceeds exponentially, a small difference in amplification efficiency between two otherwise identical samples will yield vastly different amounts of product.

A critical factor required for quantification is an exponential relationship between number of cycles and amplification of product. It is generally believed that this relationship breaks down beyond 15 and 20 cycles; after this, the amount of abundant product begins to reach a plateau, while less abundant products may continue to be amplified exponentially. The actual number of cycles before the plateau effect is seen depends on the conditions used and should be determined empirically.

Two methods are commonly used to quantify PCR product. The most effective is to use a radioactive label such as $[\alpha\text{-}^{32}\text{P}]\text{dCTP}$ in the PCR reaction. Radioactivity can be measured using a scintillation counter after removal of unincorporated label from the PCR product, or after excision of the product from a gel. Alternatively, PCR gels can be photographed and quantified by densitometry of the negative. Ideally, results should be related back to a standard curve made using different dilutions of a control RNA sample as starting material. For this purpose it is convenient to use an RNA generated *in vitro* from a template cloned into a transcription vector such as pGem (Promega) or pBlueScript (Stratagene), using the appropriate RNA polymerase (SP6, T3, or T7).

Tube-to-tube, sample-to-sample, and assay-to-assay variability can best be overcome by using an internal standard in each sample; for example, an *in-vitro*-generated transcript can be added to each sample. The transcript may be an RNA other than that being studied, or can be a version of the same RNA engineered to be slightly longer or shorter so that the two PCR products will be distinguishable on the gel (9, 10). The latter method has the advantage that both the test and the standard transcripts will amplify with similar efficiency, since both use the same pair of primers. An alternative strategy is simultaneously to amplify cDNA corresponding to a gene known to be expressed constitutively in the tissue of interest (such as *Hprt*; see Section 4). This method allows the amount of transcript detected to be related back to the amount of tissue being assayed; an important consideration with embryo samples when it may be difficult to standardize total RNA input. For all methods, multiple independent assays of each sample should be performed. Clearly this will magnify the workload of the experiment several fold, but without this extra effort attempts at quantification are likely to be inaccurate.

3.3 Using PCR to distinguish closely related transcripts

To distinguish between the gene of interest and any closely related genes (for example, a heterologous transgene versus its endogenous counterpart, or members of a multigene families), at least one nucleotide must differ between the transcripts. Normally there will be many more differences, for example in the 5'- or 3'-untranslated regions of the transcripts, which will allow the design of primers that will specifically recognize only the cDNA of interest. Where the number of nucleotide differences is limited, these differences should be incorporated into the 3' end of the primers, to reduce the possibility of false priming. The specificity-enhancer Perfect Match (Stratagene) is useful, but must be carefully titrated so as not to inhibit amplification.

Often, the sequences of closely related genes will differ at a restriction enzyme recognition site. If this is the case, primers can be designed to amplify both gene products, which can then be distinguished on an agarose gel after the appropriate restriction digest. If the sequences differ in a way which approximates to a restriction site difference, but does not match exactly, PCR primers can be designed with specific mismatches to engineer a restriction site difference between the amplification products from different genes (11). If all else fails, sequencing of PCR products can be used to identify which of two closely related genes is expressed.

4. Safeguards and controls

The sensitivity of PCR is both its greatest asset and its greatest problem. Unless painstaking care is taken, contamination is almost certain to arise, causing false positives and ruining experiments. The DNA molecules being amplified pose a great threat, since such vast quantities of these molecules are produced that it is easy for some of them to find their way into PCR reagents. Other major dangers are plasmid subclones of the gene of interest, which are likely to be wafting around the laboratory and lurking on key pieces of equipment.

Safeguards against contamination have been described elsewhere (12) but the importance of extreme care cannot be overemphasized. The PCR set-up area should be physically separated from the area where PCRs are run and analysed; dedicated equipment (in particular, pipettes, centrifuges, and freezers) should be used if possible. It is helpful to UV-irradiate pipettes between experiments, using a cross-linking chamber or trans-illuminator (but beware, these pieces of equipment tend to be contaminated with plasmids and PCR products). Reagents should be made up from stock powders and solutions reserved for PCR; buffers and primers should be split into aliquots when made up, and aliquots changed frequently; and gloves should be changed often, particularly after touching anything other than the PCR equipment.

Each assay should, of course, include a positive control (e.g. cDNA from a known expressing tissue) and a negative control (e.g. cDNA from tissue known not to express the gene of interest). The latter is not as easy as it sounds, since there is evidence that transcripts of any gene can be detected in any cell type by PCR (13). Alternatives are to use tissue from a different species, in which the homologous transcript is not detectable with the primers used, or a mock RNA extraction can be performed on an aliquot of water or an aliquot of the PBS from the dish in which the test tissue was dissected. Extremely careful workers will want a negative control for each RNA sample that gives a positive result; suitable controls would be omission of reverse transcriptase or addition of RNAase.

It is useful to include primers for an irrelevant but ubiquitously expressed gene, in addition to the primers for the gene of interest, as a control for the quality and quantity of RNA in each sample. For studies involving mouse embryos, the following primers for mouse *hypoxanthine phosphoribosyltransferase* (*Hprt*) can be used (see *Figure 1*).

Hprt.1a: 5′-CCTGC TGGAT TACAT TAAAG CACTG-3′
Hprt.1b: 5′-GTCAA GGGCA TATCC AACAA CAAAC-3′

These primers have been used at annealing temperatures from 45 to 65 °C, and so are compatible with almost any other primer set (14–16). They give a product of 354 bp, and span exons 3–8 of *Hprt*. No PCR product is obtained from genomic DNA. These primers amplify with extraordinary efficiency, and to ensure that the *Hprt* band does not obscure the signal from the gene of interest, the *Hprt* primers can be used down to 25 ng per reaction.

Peter Koopman

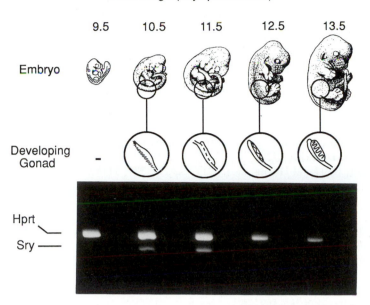

Figure 1. Time course of *Sry* expression in mouse genital ridges and testes. RNA was extracted from dissected mouse embryo torsos (prior to genital ridge formation) at 9.5 days *post coitum* (dpc), genital ridges at 10.5 and 11.5 dpc, and testes at 12.5 and 13.5 dpc, and assayed for transcripts of the Y-linked testis determining gene *Sry* (17, 18), using the methods described in this chapter. *Sry* expression is confined to a 48 h period immediately prior to overt differentiation of the testis. The simultaneous assay of *Hprt* transcripts acts as a control for the quality and quantity of RNA in each sample. Adapted from ref. 15.

Acknowledgements

The support of Robin Lovell-Badge and members of his laboratory, and of the Medical Research Council (UK), in the development of these protocols is gratefully acknowledged.

References

1. Chelly, J., Kaplan, J.-C., Maire, P., Gautron, S., and Kahn, A. (1988). *Nature*, **333**, 858.
2. Rappolee, D. A., Brenner, C. A., Schultz, R., Mark, D., and Werb, Z. (1988). *Science*, **241**, 1823.
3. Conquet, F. and Brulet, P. (1990). *Mol. Cell Biol.*, **10**, 3801.
4. Murray, R., Lee, F., and Chiu, C.-P. (1990). *Mol. Cell. Biol.*, **10**, 4953.

5. Rappolee, D. A., Wang, A., Mark, D., and Werb, Z. (1989). *J. Cell. Biochem.*, **39**, 1.

6. Shuldiner, A. R., de Pablo, F., Moore, C. A., and Roth, J. (1991). *Proc. Natl. Acad. Sci. USA*, **88**, 7679.

7. Chomczynski, P. and Sacchi, N. (1987). *Anal. Biochem.*, **162**, 156.

8. Sambrook, J., Fritsch, T., and Maniatis, T. (1989). *Molecular cloning: a laboratory manual*, 2nd ed. Cold Spring Harbor Laboratory Press, Cold Spring Harbor, New York.

9. Wang, A .M., Doyle, M. V., and Mark, D. F. (1989). *Proc. Natl. Acad. Sci. USA*, **86**, 9717.

10. Gilliland, G., Perrin, S., Blanchard, K., and Bunn, H. F. (1990). *Proc. Natl. Acad. Sci. USA*, **87**, 2725.

11. Sorscher, E. J. and Huang, Z. (1991). *Lancet*, **337**, 1115.

12. Kwok, S. and Higuchi, R. (1989). *Nature*, **339**, 237.

13. Chelly, J., Concordet, J.-P., Kaplan, J.-C., and Kahn, A. (1989). *Proc. Natl. Acad. Sci. USA*, **86**, 2617.

14. Koopman, P., Gubbay, J., Collignon, J., and Lovell-Badge, R. (1989). *Nature*, **342**, 940.

15. Koopman, P., Münsterberg, A., Capel, B., Vivian, N., and Lovell-Badge, R. (1990). *Nature*, **348**, 450.

16. Gubbay, J., Koopman, P., Collignon, J., Burgoyne, P., and Lovell-Badge, R. (1990). *Development*, **109**, 647.

17. Gubbay, J., Collignon, J., Koopman, P., Capel, B., Economou, A., Münsterberg, A., Vivian, N., Goodfellow, P., and Lovell-Badge, R. (1990). *Nature*, **346**, 256.

18. Koopman, P., Gubbay, J., Vivian, N., Goodfellow, P., and Lovell-Badge, R. (1991). *Nature*, **351**, 117.

25

Cloning genes using the polymerase chain reaction

PETER W. H. HOLLAND

1. Introduction

The polymerase chain reaction (PCR) is an *in vitro* method for amplifying regions of DNA, utilizing information based on nucleotide sequences flanking the region of interest (1–3). PCR is used in the study of DNA sequence variation, genomic organization or gene expression (the latter using cDNA as a template, reverse transcribed from RNA; Chapter 24). PCR can also be used to amplify DNA from previously uncharacterized genes, if predictions can be made regarding the sequence of part of the gene(s).

This chapter is concerned with the use of PCR to identify and clone previously uncharacterized genes potentially involved in embryonic development. This has three major applications in developmental biology:

(a) Many genes with important roles in embryonic development are members of multigene families. Hence, if one gene involved in development has been cloned, it is often possible to use PCR to amplify structurally similar genes with different (often related) functions. The amplified DNA can then be cloned; to avoid the possibility of cross-hybridization between gene family members, the PCR clones are usually used to isolate flanking DNA regions for use as hybridization probes.

(b) Many genes utilized during embryonic development have been relatively conserved through evolution. It is therefore possible to use sequence information from one species to isolate homologues from other species. PCR is often far more successful for this purpose than is reduced stringency screening of conventional DNA libraries.

(c) Several recent studies have used PCR to explore the evolution of particular developmental gene families. This application involves using PCR to survey the sequence diversity and the number of genes within a gene family, from multiple species. Such comparative surveys are essential if orthologous genes (related by speciation) are to be distinguished from paralogous genes (related by gene duplication).

2. Design of PCR-based strategy

2.1 Genomic DNA or cDNA as a target?

Previously uncharacterized genes may be amplified from either genomic DNA or cDNA. Any cDNA preparation is likely to contain only a subset of the genes in a particular gene family due to differential patterns of gene expression. This can be useful since it allows pre-selection for amplification of gene family members with particular expression characteristics. In most applications, however, differential amplification is not wanted, and genomic DNA is used. One problem with genomic DNA is that introns may be present between the PCR priming sites, and PCR is less efficient at amplifying longer DNA fragments. Furthermore, prediction of intron position is not straightforward since it can vary between taxa.

2.2 Choice of species to be analysed

If comparative data concerning developmental gene sequence or gene family composition are to be useful in evolutionary analyses, species should be chosen which have known phylogenetic relationships. If this is not known, the polarity of change in gene number or sequence cannot be determined, and no evolutionary conclusions can be made.

If PCR is to be used to attempt comprehensive surveys of gene family composition, it is preferable to restrict the choice of taxa to those that share a common ancestor with the species from which the PCR primers were designed. This is because PCR amplification will only be successful if the oligonucleotide priming sites are conserved in evolution, and predictions of conservation cannot be extrapolated safely beyond the time point of divergence of those species for which there are sequence data. For example, it is easier to amplify fish genes using primers designed from comparison of insect and mammal genes, than using primers designed from bird and mammal genes.

3. Sample preparation

Genomic DNA or cDNA prepared by most standard protocols is adequate for PCR, as long as stringent precautions are taken to avoid contamination with nucleic acid from any other source (including food items and parasites of the sample organism, and previously cloned or amplified DNA from the laboratory). Protocols for sample collection and storage are given in Section 3.1; for subsequent DNA purification in Section 3.2.

3.1 Tissue collection and storage

DNA is well preserved in any tissue if it is minced, or otherwise dispersed, in

BLB (5% SDS, 250 mM EDTA, 50 mM Tris-HCl pH 8; W. Amos, personal communication). Up to 1 cm^3 of animal tissue can be minced in 1 ml BLB. For fresh blood, DNA is adequately protected from degradation by mixing 1:1 with the more dilute RGB (2% SDS, 50 mM EDTA, 50 mM Tris-HCl pH 8; R. Griffiths, personal communication). In either case, samples can be stored at ambient temperature for many weeks, or 4 °C for many months. This is particularly convenient for collection of samples at locations with no laboratory facilities, and for transport or postage of samples at room temperature. For longer term storage, samples in BLB or RGB should be frozen (−20 °C or −70 °C). An alternative collection and storage method, convenient only when laboratory facilities are immediately available, is to snap-freeze the tissue and store frozen (preferably at −70 °C).

3.2 DNA purification

Protocols 1, *2*, and *3* describe DNA extraction from tissue stored by the methods outlined in Section 3.1.

Protocol 1. DNA purification from BLB-preserved animal tissue

1. Warm samples to 37 °C to ensure dissolution of SDS.

2. Pellet tissue by centrifugation (15 min at 15 000 *g* in a microcentrifuge; or for small cell numbers, 2 h at 230 000 *g* in an ultracentrifuge). Discard the supernatant.

3. Resuspend in 1 ml of proteinase K buffer[a], warm to 37 °C and repeat step 2.

4. Resuspend in 500 µl proteinase K buffer, add 20 µl proteinase K (from a 10 mg/ml stock stored at −20 °C), and incubate at 55 °C for 4 to 24 h.

5. Extract with phenol (i.e. add an equal volume of phenol, mix by vortexing for 20 sec, centrifuge for 20 min at 15 000 *g* and recover the aqueous phase). Extract twice with 1:1 phenol/chloroform, then twice with chloroform (these centrifugation steps can be reduced to 5 min). The phenol and phenol/chloroform should be equilibrated with proteinase K buffer (without SDS) before use.

6. To the recovered aqueous phase, add 1/10 vol. of 3 M sodium acetate pH 5.2 and 2 vol. of ethanol. Recover the DNA by centrifugation (15 000 *g*, 15 min) and wash sequentially with 1 ml 70% and 1 ml 100% ethanol (centrifuge 5 min after each wash). Vacuum dry the pellet and resuspend in 10 µl to 1000 µl TE (10 mM Tris-HCl, 1 mM EDTA pH 7.6). Assess quality of the DNA by agarose gel electrophoresis (4).

[a] Proteinase K buffer: 1% SDS, 50 mM NaCl, 10 mM EDTA, 50 mM Tris-HCl pH 8.

Protocol 2. DNA purification from RGB-preserved blood

1. Mix 30 μl RGB-preserved blood with 300 μl SET (100 mM NaCl, 1 mM EDTA, 100 mM Tris-HCl pH 8) and 30 μl 10% SDS.
2. Add 15 μl 10 mg/ml proteinase K; incubate at 55 °C for 4 to 24 h.
3. Follow steps **5** and **6** of *Protocol 1*.

Protocol 3. DNA purification from fresh frozen animal tissue

1. Grind tissue to a powder using a pestle and mortar whilst keeping it 'wet' with liquid nitrogen. Do not let the tissue thaw.
2. Add the dry powder to proteinase K buffer (*Protocol 1*), add proteinase K to a final concentration of 400 μg/ml and incubate at 55 °C for 4 to 24 h.
3. Follow steps **5** and **6** of *Protocol 1*.

3.3 Variants and troubleshooting for DNA purification protocols

(a) If purified DNA is of low molecular weight, try increasing the concentration of EDTA in proteinase K buffer to 100 mM (e.g. necessary with fish muscle).

(b) A high concentration of RNA inhibits some PCR reactions. Relative RNA and DNA concentrations can be estimated by gel electrophoresis with and without 500 μg/ml pre-boiled RNAase A added to the sample loading buffer. To remove RNA, digest the samples with 100 μg/ml pre-boiled RNAase A for 15 min at 37 °C, extract with phenol/chloroform, then with chloroform, and finally precipitate with ethanol (*Protocol 1*, steps **5–6**).

(c) Other inhibitors of PCR occasionally co-purify with DNA. These are best removed by PEG precipitation of DNA (often necessary for insect tissues). Add an equal volume of 13% (w/v) PEG-8000 in 1.2 M NaCl to the DNA solution, chill on wet ice for 30 min, centrifuge for 15 min (15 000 *g*), carefully pipette off the supernatant, wash the DNA pellet twice with 70% ethanol, vacuum dry, and resuspend in TE.

(d) An alternative strategy for inhibitor removal is overnight dialysis against TE (simple dialysis is performed using microcollodion bags; Sartorius).

(e) Last resort strategies for additional purification steps include isolation of high molecular weight DNA from an agarose gel slice; or alternatively 'CTAB' extraction. The latter is useful for invertebrates with copious slime or mucus. Add an equal volume of 2 × CTAB (2% w/v hexadecyltrimethylammonium bromide, 1.4 M NaCl, 20 mM EDTA, 100 mM Tris-HCl pH 8, plus 2% (v/v) freshly added 2-mercaptoethanol), incubate at 60 °C for 15 min, extract with chloroform and precipitate with ethanol.

(f) DNA purification is unnecessary for small planktonic or larval animals, which can be lysed by sequential freezing and boiling in H_2O, prior to PCR amplification from the supernatant.

4. Primer choice and design

4.1 General properties of PCR primers

DNA polymerases synthesize DNA unidirectionally by adding nucleotides to the 3' end of one strand of a pre-existing DNA duplex. PCR enables a particular target DNA sequence to be copied multiple times, by repeated cycles of template-directed (and primer-initiated) DNA replication followed by DNA duplex denaturation. Two oligonucleotide primers are used to hybridize, by complementary base pairing, to opposite strands of a DNA duplex, such that the 3' end of each primer is facing toward the other. In the first few copying cycles of PCR, DNA replication copies the target DNA (e.g. genomic DNA); however, as these newly synthesized copies accumulate in number, these become the dominant sites for primer annealing and subsequent template-directed replication.

For two oligonucleotide primers to function in PCR-mediated amplification of a specific DNA region, the primer sequences must be sufficiently complementary to the target DNA sequence to allow hybridization, yet sufficiently long and complex to avoid significant non-specific hybridization. In practice, both primers should be between 15 and 30 nucleotides in length, but neither need be an exact complementary copy of the target DNA. Identity to the target of 70% is sufficient for efficient amplification, if the terminal three nucleotides at the 3' end of each primer match the target exactly (5). If there is a mismatch at the 3' terminal base, PCR amplification is still possible, but it may be severely reduced in efficiency and is more dependent on other features of the primer sequence (6). Until these rules are better formulated, it is recommended that the problems of primer mismatch near the 3' terminus are overcome by using mixtures of similar sequence PCR primers, designed with mixed base degeneracies at some potentially mismatching sites.

4.2 Selection of region to be amplified

For amplification of members of gene families, two PCR primers should be synthesized to match two regions of potential sequence conservation, separated by a DNA region of between 60 and 1000 nucleotides (longer regions are amplified less efficiently). If a gene family has been well studied, choice of gene region is usually straightforward, since conserved amino acids may be identified by aligning the deduced protein sequences of previously sequenced gene family members. A useful program for performing multiple sequence alignment is CLUSTAL V (7), which will run on a range of computers including IBM-compatible personal computers. To obtain the program send an electronic mail message with the words 'help' and 'help software' (on two lines, with no quotes) to the EMBL file server at:

Netserv@EMBL-Heidelberg.DE

Alternatively, send a formatted disk for PC or Mac (specify which) to Dr D. G.Higgins, EMBL Data Library, Postfach 10.2209, 6900 Heidelberg, Germany.

From the protein sequence alignment, select two patches of at least five contiguous, absolutely conserved, amino acid residues, either side of a region of 20 to 200 amino acids. When totally conserved stretches cannot be found, broken stretches of identity can prove successful, if the internal ends of the primers are located at invariant residues. If genomic DNA is to be used as a target, intron positions should be noted for any member of the gene family already studied and avoided if possible.

If the starting point is a gene of a novel family, there is no simple way to identify potential conserved domains. Reduced stringency hybridization to genomic Southern blots can help identify conserved regions; alternatively, if weak similarity is found to a protein superfamily (e.g. kinases, cysteine proteases) it is usually possible to predict conservation of a few individual amino acid residues. Failing this, protein secondary structure prediction programmes may be useful in identifying residues more likely to be conserved. As a last resort, amino acid residues should be chosen which are rarely substituted for (e.g. Cys, His, Pro). In all cases, PCR primers should be designed which end at their 3' ends on these putatively conserved positions.

4.3 Design of PCR primers

Once conserved protein regions have been selected, degenerate oligo-nucleotide primers must be designed to match the potential DNA sequences which could encode the amino acids. A mixture of 'forward' primers (more 5' in the gene) are synthesized to represent DNA sequences potentially coding for the more N-terminal protein patch; 'reverse' primers (more 3' in the gene) are made to match the inverse complement of the DNA sequences which could encode the more C-terminal protein patch. Guidelines for the design of these primers are given below (see also *Figure 1*).

(a) Each degenerate primer should approximately match the potential encoding DNA sequences for at least 15, and preferably 20, nucleotides.

(b) Degeneracy should be introduced into the 3′-most 6–10 nucleotide positions of each primer sequence (except for the extreme 3′ position) such that all possible codons which could code for the conserved amino acids are represented. The 5′ half of each primer need not include degeneracies.

(c) The 3′-terminal base of the forward primer should match either the third position of a Met or Trp codon, or the second codon position for an amino acid encoded by 2, 3, or 4 codons (i.e. not Arg, Leu, Ser). The 3′-terminal base of the reverse primer should complement the first base of a codon specifying an amino acid encoded by few alternative codons (e.g. Met, Cys, Trp, His). Primers with C or G as the 3′ nucleotide are often most reliable.

(d) The extent of degeneracy should be kept low, by careful choice of priming sites. Degeneracies up to 1024-fold are acceptable.

(e) If attempting to clone genes more divergent than the species you have sequence data for, design primers around regions with a predominance of amino acids unlikely to be substituted (e.g. Cys, His, Pro; avoid Ser, Ala, Asp, Glu, Phe, Tyr, Lys, Arg; see *Table 1*).

(f) If the amplified DNA fragments will be longer than 600 base pairs, alter the 5′ end of each primer such that a six-cutter restriction site is present (the restriction site can be added on to the 5′ end of each primer, tailed by a further two or three bases). Not all enzymes work well near the ends of primers; we use *Bam*HI, *Eco*RI or *Xba*I.

(g) Using computer programs such as OLIGO (MedProbe), analyse the designed primer sequences to ensure that they cannot form stable hybrids with each other or amongst degenerate primer mixes. Avoid complementarity of greater than six contiguous bases, or three bases if these lie at the extreme 3′ end of a primer.

(f) Use OLIGO to ensure that the primers cannot form stable hairpin loops, and that each has a duplex melting temperature (T_m) of between 70 to 84 °C (OLIGO version 2.0 gives T_d; $T_m = T_d + 7.6$ °C). The two primers should have similar T_m; this can be adjusted by altering the length or GC content.

(i) Primers sequences should be devoid of homopolymer or dinucleotide runs, and should not end on three G or C residues at the 3′ end.

(j) Database searches should be performed with the designed primer sequences, to ensure they do not match something conserved but unwanted. In particular, identity to ribosomal RNA (and hence repetitive ribosomal DNA) sequences should be avoided.

Forward primer

```
                                  A   A
5'   ATGGATCC A C  CT C  CG C  AA A  CA C  AA   3'
              A    T G    G T   A G   A T
```

```
Mouse msx1 (Hox-7)   Cys Thr Leu Arg Lys His Lys.........//.........Glu Ala Glu Leu Glu Lys Leu Lys
                     TGC ACC CTA CGC AAG CAC AAG.........//.........GAG GCG GAG CTG GAG AAG CTG AAG
Drosophila msh       Cys Asn Leu Arg Lys His Lys.........//.........Glu Ala Glu Ile Glu Lys Ile Lys
                     TGC AAC CTG CGG AAG CAC AAG.........//.........GAG GCC GAG ATC GAG AAG ATC AAG
```

```
                                              T
                                T        T  G A G
                     3'   CT C  CG C  CT C  C A C  CTC  TTC GAA TTC   5'
                             C    G       C  A
                                            A
```

Reverse primer

Figure 1. Degenerate PCR primer design. Alignment of related homeodomain proteins from *Drosophila* and mouse is used to identify conserved amino acids; degenerate primer design takes into account amino acid conservation, amino acid properties, DNA hybridization properties and genetic code degeneracy.

Table 1. Codon degeneracy and conservative substitutions. Degenerate PCR primers should be based around those amino acids encoded by few codons or those less prone to conservative substitution (e.g. those in **bold**).

Properties	1 codon	2 codons	3 codons	4 codons	6 codons
Hydrophilic					
Acidic		Asp Glu			
Basic		Lys			Arg
Imidazole		**His**			
Amide		Asn Gln			
Hydroxy				Thr	Ser[a]
Thiol		**Cys**			
Hydrophobic					
Cyclic hydrophobic	**Trp**	Tyr Phe			
Long chain	**Met**		Ile	Val	Leu
Short chain				Gly Ala[a]	
Helix breaking				**Pro**	

[a] Ala to Ser (and vice versa) is another frequent substitution.

4.4 Alternative strategies for primer design

(a) As an alternative to mixed-base primers, the nucleotide inosine may be incorporated at variable sites. Inosine can stably base pair with A, C, G, or T.

(b) If cloning genes from mitochondrial or ciliate genomes, take into account their divergent genetic codes.

(c) A variant which dramatically increases specificity is 'nested' PCR. A third degenerate primer is synthesized, partially or completely internal to one of the two initial primers; after the first 30 PCR cycles, a 1% aliquot is amplified substituting the third primer in place of the one it overlaps or is next to.

5. PCR amplification and cloning of previously uncharacterized genes

5.1 PCR amplification

Precautions must be taken to avoid any contamination with previously cloned or amplified DNA. This is the major problem in any laboratory working on a particular gene family; strict precautions, to the verge of paranoia, must be taken. Sterile reagents, tips, and tubes should be used, latex gloves worn and changed frequently, and reaction components stored in small aliquots. The commonest source of contamination is from aerosols when using standard pipetting devices (e.g. Gilson Pipetman); this can be avoided by using positive displacement pipettes (e.g. Alpha Labs PCR Micropettor; Costar

OnePette PD-25) or by regularly acid treating Pipetman barrels and plungers (500 mM HCl for 10 min). Control reactions without template DNA should always be run in parallel; however, there is no simple control for contamination in the template DNA samples. PCR amplification is described in *Protocol 4*.

Protocol 4. PCR amplification

1. Using dedicated pipettes, prepare a stock mixture containing the following (multiply the volumes by the number of reactions to be performed, plus 10% to allow for pipette error):

   - autoclaved, 0.2 μm filtered, H_2O 17.20 μl
   - 25 mM dNTPs[a]   0.16 μl
   - *Taq* DNA polymerase (5 units/μl; Promega)[b]   0.20 μl
   - 10 × *Taq* DNA polymerase buffer[c]   2.00 μl
   - 5′ primer[d] (1 mg/ml)   0.20 μl
   - 3′ primer[d] (1 mg/ml)   0.20 μl

2. Aliquot 20 μl[e] per 0.5 ml microtube (Treff) containing 0.5 μl target genomic DNA. Try several dilutions of DNA (e.g. 1 ng/μl, 10 ng/μl, 100 ng/μl). Add 20–50 μl light mineral oil (Sigma).

3. Perform thermal cycling reaction in a programmable dri-block or water bath. A standard set of conditions is: 94 °C, 2.5 min; then 30 to 40 cycles of (94 °C, 45 sec; 50 °C, 2 min; 68 °C, 1 min; 72 °C, 1 min); then 72 °C for 5 min[f].

4. Remove the reaction from under the oil using a Pipetman and mix with 2 μl loading dye[g]. Electrophorese through a 1% to 2% agarose gel containing 200 ng/ml ethidium bromide, alongside molecular weight markers.

5. Visualize DNA band of the expected size with low intensity, 302 nm UV light. If the size cannot be predicted, blot a duplicate gel onto a Nylon filter and hybridize with a probe from a related gene (4).

---

[a] A fresh 1:1:1:1 mix of 100 mM dATP, dCTP, dGTP, dTTP solutions (Pharmacia LKB).
[b] Store stock at −70 °C and working batch at +4 °C for up to a few weeks.
[c] 10 × PCR buffer: 500 mM KCl, 100 mM Tris-HCl pH 8.8 at 25 °C, 15 mM $MgCl_2$, 1% Triton X-100; Promega. An important variable in the PCR is $MgCl_2$ concentration; 1.5 mM final concentration is usually optimal, but try a range from 0.5 mM to 3 mM final.
[d] There is no need to gel purify the primers; after elution from the column and deprotection, extract twice with phenol/chloroform, once with chloroform, and precipitate with ethanol. Store at −20 °C.
[e] Performing amplification reactions in small volumes (10 to 20 μl) economizes on reagents and optimizes heat transfer through the reaction mixtures, yet still yields sufficient product for recombinant DNA cloning.

f Optimal parameters will vary between machines; the above is for a Techne PHC-2 thermal cycler. The annealing temperature is given as 50 °C; try a range and select the highest annealing temperature which gives a clear band (usually between 45 °C and 57 °C). The extension time (at 72 °C) should be increased in proportion to the length of the target DNA sequence. 1 to 2 min is usual for 150 to 600 base pairs; 3 to 4 min is required for amplification of 2 to 3 kb.

g Sample loading dye: 50% glycerol, 10 mM EDTA, 10 mM Tris-HCl pH 7.6, 1 mg/ml Orange G.

5.2 Recombinant DNA cloning of PCR amplified bands

Many alternative cloning strategies are possible for recombinant DNA cloning of PCR amplified products; the more popular include the use of restriction sites in PCR primers and methods which utilize the additional A residue added to the 3′ ends of amplified DNA by some thermostable polymerases (e.g. TA cloning from Invitrogen). A simple blunt end cloning method (*Protocol 5*) has proved consistently efficient in our laboratory.

Protocol 5. Blunt end cloning of PCR amplified bands

1. Separate the PCR products by agarose gel electrophoresis. Using a sterile razor blade, cut slits in the agarose in front of, and behind, each band required.

2. Into each slit insert a sliver of pre-wetted NA45 DEAE cellulose paper (Schleicher and Schuell), and resume electrophoresis at 70 mA for 10 min, such that the DNA binds to the paper in front of the band.

3. Take this paper out, rinse in sterile TE or water, and immerse in 100 μl 1 M NaCl. Incubate at 70 °C for 60 min.

4. Pass the salt solution down a 1 ml Sephadex G-50 (Pharmacia LKB) spun column (equilibrated with sterile water; ref. 4). Recover the 100 μl sample.

5. Add 10 μl T4 PNK buffer (500 mM Tris-HCl pH 7.6, 100 mM MgCl$_2$, 50 mM DTT, 1 mM spermidine, 1 mM EDTA), 1 μl 10 mM ATP and 10 units T4 polynucleotide kinase (Pharmacia LKB). Incubate at 37 °C for 30 min.

6. Add 5 units either Klenow fragment or T4 DNA polymerase (Gibco BRL), 1 μl each of 10 mM dATP, dCTP, dGTP, dTTP; incubate at 37 °C for 30 min.

7. Extract once with phenol/chloroform, once with chloroform, and check 10 μl on an agarose gel to confirm DNA presence. If desired, DNA can be concentrated by passing down a Sephadex G-50 spun column followed by ethanol precipitation.

Protocol 5. *Continued*

8. Set up ligation reaction, either using a Rapid Ligation Kit (Amersham), or as follows:

   - amplified DNA band (0.1 to 1.0 ng/µl)   10.0 µl
   - 5 × ligase buffer (250 mM Tris-HCl pH 7.6, 50 mM MgCl$_2$, 5 mM ATP, 5 mM DTT, 25% (w/v) PEG-8000; Gibco BRL)   4.0 µl
   - T4 DNA ligase (2 units/µl; Gibco BRL)   1.0 µl
   - 10 mM ATP (Pharmacia LKB)   0.5 µl
   - *Sma*I-cut, phosphatased pUC18 (5 ng/µl; Pharmacia LKB)[a]   3.5 µl

   Incubate at 14 °C for 12 h.

9. Transform into competent *Escherichia coli* DH5α and select white and pale blue transformants after plating and overnight incubation on ampicillin plates (4) spread with 65 µl 2% (w/v) Bluogal (Gibco BRL) in dimethyl formamide and 65 µl 100 mM IPTG.

10. Pick potential recombinants, perform minipreps (4) from overnight cultures, and analyse by restriction digestion[b] (4).

11. Sequence double-stranded alkaline lysis minipreps (4) of the recombinants, using T7 DNA polymerase and 7-deaza-dGTP sequencing mixes, following the supplier's protocols (United States Biochemical).

12. If flanking sequence is required, either use the clones for high stringency library screening or identify gene-specific stretches of DNA sequence within the clones and design primers for RACE PCR or (more reliably) inverse PCR (2, 3).

---

[a] Ligation efficiencies are low for PCR products; the use of commercially obtained pre-digested pre-phosphatased vectors (e.g. pUC and pT7T3; Pharmacia LKB) can dramatically increase the proportion of recombinants obtained.

[b] With many plasmid vectors, including pUC18, a simple way to distinguish between recombinants, background vector religations and incomplete restriction digests is to digest with either *Pvu*II (for inserts < 1 kb) or *Dra*I (for inserts > 1 kb).

Acknowledgements

I thank members of the Molecular Zoology Laboratory, Oxford, for their contributions to these methods. Bill Amos, David Dixon, and John McVey have also given advice. The author's research in this field is funded by SERC and The Royal Society.

References

1. Saiki, R. K., Gelfand, D. H., Stoffel, S., Scharf, S. J., Higuchi, R., Horn, G. T., Mullis, K. B., and Erlich, H. A. (1988). *Science*, **239**, 487.
2. Innis, M. A., Gelfand, D. H., Sninsky, J. J., and White, T. J., ed. (1990). *PCR protocols*. Academic Press, San Diego.
3. McPherson, M. J., Quirke, P., and Taylor, G. R. (1991). *PCR: a practical approach*. IRL Press at Oxford University Press, Oxford.
4. Sambrook, J., Fritsch, E. F., and Maniatis, T. (1989). *Molecular cloning: a laboratory manual*. Cold Spring Harbor Laboratory Press, Cold Spring Harbor, New York.
5. Sommer, R. and Tautz, D. (1989). *Nucleic Acid Res.*, **17**, 6749.
6. Kwok, S., Kellogg, D. E., McKinney, N., Spasic, D., Goda, L., Levenson, C., and Sninsky, J. J. (1990). *Nucleic Acid Res.*, **18**, 999.
7. Higgins, D. G., Bleasby, A. J., and Fuchs, R. (1992). *CABIOS*, **8**, 189.

26

In situ hybridization

DAVID G. WILKINSON

1. Introduction

The visualization of patterns of gene expression in embryos is an essential technique for the molecular analysis of development. The detection of mRNA by *in situ* hybridization has found widespread use because of the speed with which specific probes can be produced from cloned DNA or by the synthesis of oligonucleotides. The most sensitive methods use antisense RNA probes and involve the following steps:

- the synthesis of a labelled RNA probe
- the fixation and pretreatment of tissue
- the hybridization of probe to the tissue
- washing to remove unhybridized probe
- the visualization of the probe

There are two major variations in the protocols for *in situ* hybridization:

(a) Probe can be labelled with a radioactive nucleotide, most commonly [^{35}S]UTP, or with a hapten-labelled nucleotide, such as digoxigenin-UTP. ^{35}S-labelled probes are detected by autoradiography and yield a signal with a resolution of about a cell diameter. The more recently developed methods for the synthesis and detection of hapten-labelled probes offer several advantages including speed, safety, and a single cell resolution of signal.

(b) Hybridization can be carried out on tissue sections or with whole embryos. Tissue sections are used for the comparison of the expression of different genes by the hybridization of adjacent sections. These can be hybridized with radioactive or hapten-labelled probes. Serial tissue sections can be used to reconstruct the spatial pattern of gene expression from serial sections (1), but a more direct and rapid visualization can be achieved by whole mount hybridization with hapten-labelled probe (2–5). This latter method also has the advantage that a large number of embryos can be hybridized simultaneously, avoiding the considerable amount of work required to prepare many tissue sections. If required, sections can be cut after signal detection.

Procedures are presented here for the *in situ* hybridization of ^{35}S-labelled probes to tissue sections and the hybridization of digoxigenin-labelled probes to whole embryos. The method for tissue sections works well for all of the vertebrate species that we have tested (*Xenopus*, chick, mouse); others have also used the protocol succesfully on invertebrates (*Drosophila*, amphioxus). However, different methods have proved to be optimal for the whole mount hybridization of different species, and we present here protocols for the analysis of mouse embryos (5), *Drosophila* embryos, and *Xenopus* embryos (3). By a simple adaptation of these procedures, digoxigenin-labelled probes can be used for the hybridization of tissue sections. Discussion of the theory of *in situ* hybridization and the rationale of the various steps is not presented here, but can found in ref. 6.

2. General precautions

It is important to avoid the degradation of cellular RNA prior to hybridization, so the PBS, saline, and triethanolamine solutions should be autoclaved and the glassware-baked. It is essential to keep containers used for post-hybridization steps (when ribonuclease is used) separate from those used for prehybridization steps.

3. *In situ* hybridization of ^{35}S-labelled probes to tissue sections

3.1 Preparation of ^{35}S-labelled single-stranded RNA probes

^{35}S-labelled RNA probes are synthesised as described in *Protocol 1* by the transcription of the required sequences cloned into a vector, such as pBlueScript (Stratagene) or pGEM (Promega), containing RNA polymerase binding sites (7). The vector is first digested with a restriction enzyme such that transcription yields an RNA probe that is complementary (antisense) to the target mRNA and lacks plasmid sequences. Transcription is carried out in the presence of unlabelled CTP, GTP, and ATP plus ^{35}S-labelled UTP. The amount of radioactive ribonucleotide limits the transcription reaction and the maximum yield is about 200 ng.

The length of the probe influences the signal strength: probes of up to 1 kb in length give optimal results; longer probes penetrate the tissue inefficiently, so should be partially degraded by limited alkaline hydrolysis as described in step **5**.

Protocol 1. Synthesis of ^{35}S-labelled probe

1. Mix in the following order at room temperature:

• sterile distilled water	2.5 µl
• nucleotide mix[a]	2 µl
• 10 × transcription buffer[b]	2 µl
• 0.2 M DTT	1 µl
• linear template (1 µg/µl)	1 µl
• ^{35}S-UTP (> 1000 Ci/mmol)	10 µl
• placental ribonuclease inhibitor (100 U/µl)	0.5 µl
• SP6, T7 or T3 RNA polymerase (10 U/µl)	1 µl

2. Incubate at 37 °C for 2 h.

3. Add 2 µl RNAase-free DNAase I (20 U/µl).

4. Incubate at 37 °C for 15 min.

5. If the transcript is greater than 1 kb in length, hydrolyse it to an average length of 750 bases by adding an equal volume of 80 mM NaHCO$_3$, 120 mM Na$_2$CO$_3$, and incubating at 60 °C for a time (t, in minutes) given by:

$$t = (L - 0.75)/0.08L \text{ (where } L \text{ is starting length in kb).}$$

6. Remove the unincorporated ribonucleotides by fractionation on a Sephadex G-50 column (e.g. in a 1 ml syringe or a Pasteur pipette) in 50 mM Tris-HCl pH 8.0, 1 mM EDTA, 0.1% SDS and then ethanol precipitate the RNA. Redissolve at 2 × 10^6 c.p.m./µl in 100 mM dithiothreitol, then add 9 vol. hybridization mix[c] and store at −70 °C.

[a] Nucleotide mix: 2.5 mM GTP, 2.5 mM ATP, 2.5 mM CTP.
[b] 10 × transcription buffer: 400 mM Tris-HCl, pH 8.25, 60 mM MgCl$_2$, 20 mM spermidine.
[c] Hybridization mix: 50% deionized formamide, 0.3 M NaCl, 20 mM Tris-HCl, 5 mM EDTA, pH 8.0, 10% dextran sulphate, 0.02% Ficoll 400, 0.02% polyvinylpyrrolidone, 0.02% BSA, 0.5 mg/ml yeast RNA.

3.2 Preparation of tissue sections

Tissue sections of embryos are prepared by fixation, embedding in paraffin wax and sectioning as described in *Protocol 2*. The sections are then dried on to subbed slides that are prepared as described in *Protocol 3*. Embryos can be stored indefinitely in 70–100% ethanol (step **2**) at −20 °C, in paraffin wax at 4 °C (after step **3**), or as sections if dessicated at 4 °C (after step **6**).

Protocol 2. Fixation, embedding, and sectioning of embryos

1. Collect embryos as appropriate for the species. Vertebrate embryos are fixed in 4% paraformaldehyde in PBS[a] at 4 °C overnight. *Drosophila* embryos are collected and fixed as descibed in Chapter 1, *Protocols 3* and *4*.

2. Successively replace solution with: saline (0.83% NaCl) at 4 °C (twice), 70% ethanol (twice), 85% ethanol, 95% ethanol, absolute ethanol (twice), toluene (twice), paraffin wax (e.g. BDH pastillated Fibrowax) at 60 °C (three times). For vertebrate embryos, use volumes of 10 ml and incubation times of at least 30 min, with occasional agitation.

3. Transfer the embryos to an embryo dish (BDH) at 60 °C, then allow to cool, orientating the embryos with a warmed needle if desired.

4. Cut sections of 6 μm on a microtome.

5. Float the ribbons of sections on a bath of distilled water at 40–50 °C until the creases disappear. Collect on subbed slides (from *Protocol 3*).

6. Dry the slides at 37 °C overnight.

[a] Fixative is prepared on the day of use by adding 4 g paraformaldehyde per 100 ml phosphate-buffered saline (PBS), heating at 65 °C until the paraformaldehyde has dissolved, and cooling on ice. If this solution is not isotonic for the embryos, the concentration of PBS should be adjusted; for example, for *Xenopus* embryos use 70% PBS. Paraformaldehyde is toxic; handle solutions within a fume hood.

Protocol 3. Preparation of subbed slides

1. Dip the slides for 1 min each in 10% HCl/70% ethanol, followed by distilled water and 95% ethanol, and then air-dry.

2. Dip the slides in 2% TESPA (3-aminopropyltriethoxysilane)[a] in acetone for 10 sec.

3. Wash the slides twice with acetone, and then with distilled water.

4. Dry the slides at 37 °C.

[a] TESPA is toxic.

3.3 Prehybridization treatments and hybridization of tissue sections

Before hybridization sections are dewaxed and subjected to a series of treatments as described in *Protocol 4*. The proteinase K treatment increases signal by digesting cellular protein, thus rendering the target RNA more accessible to the probe. It is advisable to optimize the digestion time for each batch of proteinase K. The post-fixation of the sections after this step is essential, otherwise the tissue will later disintegrate. The acetic anhydride step acetylates amino residues and is intended to prevent the non-specific binding of probe. This step is optional.

Protocol 4. Prehybridization treatments and hybridization

1. De-wax the slides in 250 ml Histoclear (National Diagnostics), twice for 10 min, and then place them in 250 ml 100% ethanol for 2 min to remove most of the Histoclear.

2. Transfer the slides quickly through 250 ml 100% ethanol (twice), 95%, 85%, 70%, 50%, and then 30% ethanol. Wash the slides with 250 ml saline, and then PBS for 5 min each.

3. Immerse the slides in 40 ml fresh 4% paraformaldehyde in PBS for 20 min.

4. Wash the slides with 250 ml PBS, twice for 5 min.

5. Drain the slides and place horizontally on a sheet of filter paper on the bench, with the sections uppermost. Overlay the sections with 20 µg/ml proteinase K (freshly diluted from a 10 mg/ml stock) in 50 mM Tris-HCl, 5 mM EDTA pH 8.0, and leave for 5 min.

6. Shake off excess liquid and wash the slides with 250 ml PBS for 5 min.

7. Repeat the fixation of step 3; the same solution can be used.

8. Place the slides in a slide dish containing 250 ml 0.1 M triethanolamine HCl pH 8.0, set up with a rapidly rotating stir bar and in a fume hood. Add 0.63 ml acetic anhydride (caution: this is toxic and volatile) and leave for 10 min.

9. Wash the slides with 250 ml PBS, and then saline for 5 min each. Dehydrate by passing through 250 ml 30%, 50%, 70%, 85%, 95%, 100%, 100% ethanol, and air-dry. To avoid salt deposits on the slides, the slides should be left in 70% ethanol for 5 min; the other dehydration steps can be carried out quickly.

10. Apply the hybridization mix to the slide adjacent to the sections ($\sim$2.5 µl per cm^3 of coverslip is sufficient) and gently lower a clean coverslip so that the mix is spread over the sections.

11. Place the slides horizontally in a box containing tissue paper soaked in 50% formamide, 5 × SSC, seal the box and incubate overnight at 60 °C.

3.4 Posthybridization washing and autoradiography

Following hybridization, slides are washed at high stringency and treated with ribonuclease to remove probe that has not annealed to the target RNA as described in *Protocol 5*. The location of the radioactive probe is then revealed by autoradiography as described in *Protocol 6*.

Protocol 5. Posthybridization washing

1. Remove the slides from the hybridization box and place in a slide rack in 250 ml 5 × SSC, 10 mM DTT (pre-warmed) at 60 °C for 30–60 min until the coverslips fall off.

2. Place the slides in 40 ml 50% formamide, 2 × SSC, 20 mM DTT, at 65 °C for 30 min[a].

3. Wash the slides in 250 ml NTE buffer (0.5 M NaCl, 10 mM Tris-HCl, 5 mM EDTA pH 8.0) at 37 °C, three times for 10 min each.

4. Immerse the slides in 40 ml 20 μg/ml ribonuclease A (diluted from a 10 mg/ml stock) in NTE buffer at 37 °C for 30 min.

5. Wash the slides in 250 ml NTE buffer at 37 °C for 15 min.

6. Repeat the high stringency wash of step **2**[a].

7. Wash the slides in 250 ml 2 × SSC, then 0.1 × SSC for 15 min each.

8. Dehydrate the slides by quickly putting them through 250 ml 30%, 60%, 80%, and 95% ethanol, all including 0.3 M ammonium acetate, followed by absolute ethanol, twice. Air dry.

[a] A lower stringency wash at 50 °C gives satisfactory results for many probes, and should be used for AT-rich probes.

Protocol 6. Autoradiography

1. Under safelight conditions (e.g. Kodak safelight filter 6B) in a darkroom melt the emulsion[a] at 43 °C and mix gently with an equal volume of 2% glycerol that has been pre-warmed to 43 °C. Aliquot into dipping chambers (for example, 15 ml per slide mailer) with a wide-mouthed pipette, and wrap in foil. Store at 4 °C. When required, melt an aliquot at 43 °C.

2. To remove bubbles, dip clean slides in the emulsion until a bubble-free coating is obtained. Then dip the experimental slides quickly into the emulsion, drain vertically for 2 sec, wipe the back of the slide and place it horizontally in a light-tight (but not air-tight) box.

3. Leave the slides for 2 h, then add a sachet of desiccant, and leave for a further 2 h.

4. Transfer the slides to a slide box containing a sachet of desiccant, seal with tape, and place at 4 °C to expose. It is important that the slides are completely dry during the exposure. Typical exposure times are between 1–2 weeks.

5. 1–2 h before developing remove the box of slides from 4 °C and warm to room temperature.

6. Under safelight conditions in a darkroom, transfer the slides through Kodak D19 developer (80 g/litre) for 2 min, 1% glycerol/1% acetic acid for 1 min, and then 30% sodium thiosulphate (Sigma) for 2 min. The lights can now be turned on. At least 5 ml per slide should be used for each of these reagents.

7. Wash the slides twice in 250 ml distilled water for 10 min, quickly transfer through 70%, then 100% ethanol, and air-dry.

8. Stain in 0.02% toluidine blue for 1 min, then pass the slides quickly through water, 70% ethanol, and absolute ethanol. Repeat this after air-drying if the staining is too weak, but do not overstain since this can obscure the signal. Once appropriate staining has been achieved, immerse the slides in Histoclear for several minutes, then briefly drain, and mount the sections under a coverslip using DPX mounting agent (BDH).

---

[a] We use Ilford K5; other emulsions also give excellent results, but the developing conditions may need to be altered as recommended by the manufacturer.

Autoradiography produces silver grains which can be observed most sensitively under a microscope with dark-field illumination. For black-and-white photography, bright- and dark-field images of the stained tissue and silver grains, respectively, are photographed (*Figure 1*), but this can have the drawback of it being difficult to relate precisely the signal to the morphology. With colour photography, it is possible to present the signal and stained tissue simultaneously by double-exposure; a dark-field image of the silver grains, coloured by the use of a filter, is superimposed upon a bright-field image of the stained tissue. Another alternative is the use of epipolarization, but the sensitivity of signal detection is much lower than with dark-field illumination.

4. Probe preparation for whole mount *in situ* hybridization

Single-stranded RNA probes are prepared by the transcription of linearized plasmid containing RNA polymerase binding sites as described in Section 3.1, but with the inclusion of digoxigenin-UTP (Boehringer) in the nucleotide mixture (*Protocol 7*). A significant difference from the preparation of ^{35}S-labelled probes is that the concentration of the labelled nucleotide is not limiting and therefore relatively large amounts of probe (up to 10 μg) are synthesized. The amount of probe made can most easily be estimated by electrophoresis of an aliquot of the transcription reaction product on an agarose gel and staining with ethidium bromide. It is important that the aliquot of probe is not degraded during this electrophoresis, so precautions should be taken to avoid any contamination of the electrophoresis buffer and apparatus with ribonucleases.

Protocol 7. Synthesis of digoxigenin-labelled RNA probe

1. Mix these reagents in the following order at room temperature:

- sterile distilled water — 13 μl
- 10 × transcription buffer[a] — 2 μl
- 0.2 M DTT — 1 μl
- nucleotide mix[b] — 2 μl
- linearised plasmid (1 μg/μl) — 1 μl
- placental ribonuclease inhibitor (100 U/μl) — 0.5 μl
- SP6, T7, or T3 RNA polymerase as appropriate (10 U/μl) — 1 μl

2. Incubate at 37 °C for 2 h.

3. Remove a 1 μl aliquot and electrophorese on a 1% agarose gel containing 0.5 μg/ml ethidium bromide. An RNA band ~10-fold more intense than the plasmid band should be seen, indicating that ~10 μg probe has been synthesized.

4. Add 2 μl DNAase I (ribonuclease-free).

5. Incubate at 37 °C for 15 min.

6. Add 100 μl TE (50 mM Tris-HCl, 1 mM EDTA, pH 8.0), 10 μl 3 M sodium acetate, 300 μl ethanol, mix, and incubate at −20 °C for 30 min.

7. Spin in a microcentrifuge at 4 °C for 10 min, wash the pellet twice with 70% ethanol and air-dry.

8. Redissolve the pellet in TE at 0.1 μg/μl and store at −20 °C. Use between 5 and 10 μl of this for each 1 ml of hybridization mix.

[a] 10 × transcription buffer: 400 mM Tris-HCl pH 8.25, 60 mM MgCl$_2$, 20 mM spermidine.
[b] Nucleotide mix: 10 mM GTP, 10 mM ATP, 10 mM CTP, 6.5 mM UTP, 3.5 mM digoxigenin-UTP, pH 8.0.

---

5. Whole mount *in situ* hybridization of mouse embryos

5.1 Fixation and prehybridization treatment

The preparation and prehybridization treatment of mouse embryos is described in *Protocol 8*. A common problem with whole-mount hybridization is the presence of high background staining in enclosed cavities, such as the mouse extra-embryonic membranes or neural tube, presumably because of the trapping of reagents. It is therefore important to dissect open any such cavities. For embryos of less than 2 mm in length, a fixation time of 2 h is adequate, but overnight fixation gives the same results and can be more convenient.

As for tissue sections, proteinase K is used to increase the accessibility of target RNA, but it is important to optimize the digestion conditions for different-sized embryos and tissue thicknesses. In addition, it is essential to post-fix the embryos after this treatment or they will disintegrate.

Since 9-day or younger mouse embryos are delicate, it is important to take precautions to avoid damaging them during these and the subsequent treatments. Damage most easily occurs on transferring embryos between containers; we minimize this by using 30 ml Universal tubes (Sterilin) which have conical bottoms that allow small volumes to be used for the hybridization and antibody steps, and larger volumes for washing. The embryos are especially fragile during the proteinase treatment, and it is important to allow the embryos to settle to the bottom of the tube and leave some liquid above them when removing the wash solutions or the embryos will be flattened. The washes need to be thorough but gentle, and this can be achieved by rocking the tubes placed on their side on a Denley A600 rocker.

Protocol 8. Fixation and pretreatment of mouse embryos

1. Dissect the embryos out in PBS, opening any cavities to avoid the trapping of reagents.
2. Fix in 4% paraformaldehyde in PBS[a,b], rocking at 4 °C for from 2 h to overnight.
3. Wash the embryos twice for 5 min each in PTW[c] at 4 °C.
4. Wash for 5 min each with 1:3, 1:1, and then 3:1 methanol/PTW, then twice with 100% methanol.
5. Rehydrate by taking the embryos through this methanol/PTW series in reverse and then wash three times with PTW.
6. Treat with 10 µg/ml proteinase K in PTW for 15 min.

Protocol 8. *Continued*

7. Wash for 5 min each with freshly-prepared 2 mg/ml glycine[d] in PTW and then twice with PTW.

8. Refix the embryos with fresh 0.2% glutaraldehyde[e]/4% para-formaldehyde in PTW for 20 min.

9. Wash twice for 5 min with PTW.

10. Add 1 ml prehybridization mix[f] and transfer the embryos to a 1.5 ml microtube.

12. Replace 1 ml prehybridization mix and incubate at 60 °C for 3–8 h. The embryos can be stored indefinitely in this mix at −20 °C.

[a] 5–10 ml of solutions are used for steps **2–10**. Unless otherwise stated, the washes are at room temperature.
[b] 4% paraformaldehyde dissolved in PBS at 65 °C, cooled, and used on the same day.
[c] PTW: PBS, 0.1% Tween-20.
[d] Made from a 10 mg/ml glycine stock kept at −20 °C.
[e] A stock of 25% glutaraldehyde (Sigma) is stored in aliquots at −20 °C. Thaw out an aliquot just before use.
[f] Prehybridization mix: 50% formamide, 5 × SSC pH 5 (use citric acid to pH),
50 µg/ml yeast RNA, 1% SDS, 50 µg/ml heparin. For hybridization, add probe to 1 µg/ml.

5.2 Hybridization and posthybridization washes for mouse embryo whole mount *in situ* hybridizations

Hybridization with labelled probe is carried out overnight, and then probe that has not annealed to target RNA is removed by ribonuclease treatment and high stringency washing (*Protocol 9*). We routinely carry out the hybridization and high stringency washing at 60 °C, but for probes in conserved regions of gene families cross-hybridization can be avoided by hybridizing and washing at 70 °C. The washes can be carried out with gentle rocking, or by occasionally agitating the tube.

Protocol 9. Hybridization and washes for mouse embryo whole mounts

1. Replace the prehybridization solution (*Protocol 8*, step **12**) with 0.4 ml hybridization mix including 1 µg/ml digoxigenin-labelled RNA probe (from *Protocol 7*, step **8**).

2. Incubate at 60 °C overnight.

3. Wash[a] the embryos twice with solution 1[b] for 30 min at 60 °C.

4. Wash with 1:1 solution 1/solution 2[c] for 10 min at 60 °C.

5. Wash three times for 5 min with solution 2 at 37 °C.

6. Wash twice with 20 µg/ml ribonuclease A in solution 2 for 30 min at 37 °C.

7. Wash the embryos for 5 min with solution 2, then with solution 3[d].

8. Wash twice with solution 3 for 30 min at 65 °C.

9. Wash three times for 5 min with TBST[e] then proceed to *Protocol 10*.

[a] All washes are with 10 ml solution. If the embryos are not rocked during the washes, they should periodically be gently agitated.
[b] Solution 1: 50% formamide, 5 × SSC pH 5, 1% SDS.
[c] Solution 2: 0.5 M NaCl, 10 mM Tris-HCl pH 7.5, 0.1% Tween-20.
[d] Solution 3: 50% formamide, 2 × SSC pH 5.
[e] 10 × TBST: 1.4 M NaCl, 27 mM KCl, 0.25 M Tris-HCl pH 7.5, 1% Tween-20.

5.3 Immunocytochemical detection of probe in mouse whole mount *in situ* hybridization

After the removal of unbound probe, the embryos are incubated with anti-digoxigenin antibody conjugated to alkaline phosphatase, followed by washing and the detection of alkaline phosphatase (*Protocol 10*). To prevent the non-specific binding of antibody, the embryos are preblocked with sheep serum. In addition, to remove any components of the polyclonal antibody that can bind non-specifically, preabsorption with embryo powder (prepared as described in *Protocol 11*) is carried out; this step may not be necessary for all systems. The histochemical reaction should be carried out in a glass container, since a precipitate can form if a plastic container is used. Levamisole is included to inhibit endogenous alkaline phosphatases, but can be omitted if the latter do not cause a problem. Typically, for a transcript of approximately 100 copies per cell, signal is detectable after an incubation time of several hours, and with the low backgrounds obtained by this method, the histochemical reaction can be allowed to proceed for 1–3 days to produce a very intense signal.

Protocol 10. Washes and histochemistry for mouse embryos[a]

1. Preblock the embryos (from *Protocol 9*, step **9**) by incubating with 10% sheep serum[b] in TBST for 60–90 min.

2. During this, preabsorb the antibody as follows. Weigh out 3 mg embryo powder[c] into a microtube, add 0.5 ml TBST, 5 μl sheep serum, and 1 μl anti-digoxigenin antibody coupled to alkaline phosphatase (Boehringer). Shake gently at 4 °C for 1 h, then spin in a microcentrifuge at 4 °C for 1 min. Recover the supernatant and dilute it to 2 ml with 1% sheep serum in TBST.

3. Remove the 10% serum from the embryos, replace with preabsorbed antibody and rock overnight at 4 °C.

4. Wash the embryos three times for 5 min, then five times for 1 h with TBST.

Protocol 10. *Continued*

5. Wash three times for 10 min each with NTMT[d].

6. Incubate the embryos with NTMT including freshly-added 4.5 μl NBT[e] and 3.5 μl BCIP[e] per ml. Transfer the solution and embryos to a glass embryo dish for easier observation and rock for the first 20 min. Keep in the dark as much as possible.

7. When the colour has developed to the desired extent, wash the embryos twice for 10 min with NTMT.

8. Fix the signal by incubation in 4% formaldehyde in PBS. Embryos can be stored in this solution at 4 °C, but the fixative should be removed by washing with PBS prior to photography or embedding.

[a] All washes are in 10 ml solution with rocking. If required, 2 mM levamisole is added on the day of use to the washes and the colour developing solution.
[b] Sheep serum is heat-treated at 70 °C for 30 min before use.
[c] Embryo powder is prepared as described in *Protocol 11*.
[d] NTMT: 100 mM NaCl, 100 mM Tris-HCl pH 9.5, 50 mM $MgCl_2$, 0.1% Tween-20. Make from stocks on the day of use, since otherwise the pH can change due to the absorption of CO_2.
[e] Stock NBT (Nitro Blue Tetrazolium salt) is 75 mg/ml in 70% dimethylformamide. Stock BCIP (5-bromo-4-chloro-3-indolyl phosphate) is 50 mg/ml in dimethylformamide. These are stored at −20 °C in the dark.

Protocol 11. Preparation of embryo powder

1. Homogenize embryos[a] (for example, 14.5-day mouse embryos) in a minimum volume of ice-cold PBS.

2. Add 4 vol. ice-cold acetone, mix, and incubate on ice for 30 min.

3. Centrifuge at 10 000 g for 10 min, remove the supernatant, and then wash the pellet with ice-cold acetone and spin again.

4. Spread out the pellet and grind it into a fine powder on a sheet of filter paper.

5. Air-dry the powder and store at 4 °C.

[a] It is important to use embryo powder prepared from the system that you are studying e.g. for the analysis of *Xenopus* embryos, prepare *Xenopus* embryo powder.

5.4 Clearing and embedding of mouse embryos

Embryos can be photographed in whole mount (see *Figure 1c*), or to see some tissues, such as the neural plate, the tissue may be dissected and flattened under a coverslip (*Figure 1d*). It can also be useful to embed and section the embryos. Although the reaction product using NTB/BCIP is soluble in many

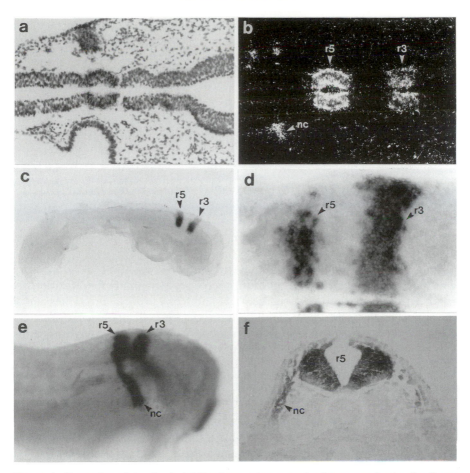

Figure 1. Examples of *in situ* hybridizations using a probe for a vertebrate zinc-finger gene, *Krox-20*. This gene is expressed in two rhomobmeres, r3 and r5, and a subpopulation of neural crest (nc) cells in the hindbrain. (a,b) Bright- and dark-field images, respectively, of a coronal section through the 9-day mouse embryo hindbrain, hybridized with ^{35}S-labelled probe. (c) Whole mount *in situ* hybridization of a 9-day mouse embryo with digoxigenin-labelled probe. (d) Flat mount of a microdissected hindbrain from an 8.5-day mouse embryo following whole mount *in situ* hybridization. (e) Whole mount *in situ* hybridization of a stage 28 *Xenopus* embryo. (f) Transverse section through a stage 28 *Xenopus* embryo following whole mount *in situ* hybridization.

organic solvents (e.g. methanol), it is stabilized by fixation with formaldehyde and the embryos can then be cleared and embedded in paraffin wax as follows.

In a fume hood, successively replace the solution from *Protocol 10*, step **8** with 5–10 ml methanol for 10 min, isopropanol for 15 min, and tetrahydro-naphthalene for 15 min. The cleared embryos can then be stored indefinitely in fresh tetrahydronaphthalene. Since the latter is an irritant and dissolves certain plastics, it is advised to use a glass container that can be sealed, for example, a glass scintillation vial. The cleared embryos can be photographed, although precautions should be taken against excessive fumes. To embed, transfer the embryos through three changes of paraffin wax at 60 °C, then place in a mould and allow the wax to set. Sections can then be cut, dried onto slides, and then de-waxed with Histoclear and mounted under a coverslip with DPX mounting agent (BDH) for photography (*Figure 1f*).

6. Whole mount *in situ* hybridization of *Drosophila* embryos

(Protocol provided by P. W. Ingham; based on the method of Tautz and Pfeifle, ref. 2).

Whole mount *in situ* hybridization to embryos was first developed for *Drosophila* embryos (2); the other protocols are essentially modifications of the *Drosophila* method. Digoxigenin-labelled RNA probes are synthesized as described in *Protocol 7*. *Drosophila* embryos are collected, dechorionated, fixed, devitellinized, and stored as described in Chapter 1. Prehybridization treatments, hybridization, and washing is described in *Protocol 12*; detection is described in *Protocol 13*.

Protocol 12. Prehybridization, hybridization, and washing of *Drosophila* embryo whole mounts

1. Rehydrate fixed *Drosophila* embryos (from Chapter 1, *Protocol 4*) through an ethanol series and refix in 4% formaldehyde[a] in 1 × PBS for 20 min at room temperature.

2. Rinse three times for 2 min in PTW[b].

3. Incubate for 4 min in PTW plus 50 µg/ml proteinase K.

4. Stop proteinase K digestion with two rinses in 2 mg/ml glycine in PTW, and rinse the embryos two more times in PTW.

5. Fix again in 4% formaldehyde in PTW for 5–10 min at room temperature. Longer fixation seems to increase the background.

6. Rinse 5 times in PTW, once in 1:1 PTW/hybridization buffer, and once in hybridization buffer[c] alone.

7. Prehybridize embryos by incubating in hybridization buffer for at least 1 hour at 55 °C.

8. Hybridize overnight in 200-400 µl hybridization buffer containing heat-denatured probe[d].

9. Wash embryos for 30 min at 65 °C in hybridization buffer, and then for another 30 min in a 1:1 mix of PTW/hybridization buffer at 65 °C. Wash a further 5 times in PTW (20 min each) at room temperature.

  [a] Many laboratories working with *Drosophila* find that 4% formaldehyde (made by dilution from stock 40% formaldehyde; BDH) is a satisfactory alternative to freshly dissolved paraformaldehyde.
  [b] PTW: 1 × PBS, 1% Tween 20.
  [c] Hybridization buffer: 50% formamide, 5 × SSC, 100 µg/ml yeast RNA, 0.1% Tween 20, 50 µg/ml heparin.
  [d] Probe concentration is usally 0.2 to 1 µg/ml, but may be varied. As a rough guide, a 1:500 dilution of the product of a normal probe synthesis resuspended in 50 µl should give good results.

Protocol 13. Immunocytochemical detection of probe in *Drosophila* embryo whole mounts

1. To 50 µl of spare fixed embryos (not the experimental embryos), add 800 µl PTW containing 2 µl anti-digoxigenin antibody coupled to alkaline phosphatase (Boehringer). Incubate for at least 4 h at room temperature or up to 5 days at 4 °C, to preabsorb the antibody.

2. Incubate the hybridized embryos (from *Protocol 12*, step **9**) with a 1:5 dilution of the preabsorbed antibody (from above) in PTW, for 1 h at room temperature.

3. Wash 4 × 20 min in PTW, then 2 × 5 min in fresh NTMT (see *Protocol 10*, footnote *d*).

4. Resuspend the embryos in 1 ml of NTMT buffer containing 4.5 µl NBT and 3.5 µl BCIP (see *Protocol 10*, footnote *e*) and allow the reaction to proceed in the dark for until the purple/brown colour appears (usually at least 1 h).

5. Stop the reaction by rinsing repeatedly in PTW. Embryos can be stored at 4 °C in PTW + 0.01% sodium azide until required for analysis.

---

7. Whole mount *in situ* hybridization of *Xenopus* embryos

For the whole mount hybridization of *Xenopus* embryos a different method (3) is superior to the protocols described above. We have not tested which of the differences affect the results, but the use of the detergent CHAPS (Sigma C-3023) in the prehybridization buffer and the posthybridization washes is probably the most significant change.

7.1 Fixation and prehybridization treatment of *Xenopus* embryos

Xenopus embryos are fixed with paraformaldehyde and pretreated with proteinase K and acetic anhydride (*Protocol 14*). As discussed in Section 5.1, the proteinase K treatment should be optimized. Overdigestion should be avoided since it leads to the stripping of epidermis and underlying mesenchyme. Since *Xenopus* embryos are quite brittle they should be treated gently.

Protocol 14. Fixation and prehybridization treatments for *Xenopus* embryos[a]

1. Dissect away the perivitelline membrane and fix the *Xenopus* embryos overnight at 4 °C in 4% paraformaldehyde in 70% PBS.

2. Remove the fixative and replace with 5–10 ml methanol. Agitate gently for 10–15 min, then replace solution with fresh methanol. The embryos can be stored indefinitely at −20 °C at this stage.

3. Rehydrate the embryos by washing for 5 min each with 5–10 ml 3:1, 1:1, and 1:3 methanol/PTW, then three times with PTW.

4. Incubate with 10 µg/ml proteinase K for 5–20 min.

5. Wash twice for 5 min with 5 ml 0.1 M triethanolamine, then add 12.5 µl acetic anhydride and rock gently for 5 min. Add a further 12.5 µl acetic anhydride and rock for 5 min.

6. Wash twice for 5 min with PTW.

7. Fix embryos for 20 min in 4% paraformaldehyde in PTW.

8. Remove most of the buffer and add 0.5 ml prehybridization (PH) buffer[b]. Allow the embryos to settle through this, then replace with fresh 0.5 ml PH buffer.

9. Incubate at 60 °C for 3–8 h. Embryos can be stored indefinitely at −20 °C at this point.

[a] See the footnotes to *Protocol 8* for the recipes of solutions.
[b] PH buffer: 50% formamide, 5 × SSC, 5mM EDTA, 100 µg/ml heparin, 1 × Denhardt's, 1 mg/ml yeast total RNA, 0.1% Tween-20, 0.1% CHAPS.

7.2 Hybridization and washing of *Xenopus* embryos

Digoxigenin-labelled RNA probe is synthesized as described in *Protocol 7* and mixed at 1 µg/ml in PH buffer (*Protocol 14*, footnote *b*). Hybridization is carried out overnight and then the embryos treated with ribonuclease and washed at high stringency (*Protocol 15*).

Protocol 15. Hybridization and washing for *Xenopus* embryo whole mounts[a]

1. Replace PH with 0.5 ml PH containing 1 µg/ml digoxigenin-labelled RNA probe and incubate at 60 °C overnight.

2. Wash the embryos at 60 °C for 10 min with PH buffer, then with 1:1 PH buffer/2 × SSC, 0.3% CHAPS, and finally 1:3 PH buffer/2 × SSC, 0.3% CHAPS.

3. Wash the embryos twice at 37 °C for 20 min with 2 × SSC, 0.3% CHAPS.

4. Treat at 37 °C for 30 min with 20 µg/ml ribonuclease A in 2 × SSC, 0.3% CHAPS.

5. Wash at 37 °C for 10 min with 2 × SSC, 0.3% CHAPS.

6. Wash the embryos at 60 °C twice for 30 min with 0.2 × SSC, 0.3% CHAPS, then twice for 10 min with PTW, 0.3% CHAPS.

7. Wash for 10 min at room temperature with PTW.

[a] All washes are with 5–10 ml of solution.

7.3 Immunocytochemical detection of probe in *Xenopus* embryo whole mounts

Anti-digoxigenin antibody coupled to alkaline phosphatase is used to detect probe exactly as described in *Protocol 10*, except that PBT (PBS, 2 mg/ml BSA, 0.1% Triton-X100) is used instead of TBST. With the low backgrounds that can be achieved (*Figure 1e*), the colour reaction can be left at room temperature for up to several days. The embryos can then be fixed, and, if desired, embedded and sectioned (*Figure 1f*) as described in Section 5.4.

8. Alternative protocols

The methods for whole mount *in situ* hybridization are still being optimized, and it may be worth testing each of the above protocols, or a combination of them, for your developmental system. It should also be noted that a considerably less stringent method works very well for zebrafish embryos (8). In addition, other methods can be used for the immunocytochemical detection of probe, for example using streptavidin coupled to β-galactosidase to detect a biotin-labelled probe (4). Modification of these methods should allow the simultaneous detection of several RNAs.

References

1. Wilkinson, D. G. and Green, J. (1990). In *Postimplantation mammalian embryos: a practical approach.* (ed. A. J. Copp and D. L. Cockroft), p. 155. IRL Press at Oxford University Press, Oxford.
2. Tautz, D. and Pfeifle, C. (1989). *Chromosoma*, **98**, 81.
3. Harland, R. M. (1991). *Methods Cell Biol*, **36**, 675.
4. Herrmann, B. G. (1991). *Development*, **113**, 913.
5. Wilkinson, D. G. (1992). In *In situ hybridization: a practical approach* (ed. D. G. Wilkinson), p. 75. IRL Press at Oxford University Press, Oxford.
6. Wilkinson, D. G. (1992). In *In situ hybridization: a practical approach* (ed. D. G. Wilkinson), p. 1. IRL Press at Oxford University Press, Oxford.
7. Melton, D. A., Krieg, P. A., Rebagliati, M. R., Maniatis, T., Zinn, K., and Green, M. R. (1984). *Nucleic Acids Res.*, **12**, 7035.
8. Krauss, S., Johansen, T., Korzh, V., and Fjose, A. (1991). *Development*, **113**, 1193.

Appendices

A1

Stage tables

1a. Westerfield (1993) stages of zebrafish development

Modified with permission from Westerfield, M. (1993). *The zebrafish book.* (2nd edn). University of Oregon Press, Eugene, USA.

Stage[a]	h[b]	HB[c]	Description
Zygote			
1-cell	0	1,2	Cytoplasm streams to animal pole to form blastodisc
Cleavage			
2-cell	¾	3	2 blastomeres
4-cell	1	4	2 × 2 array of blastomeres
8-cell	1¼	5	2 × 4 array of blastomeres
16-cell	1½	6	4 × 4 array of blastomeres
32-cell	1¾	7	2 regular tiers of blastomeres
64-cell	2	8	3 regular tiers of blastomeres
Blastula			
128-cell	2¼	9	5 irregular tiers of blastomeres
256-cell	2½		7 blastomere tiers
512-cell	2¾		9 blastomere tiers; YSL[d] formation
1k-cell	3	10	11 blastomere tiers; single row of YSL nuclei
High	3⅓		> 11 blastomere tiers; two rows of YSL nuclei
Oblong	3⅔	11	Multiple rows of YSL nuclei; elliptical shape
Sphere	4	12	Continued flattening produces spherical shape
Dome	4⅓	13	Blastula spherical; yolk cell bulging towards animal pole
30%-epiboly[e]	4⅔	14	Blastoderm shaped like cup of uniform thickness
Gastrula			
50%-epiboly	5¼		Blastoderm still uniform thickness
Germ-ring	5⅔		Margin of blastoderm thicker (germ ring); 50% epiboly

Shield	6	15	Dorsal germ ring thickest (embryonic shield); 50% epiboly
75%-epiboly	8	16	From 60–100% epiboly, extent of epiboly[e] is staging index
90%-epiboly	9		
Bud	10	17	Tail bud; groove in anterior neural keel; 100% epiboly

Segmentation

1-somite	10⅓		First somite furrow forms, posterior to somite 1
5-somite	11⅔	18	Staging index is number of somites, from 1–26; optic vesicle
14-somite	16	19	Otic placode; brain subdivided; somites V-shaped
20-somite	19	20[f]	Rhombic flexure present; rhombomeres often prominent
26-somite	22		2 otoliths in otic capsules; Prim-3[g]; active lateral flexures

Straightening

Prim-5	24	$HTA^h = 135°$; $OCL^i = 5$; about 30 somites
Prim-11	30	$HTA = 100°$; $OCL = 3$; straight tail
Prim-23	36	$HTA = 85°$; $OCL = 1–2$
Pec-ridge	42	$HTA = 75°$; $OCL = < 1$; prominent bend in chambered heart

Hatching

| Long bud | 48 | $HTA = 45°$; pointed pectoral buds; yellowish head |
| Pec-fin | 60 | $HTA = 30°$; mouth open; pectoral fin flattened distally |

Early larva

| Early larva | > 72 | Hatched; swim bladder develops; larvae seek food |

[a] The staging series is for living, not fixed, embryos and is not necessarily applicable to other fish species.
[b] h: hours after fertilization at 28.5°C.
[c] HB: Stage number of Hisoaka, K. K. and Battle, H. I. (1958) *J. Morphol.*, **102**, 311-323.
[d] YSL: yolk syncytial layer. Nomarski optics are required to see YSL nuclei.
[e] Margin of blastoderm moving; % epiboly is position of margin between animal and vegetal pole.
[f] HB descriptions are considered inaccurate after stage 20.
[g] Prim: the myotome to which the posterior lateral line has advanced. Nomarski required.
[h] HTA: head-trunk angle, lower values denote a straighter axis.
[i] OCL: the distance between otic capsule and eye, in 'otic capsule lengths'.

Figures 1 and 2. Stages of normal development of the zebrafish, *Brachydanio rerio*. Reproduced, with kind permission, from Westerfield, M. (1993). *The zebrafish book* (2nd edn) University of Oregon Press, Eugene.

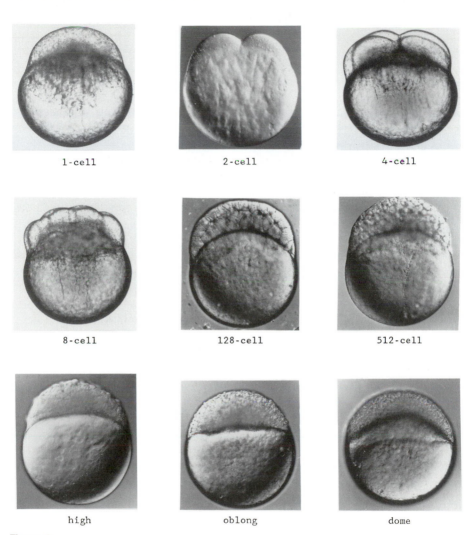

Figure 1

Appendix 1

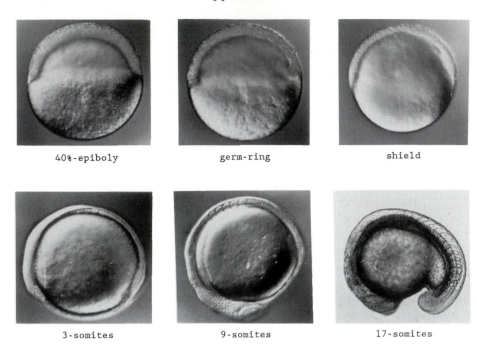

40%-epiboly

germ-ring

shield

3-somites

9-somites

17-somites

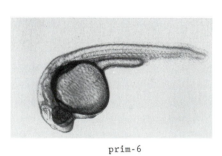

prim-6

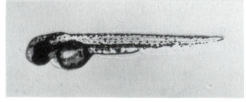

48-h

Figure 2

1b. Nieuwkoop and Faber (1967) table of *Xenopus laevis* development (very abridged)

Stage	Brief description
1	(0 h). 1-cell stage. Length 1.4–1.5 mm.
2	(1.5 h). Advanced 2-cell stage.
3	(2 h). Advanced 4-cell stage.
4	(2.25 h). Advanced 8-cell stage.
5	(2.75 h). Advanced 16-cell stage.
6	(3 h). Advanced 32-cell stage.
7	(4 h). Large-cell blastula stage; 10 micromeres along meridian in animal view.
8	(5 h). Medium-cell blastula. Gradual transition in cell size from animal to vegetal.
9	(7 h). Fine-cell blastula. Animal cells smaller dorsal than ventral. Border between marginal zone and vegetal part distinct, particularly dorsally.
10	(9 h). Initial gastrula. Blastopore appears; no groove—only pigment.
11	(11.75 h). Horse-shoe blastopore. Groove surrounds half of future yolk plug and indicated ventrally.
12	(13.25 h). Medium yolk plug stage. Diameter of yolk plug less than one-quarter the diameter of egg.
13	(14.75 h). Slit-blastopore stage. Neural plate faintly delimited. Caudal part of median groove formed.
14	(16.25 h). Neural plate stage. Initial elevation of neural folds. Length 1.5–1.6 mm.
15	(17.5 h). Early neural fold stage. Cement gland faint. Anterior neural plate roundish. Neural folds distinct except medio-rostrally.
16	(18.25 h). Mid-neural fold stage. Anterior neural plate rectangular. Neural plate sharply constricted in the middle.
17	(18.75 h). Late neural fold stage. Anterior neural plate oblong triangular.
18	(19.75 h). Neural groove stage. Anterior neural plate narrow, club-shaped. Parallel neural folds in trunk region very close. Anterior 3–4 somites forming.
19	(20.75 h). Initial neural tube stage. Neural folds touching. 4–6 somites segregating.
20	(21.75 h). Neural folds fused, suture still present. Length 1.7–1.8 mm.
21	(22.5 h). Suture of neural tube completely closed. Delimitation of frontal field by pigment lines. Beginning of eye protrusion. Length 1.9–2 mm.
22	(24 h). Distinct protrusion of eyes. Initial groove between jaw and gill areas. Length 2–2.2 mm.

23   (24.75 h). Jaw and gill areas completely separated by groove. 12 somites formed. Length 2.2–2.4 mm.

24   (1 day, 2.25 h). Eyes protrude less far than gill area. Gill area more prominent than jaw area; 15 somites. Length 2.5–2.7 mm.

25   (1 day, 3.5 h). Eyes protrude equally far or further than gill area. Gill area grooved. Invagination of ear vesicle indicated by pigment spot. Beginning of fin formation. 16 somites. Length 2.8–3 mm.

26   (1 day, 5.5 h). Ear vesicle protruding. Pronephros clearly visible. Myotomes (somites) visible from outside for the first time. Fin somewhat broadened dorso-caudally. Movements start. 17 somites formed; length 3–3.3 mm.

27   (1 day, 7.25 h). Fin translucent, except just behind anus. Tail bud formation accentuated laterally; 19 somites. Length 3.4–3.7 mm.

28   (1 day, 8.5 h). Fin broadened and distinctly divided into outer transparent and inner translucent band, extending to anus. 20–22 somites. Length 3.8–4 mm.

29/30   (1 day, 11 h). Fin transparent to base over its whole length. Grey eye cup shows through for the first time. 24–25 somites formed, reaching tail. Length 4–4.5 mm.

31   (1 day, 13.5 h). Tail bud as long as broad. Mushroom-shaped pineal (epiphysis) evagination. 22–23 post-otic somites. Length 4.2–4.8 mm.

32   (1 day, 16 h). Length of tail bud about 1.5 times its breadth. Eye cup horseshoe shaped. ~26 post-otic somites. Length 4.5–5.1 mm.

33/34   (1 day, 20.5 h). Dorsal part of eye more pigmented than ventral. Melanophores appearing dorsally on head and laterally in a row from below the pronephros backwards. Length of tail bud about twice its breadth. Heart starts to beat. Length 4.7–5.3 mm.

35/36   (2 days, 2 h). Formation of two gill rudiments. Eye entirely black, choroid fissure nearly closed. Melanophores appear on back. Tail bud length three times its breadth. Hatching starts; length 5.3–6 mm.

37/38   (2 days, 5.5 h). Both gill rudiments nipple-shaped, a branch of the anterior one indicated. Posterior outline of proctodeum straight, forms very obtuse angle to ventral border of tail myotomes. Melano-phores over tail. Length 5.6–6.2 mm. Tail with fin ~1.8 mm long.

39   (2 days, 8.5 h). Outline of proctodeum and tail myotomes form angle of 135°; Melanophores along ventral edge of tail musculature. Length 5.9–6.5 mm. Tail with fin ≤ 2.6 mm.

40   (2 days, 18 h). Mouth broken through. Beginning of blood circulation in gills. Gills two times longer than broad, posterior sometimes branched. Length 6.3–6.8 mm.

41   (3 days, 4 h). Formation of left rostral and right caudal furrow in yolk mass. Conical proctodeum, 60° to tail myotomes. Formation of fin rostral to proctodeum. Length 6.7–7.5 mm.

42   (3 days, 8 h). Beginning of formation of opercular folds. Torsion of

intestine about 90°. Proctodeum connected with yolk mass by short horizontal intestinal tube. Length 7–7.7 mm.

43   (3 days, 15 h). Lateral line system becoming visible. Cement gland losing pigment. 180° torsion of intestine. Length 7.5–8.3 mm.

44   (3 days, 20 h). Appearance of tentacle rudiments. Coiling part of intestine shows S-shaped loop. Torsion 360°. Usually no blood circulation in gills. Length 7.8–8.5 mm.

45   (4 days, 2 h). Intestine spiral in ventral aspect, showing 1.5 revolutions. Feeding starts. Length 8–10 mm.

46   (4 days, 10 h). Edge of operculum convex. Xanthophores appear on eye and abdomen. Intestine with 2–2.5 revolutions. Hindlimb bud visible. Length 9–12 mm.

47   (5 days, 12 h). Edge of operculum forms quarter-circle. Xanthophores form opaque layer on abdomen. Intestine shows 2.5–3.5 revolutions. Length 12–15 mm.

48   (7.5 days). Forelimb bud visible. Shining gold-colour abdomen. Length 14–17 mm.

49   (~12 days). Forelimb bud distinct. Hindlimb larger, distally still circular, no constriction at base. Melanophores on dorsal and ventral fins. Length 17–23 mm.

50   (~15 days). Forelimb bud oval shaped dorsally. Hindlimb bud longer than broad, constricted at base, distal outline conical. Length 20–27 mm.

51   (~17 days). Forelimb bud oval in lateral aspect. Hindlimb conical, length about 1.5-times its breadth. Melanophores on hindlimb bud. Length 28–36 mm.

52   (~21 days). Forelimb bud irregularly conical. Hindlimb shows first indication of ankle constriction and flattening of foot. Length 42–56 mm.

53   (~24 days). Paddle stage in all limbs. Hindlimb (without foot) longer than broad, toes 4 and 5 indicated. Length 50–60 mm.

54   (~26 days). All four fingers indicated. Length of hindlimb without foot two times breadth, all five toes indicated, second only slightly. Length 58–65 mm.

55   (~32 days). Hand pronated 90°. Length of hind-limb without foot about three times breadth. Length 70–80 mm.

56   (~38 days). Elbow and wrist indicated. Length 70–100 mm.

57   (~41 days). Angle of elbow more than 90°, length of fingers about seven times their breadth. Length 75–105 mm.

58   (~44 days). Forelimbs broken through. All three claws present on hindlimb. Length 80–110 mm.

59   (~45 days). Stretched forelimb reaches base of hindlimb.

60   (~46 days). Distal half of fingers of stretched forelimb extend beyond base of hindlimb.

61   (~48 days). 4th arterial arch seen just in front of adult skin area of forelimb. Forelimb at level of posterior half of heart.

62   (~49 days). Head somewhat broader than cranial part of trunk. Forelimb at middle of heart.

63   (~51 days). Head narrower than trunk. Tentacles mostly disappeared. Corner of mouth at level of caudal border of eye. Operculum closed. Tail still slightly longer than body.

64   (~53 days). Corner of mouth behind eye. Tail about one-third of body length.

65   (~54 days). Tail length about 1/10 body length.

66   (~58 days). Tail no longer visible from ventral side. Border lines between adult skin areas disappeared.

Appendix 1

Figures 1–10. Stages of normal development of the South African clawed frog, *Xenopus laevis*. Reproduced, with kind permission of Prof. Nieuwkoop and Elsevier/North Holland, from Nieuwkoop, P.D. and Faber, J. (1967). *Normal table of Xenopus laevis (Daudin).* Elsevier/North Holland, Amsterdam.

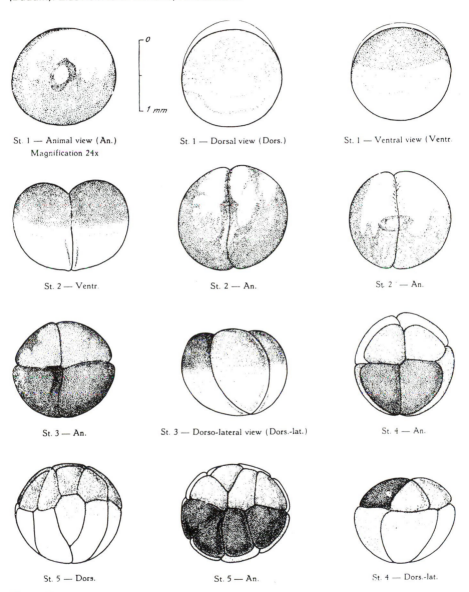

St. 1 — Animal view (An.)
Magnification 24x

St. 1 — Dorsal view (Dors.)

St. 1 — Ventral view (Ventr.

St. 2 — Ventr.

St. 2 — An.

St. 2 — An.

St. 3 — An.

St. 3 — Dorso-lateral view (Dors.-lat.)

St. 4 — An.

St. 5 — Dors.

St. 5 — An.

St. 4 — Dors.-lat.

Figure 1

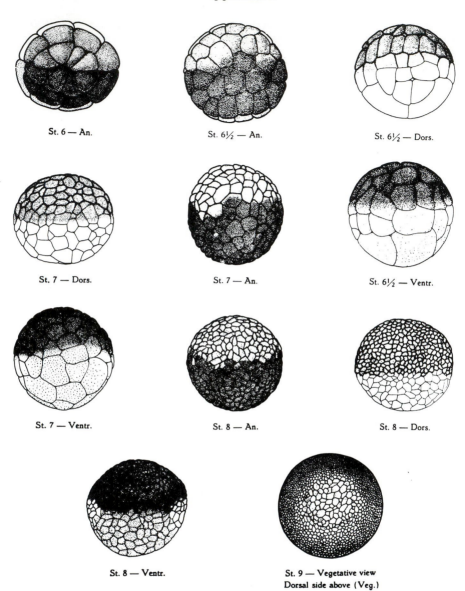

St. 6 — An.

St. 6½ — An.

St. 6½ — Dors.

St. 7 — Dors.

St. 7 — An.

St. 6½ — Ventr.

St. 7 — Ventr.

St. 8 — An.

St. 8 — Dors.

St. 8 — Ventr.

St. 9 — Vegetative view
Dorsal side above (Veg.)

Figure 2

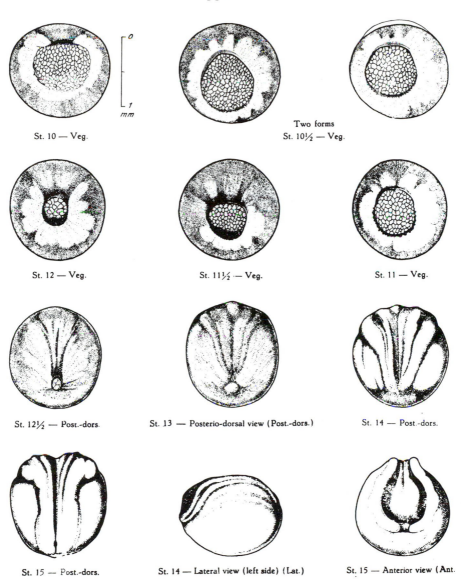

St. 10 — Veg.

Two forms
St. 10½ — Veg.

St. 12 — Veg.

St. 11½ — Veg.

St. 11 — Veg.

St. 12½ — Post.-dors.

St. 13 — Posterio-dorsal view (Post.-dors.)

St. 14 — Post.-dors.

St. 15 — Post.-dors.

St. 14 — Lateral view (left side) (Lat.)

St. 15 — Anterior view (Ant.

Figure 3

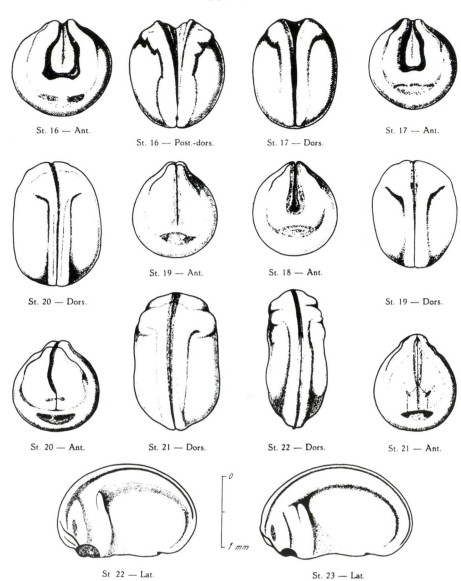

St. 16 — Ant.

St. 16 — Post.-dors.

St. 17 — Dors.

St. 17 — Ant.

St. 19 — Ant.

St. 18 — Ant.

St. 20 — Dors.

St. 19 — Dors.

St. 20 — Ant.

St. 21 — Dors.

St. 22 — Dors.

St. 21 — Ant.

St. 22 — Lat.

St. 23 — Lat.

Figure 4

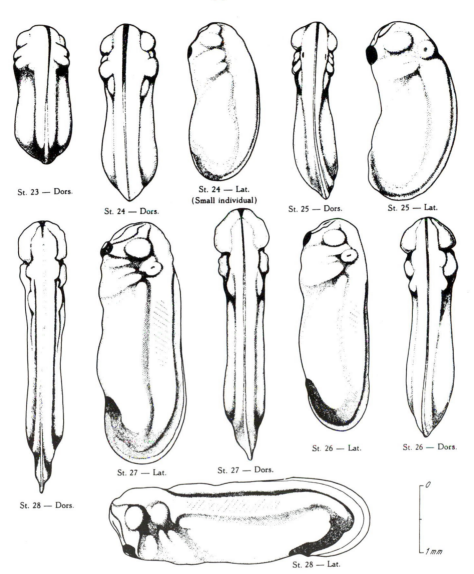

St. 23 — Dors.

St. 24 — Dors.

St. 24 — Lat.
(Small individual)

St. 25 — Dors.

St. 25 — Lat.

St. 28 — Dors.

St. 27 — Lat.

St. 27 — Dors.

St. 26 — Lat.

St. 26 — Dors.

St. 28 — Lat.

0

1 mm

Figure 5

289

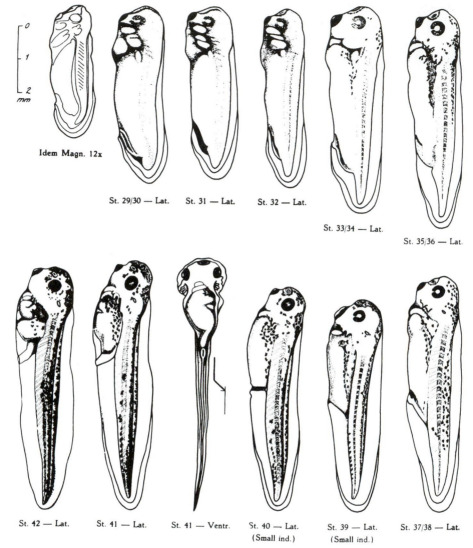

Idem Magn. 12x

St. 29/30 — Lat.   St. 31 — Lat.   St. 32 — Lat.

St. 33/34 — Lat.

St. 35/36 — Lat.

St. 42 — Lat.   St. 41 — Lat.   St. 41 — Ventr.   St. 40 — Lat.   St. 39 — Lat.   St. 37/38 — Lat.
(Small ind.)   (Small ind.)

Figure 6

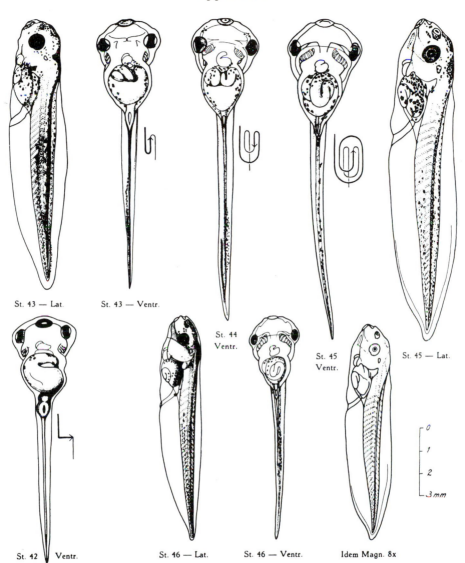

St. 43 — Lat.   St. 43 — Ventr.   St. 44 Ventr.   St. 45 Ventr.   St. 45 — Lat.

St. 42 — Ventr.   St. 46 — Lat.   St. 46 — Ventr.   Idem Magn. 8x

Figure 7

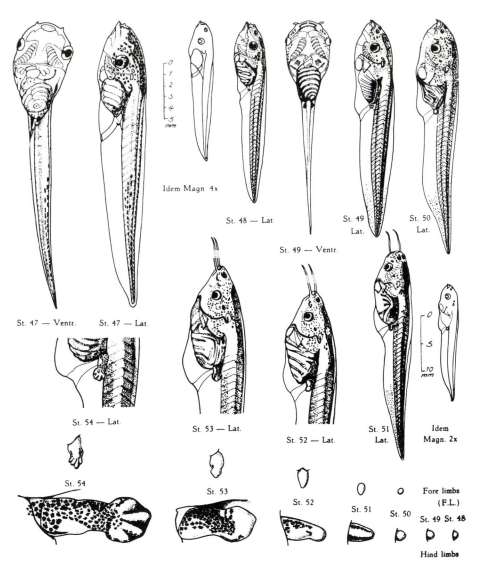

Figure 8

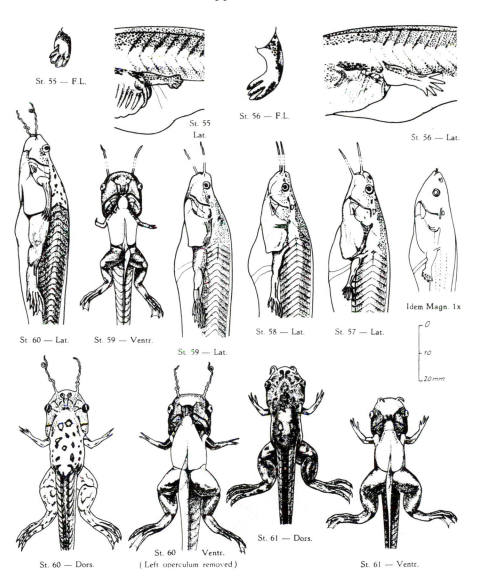

Figure 9

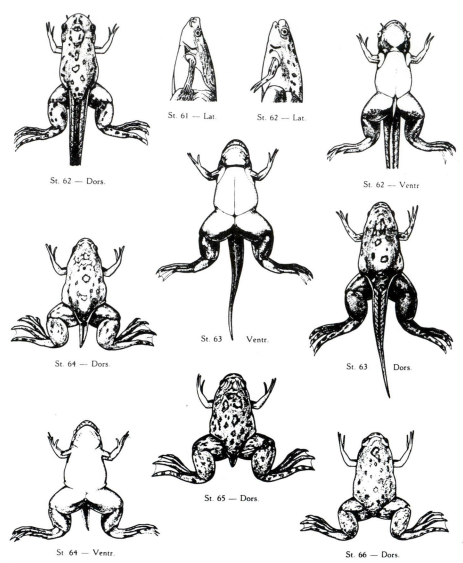

St. 62 — Dors.

St. 61 — Lat.   St. 62 — Lat.

St. 62 — Ventr

St. 63   Ventr.

St. 64 — Dors.

St. 63   Dors.

St. 65 — Dors.

St 64 — Ventr.

St. 66 — Dors.

Figure 10

1c. Eyal-Giladi and Kochav (1976) table of normal stages of chick development before formation of the primitive streak (very abridged)

Stage   **Brief description**
I   (0–1 h uterine age). Flaccid shell membrane. Germinal disc 3.5–4 mm diameter, sometimes with cleavage furrows. 5–6 cell divisions during first 2 h. Cells still open peripherally.
II   (2 h uterine age). 14–16 lateraly closed cells enclosed by vertical cleavage furrows.
III   (3–4 h uterine age). Smaller diameter disc, but thicker than stage I. 80–90 blastomeres closed laterally on upper surface. 10–16 cells on lower surface.
IV   (5 h uterine age). Upper surface has 250–300 closed cells. Centre of lower surface has 80–90 ventrally closed cells.
V   (8–9 h uterine age). Sub-germinal cavity stretches towards the periphery of the blastodisc. Cleavage much more advanced than stage IV.
VI   (10–11 h uterine age). Entire cytoplasmic mass of disc now cleaved on both surfaces. Cells form epithelial sheet of uniform thickness. Embryo now called *blastoderm*.
VII   (12–14 h uterine age). Upper layer cells smaller than lower layer. Shedding of cells from (posterior half of) upper layer. The region of upper layer shedding cells becomes thinner and is first sign of *area pellucida*.
VIII   (15–17 h uterine age). Thinner (transparent) area becomes sickle-shaped. Margin of ring clearly not included in thinning and does not change: *area opaca*. Ventrally there are still very large cells.
IX   (17–19 h uterine age). *Area pellucida* spreads in anterior direction. Anteriorly the border between *area opaca* and *area pellucida* is not yet sharp.
X   (20 h uterine age or freshly laid egg). Formation of *area pellucida* completed. On lower surface, clusters of cells smaller than those shed earlier form mesh-like layer posteriorly. Transparent sickle-shaped belt seen posteriorly.
XI   (from this stage, eggs need to be incubated but period depends on stage at laying and other factors). Posterior sickle visible from ventral side (Koller's sickle), anterior to which the hypoblast sheet begins to form.
XII   Hypoblast covers half of lower surface of *area pellucida*, extending anteriorly from Koller's sickle.
XIII   Hypoblast formation complete; sheet covers all of ventral surface of

area pellucida. Posterior, but not anterior, border of hypoblast has well defined edge. Upper surface still smooth.

XIV   Anterior part of hypoblast now also has well defined edge. Posterior to the posterior margin of the hypoblast (Koller's sickle), a 'cellular bridge' can be seen.

(followed by Hamburger and Hamilton, 1951 stage 2).

Figures 1–5. The early stages of development of the chick. Reproduced, with kind permission of Prof. Eyal-Giladi and Academic Press, from Eyal-Giladi, H. and Kochav, S. (1976) *Dev. Biol.*, **49**, 321.

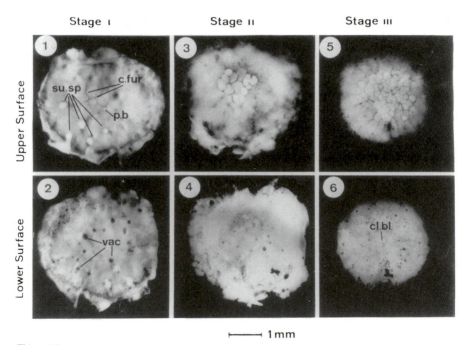

Figure 1

Appendix 1

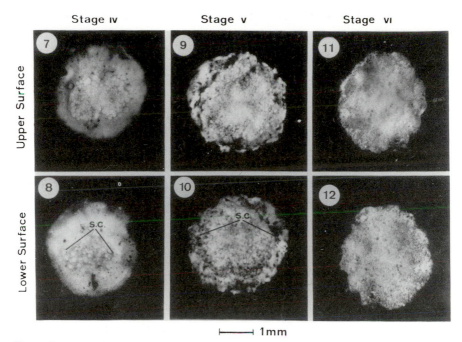

Figure 2

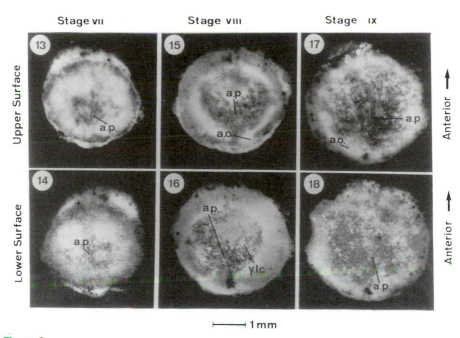

Figure 3

297

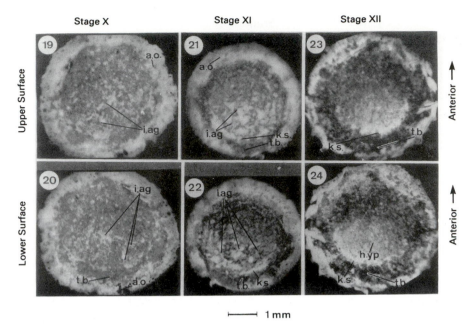

Figure 4

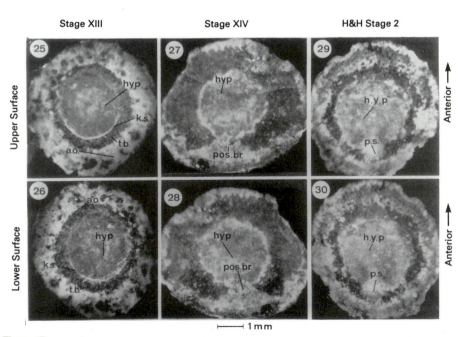

Figure 5

1d. Hamburger and Hamilton (1951) table of normal stages of the chick embryo after appearance of the primitive streak (very abridged)

Stage Brief description

2 (~6–7 h; brief stage) Primitive streak is short conical thickening, 0.3–0.5 mm long.

3 (12–13 h) Primitive streak extends to centre of embryo; streak broad throughout. No groove.

4 (18–19 h) Primitive streak grooved, ~1.88 mm long. Hensen's node present. *Area pellucida* pear-shaped.

5 (19–22 h). Notochord (head-process) visible as a rod of mesoderm extending forwards from Hensen's node. No head fold.

6 (23–25 h; brief stage) A fold of the blastoderm anterior to the notochord marks anterior end of embryo proper. No somites.

7 (23–26 h). 1 somite. Neural folds visible in head region.

8 (26–29 h). 4 somites. Neural folds meet at level of midbrain. Blood islands present posteriorly.

9 (29–33 h). 7 somites. Primary optic vesicles present. Paired primordia of heart begin to fuse.

10 (33–38 h). 10 somites. First somite becoming dispersed. First indication of cranial flexure. 3 brain vesicles clearly visible. Optic vesicles not constricted at bases. Heart bent slightly to right.

11 (40–45 h). 13 somites. Slight cranial flexure. 5 distinct neuromeres in hindbrain. Anterior neuropore closing. Optic vesicles constricted at bases. Heart bent to right.

12 (45–49 h). 16 somites. Head turning onto left side. Anterior neuropore closed. Primary optic vesicles and optic stalk well established. Auditory pit deep, but open. Heart slightly S-shaped . Head fold of amnion covers entire forebrain.

13 (48–52 h). 19 somites. Head turning onto left side. Telencephalon enlarged. Atrio-ventricular canal indicated by constriction. Head fold of amnion extends to anterior hindbrain.

14 (50–53 h). 22 somites. Forebrain and hindbrain form about a right angle. Visceral (branchial) arches 1 and 2, and clefts 1 and 2 distinct. Primary optic vesicle invaginating; lens placode formed. Opening of auditory pit constricted. Rathke's pouch can be recognized. Ventricular loop of heart now ventral to atrio-ventricular canal. Amnion extends to somites 7–10.

15 (50–55 h). 24–27 somites. Lateral body folds extend to anterior end of wing level (somites 15–17). Prospective limb areas not yet demarcated. Amnion extends to somites 7–14. Axes of forebrain and hindbrain nearly

parallel. Trunk rotation extends to somites 11–13. Visceral arch 3 and cleft 3 distinct. Cleft 3 shorter than cleft 2 and usually oval shape. Optic cup completely formed; double contour distinct in region of iris.

16 (51–56 h). 26–28 somites. Lateral body folds extend to somites 17–20, between wings and legs. Wing lifted off blastoderm, represented by thickened ridge. Primordium of leg still flat. Amnion extends to somites 10–18. Rotation of trunk extends to somites 14–15. Tail bud a short, straight cone, delimited from blastoderm. Pineal gland (epiphysis) indistinct or not yet formed.

17 (52–64 h). 29–32 somites. Lateral body folds extend around entire circumference of body. Both wing and leg buds lifted off blastoderm, both approximately equal size. Trunk rotation to somites 17–18. Tail bud bent ventrally, unsegmented. Pineal gland (epiphysis) is distinct. Indication of nasal pits.

18 (3 days). 30–36 somites, extending beyond leg bud. Leg buds slightly larger than wing buds. L/W of wing $\leqslant 6.0$ (L = anterior-posterior length measured along body wall; W = distance from body wall to apex). Axis of medulla at cervical flexure about 90° to posterior trunk. Trunk flexure shifted to lumbar region. Rotation extends to posterior part of body. Leg buds no longer in horizontal plane. Tail bud turned to right, about 90° to axis of posterior trunk. Allantois short, thick-walled, not yet vesicular.

19 (3–3½ days). 37–40 somites, extending into tail but end of tail unsegmented. Leg buds slightly larger than wing buds. L/W of wing buds = 4–6. Tail bud curved, tip pointing forward. Maxillary process a distinct swelling approximately same length as mandibular process.

20 (3–3½ days). 40–43 somites; tip of tail unsegmented. Leg buds distinctly larger than wing buds. Wing buds approximately symmetrical, leg buds slightly asymmetrical. L/W of wing: 3–3.9; leg: 3–3.3 (error in original?). Tail bend extending forward into lumbo-sacral region. Rotation completed. Maxillary process equal or larger than mandibular process. 2nd arch projects over surface. 4th arch less prominent and smaller than 3rd. 4th cleft shorter than 3rd. Allantois vesicular, variable in size; about the size of midbrain. Eye pigment: faint greyish hue.

21 (3½ days). 43–44 somites; extreme tip of tail unsegmented. Proximo-distal axes of wing and leg directed caudally. Posterior contours of wing and leg buds meet the baseline at an angle of ~90°. L/W of wing = 2.3–2.7; leg = 2.0–2.5. The posterior curvature includes lumbo-sacral region. Maxillary process definitely longer than mandibular, extending approx. to middle of eye. 2nd arch overlaps 3rd ventrally. 4th arch distinct; 4th cleft visible as a slit. Eye pigmentation: faint.

22 (3½–4 days). Tail segmented to tip. Elongated limb buds, pointing caudally. Anterior and posterior contours of limbs nearly parallel at their bases. L/W of wing = 1.5–2; leg = 1.3–1.8. Allantois extends to head. Eye pigmentation: distinct.

23 (4 days). At proximal parts, anterior and posterior contours of limbs parallel and lengthened; wing and leg buds about as long as wide. 1st visceral cleft represented by broken line. Slight protuberance ('a' in *Figure 8*) noticeable anterior to the dorsal slit. Caudal part of 2nd arch distinctly elevated over surface. Arches 3 and 4 still completely exposed. 3rd cleft a distinct groove, and 4th cleft reduced to narrow oval pit dorsally.

24 (4½ days). Wing and leg buds distinctly longer than wide. Digital plate in wing not yet demarcated; toe plate in leg bud distinct. Toes not yet demarcated. 1st visceral cleft a distinct curved line. Slight indication of two protuberances ('a,' 'b' *in Figure 8* on mandibular process) and of three ('d,' 'e,' 'f', in *Figure 8*) on 2nd arch. Part 'c' of mandibular process receding. Second arch longer ventrally (at 'f', *Figure 8*) and much wider than mandibular process. 3rd arch reduced and partly overgrown by 2nd. 4th arch flattened. Both sunk beneath the surface. 3rd cleft an elongated groove. 4th cleft reduced to a small pit.

25 (4½–5 days). Elbow and knee joints distinct. Digital plate in wing distinct, but no digits. Faint grooves demarcate the third toe on leg. Maxillary process lengthened, meeting wall of nasal groove (notice the notch at point of fusion). Three protuberances on each side of 1st visceral cleft ('a' to 'f', in *Figure 8*). In dorsal view, 'a,' 'b,' and 'd' appear as round knobs, and 'c' as a flat ridge. Part 'f' ('collar') conspicuous and projects distinctly over the surface. Dorsal part of 3rd arch still visible. 3rd and 4th clefts reduced to small pits.

26 (*ca.*5 days). Contour of digital plate rounded. Faint groove between second and third wing digit; first three toes distinct. Contour of maxillary process a broken line. Mandibular process lengthened ventrally. 'a' and 'b' (*Figure 8*) project over surface. 'b' subdivided by a shallow groove. Small knob distinct at dorsal edge of 'c'. On 2nd arch, 'd' and 'e' are only slightly elevated over the surface. The 'collar' ('f') has broadened and overgrown visceral arches III and IV. Deep groove separates 'f' from 'c'. 3rd and 4th clefts no longer visible.

27 (5–5½ days). Contour of digital plate angular in region of first digit. Digits 1, 2, 3 separated by grooves. Grooves between toes distinct on outer and inner surfaces of toe plate. First toe projects over tibial part at an obtuse angle. Tip of third toe not yet pointed. Mandibular process has broadened ventrally (at 'c', *Figure 8*) and grown forward. 'a' and 'b' project over surface. 'd' and 'e' flat. 'b' and 'e' close to fusion, but separating line still distinct. The 'collar' ('f') has broadened and rises above the surface. Groove between 'c' and 'f' has widened. Beak: barely recognizable.

28 (5½–6 days). Second digit and third toe slightly longer than others. Three digits and four toes distinct. No fifth toe. Protuberance 'a' still projects over surface. Mandibular process has lengthened and grown forward.

301

Parts 'b' and 'e' have fused. 'b', 'd' and 'e' no longer project above surface. External auditory opening now very distinct between 'a', 'b' and 'd.' 'Collar' ('f') projects distinctly over surface. Neck between 'collar' and mandible lengthened. Beak: distinct outgrowth visible in profile.

29 (6–6½ days). Wing bent in elbow, second digit distinctly longer than others. Shallow grooves between digits 1, 2, 3. Toes 2–4 stand out as ridges separated by distinct grooves, with webs between them. Distal contours of webs are straight lines. Rudiment of fifth toe visible. Mandibular process and 2nd arch broadly fused. Auditory meatus distinct at dorsal end of fusion. All protuberances have flattened. 'Collar' stands out conspicuously. Beak more prominent than in stage 28. No egg tooth visible.

30 (6½–7 days). The three segments of wing and leg clearly demarcated, bent at elbow and knee joints. Distinct grooves between digits 1 and 2. Webs between digits 1/2 and between all toes slightly curved, concave lines. Mandibular process approaches beak, but gap between the two still conspicuous. Lengthening of neck between 'collar' and mandible very conspicuous. 'Collar' begins to flatten. Two dorsal rows of feather germs to either side of spinal cord at brachial level. Three rows at level of legs; indistinct at thoracic level. None on thigh. Scleral papillae: one on either side of choroid fissure; never more than two. Egg tooth distinct, slightly protruding.

31 (7–7½ days). Indication of web between digits 1/2. Rudiment of fifth toe still distinct. Gap between mandible and beak narrowed to a small notch. 'Collar' inconspicuous or absent. Feather germs: on dorsal surface, continuous from brachial to lumbo-sacral level. ~7 rows at lumbo–sacral level. Distinct feather papillae on thigh. One indistinct row on each lateral edge of tail. 6 scleral papillae: 4 on dorsal side near choroid fissure, 2 on opposite side.

32 (7½ days). All digits and 4 toes have lengthened conspicuously. 5th toe has disappeared. Webs between digits and toes thin, contours concave. Differences in size of individual digits and toes conspicuous. Anterior tip of mandible reaches beak. ≥ 11 rows of feather germs on dorsal surface at level of legs. 1 row on tail distinct, 2nd row indistinct. Scapular and flight feather germs barely perceptible or absent. 6–8 scleral papillae in two groups; one dorsal and one ventral. Circle not yet closed.

33 (7½–8 days). Web on radial margin of arm and digit 1 discernible. Mandible and neck lengthened. 3 rows of feather germs distinct in tail, middle row larger than others. 13 scleral papillae, forming an almost complete circle, with gap for one missing papilla ventrally near middle of jaw.

34 (8 days). Differential growth of second digit and third toe conspicuous; contours of webs between digits and toes concave and arched. Feather germs visible on scapula, ventral neck, procoracoid, and posterior (flight)

edge of wing. Feather germs next to dorsal midline, particularly at lumbo–sacral level, extend slightly over surface when viewed in profile. Feather germs on thigh protrude conspicuously. One row on inner side of each eye. None around umbilical cord. 13 or 14 scleral papillae. Nictitating membrane extends halfway between outer rim of eye (eyelid) and scleral papillae.

35 (8½–9 days). Webs between digits and toes inconspicuous. Phalanges in toes distinct. Feather germs more conspicuous. Mid-dorsal line stands out distinctly in profile view. ≥ 4 rows in inner side of each eye. New feather germs near mid-ventral line, close to sternum, and extending to both sides of umbilical cord. Nictitating membrane approaches outer scleral papillae. Eyelids (external to nictitating membrane) extend towards beak and begin to overgrow eyeball. Circumference of eyelids ellipsoidal. 4 rows on inner side of each eye.

36 (10 days). Distal segments of limb proportionately much longer. Length of third toe, from tip to middle of metatarsal joint = 5.4 ± 0.3 mm. Tapering primordia of claws just visible on termini of toes and on digit 1 of wing. Primordium of the comb appears as prominent ridge with slightly serrated edge along dorsal mid-line of beak. 'Labial groove' clearly visible at tip of upper jaw, but barely indicated on tip of mandible. Nostril narrowed to a slit. Length of beak from anterior angle of nostril to tip of bill = 2.5 mm. Flight feathers conspicuous; coverts just visible in web of wing. Feather germs cover tibio-fibular portion of leg. ≥ 9–10 rows of feather germs between each upper eyelid and dorsal midline. Sternal tracts prominent, with 3–4 rows on each side of ventral midline when counted in anterior part of sternum, merging into many rows around umbilicus. Nictitating membrane covers anteriormost scleral papillae and approaches cornea. Lower lid has grown to level of cornea.

37 (11 days). Claws of toes flattened laterally and curved ventrally, dorsal tips opaque. Tip of claw on wing opaque. Pads on plantar surface of foot conspicuous and smooth. Transverse ridges along superior surfaces of metatarsus and phalanges (future scales). Length of third toe: 7.4 ± 0.3 mm. Labial groove on mandible is clearly marked off. Comb clearly serrated. Beak (from anterior angle of nostril to tip of bill) = 3.0 mm. Feather germs: much more numerous, and in most advanced tracts (e.g. back, tail) elongated into long, tapered cones. External auditory meatus nearly surrounded by feather germs. Circumference of eyelids bordered by single row of just-visible primordia; none on remainder of lids. Sternal tracts contain 5–6 prominent rows when counted at anterior end of sternum. Nictitating membrane has reached anterior edge of cornea. Upper lid has reached dorsal edge of cornea. Lower lid has covered ⅓–½ of cornea.

38 (12 days). Primordia of scales marked off over entire surface of leg; ridges not yet grown out to overlap surface. Tips of toes with ventral centre of

cornification and more extensive dorsal one. Main plantar pad ridged in profile. Length of 3rd toe: 8.4 ± 0.3 mm. Labial groove marked off by deep furrow at end of each jaw. Length of beak (anterior angle of nostril to tip of bill): 3.1 mm. Feather germs: coverts of web of wing becoming conical. External auditory meatus surrounded by germs. Sternum covered except midline. Upper eyelid covered with newly formed germs; lower lid naked except for 2–3 rows at its edge. Lower eyelid covers ⅔–¾ of cornea. Opening between lids much reduced.

39 (13 days). Scales overlapping on superior surface of leg. Major pads of phalanges covered with papillae; minor pads smooth. Length of third toe: 9.8 ± 0.3 mm. Mandible and maxilla opaque (cornified) back to level of proximal edge of egg tooth. Channel of auditory meatus visible only at posterior edge of its shallow external opening. Length of beak (anterior angle of nostril to tip of bill): 3.5 mm. Feather germs: Coverts of web of wing are very long tapering cones. Note great increase in length of feather germs in major tracts. 4–5 rows of germs at edge of lower eyelid. Opening between eyelids reduced to a thin crescent.

40 (14 days). Length of beak (anterior edge of nostril to tip of bill): 4 mm. Length of third toe: 12.7 ± 0.5 mm. Scales overlapping on inferior as well as superior surfaces of leg. Dorsal and ventral cornification extends to base of exposed portion of claw. Entire plantar surface of phalanges is covered with well-developed papillae.

41 (15 days). Length of beak (anterior angle of nostril to tip of upper bill): 4.5 mm. Third toe length: 14.9 ± 0.8 mm.

42 (16 days). Length of beak (anterior angle of nostril to tip of upper bill): 4.8 mm. Third toe length: 16.7 ± 0.8 mm.

43 (17 days). Length of beak (anterior angle of nostril to tip of upper bill): 5 mm. 'Labial grooves' reduced to a white granular crust at edge of jaw. Third toe length: 18.6 ± 0.8 mm.

44 (18 days). Length of beak (anterior angle of nostril to tip of upper bill): 5.7 mm. Translucent peridermal covering of beak starting to peel off proximally. Third toe length = 20.4 ± 0.8 mm.

45 (19–20 days). Beak shiny all over and more blunt at tip. Both labial grooves have disappeared with periderm. Yolk sac half enclosed in body cavity. Chorio-allantoic membrane contains less blood and is 'sticky' in the living embryo.

46 Newly hatched chick (20–21 days).

Figure 1–16. Stages of chick development after the initial appearance of the primitive streak. Reproduced, with kind permission of Professor Hamburger and John Wiley & Sons, from Hamburger, V. and Hamilton, H.L. (1951). *J. Morphol.* **88**, 49.

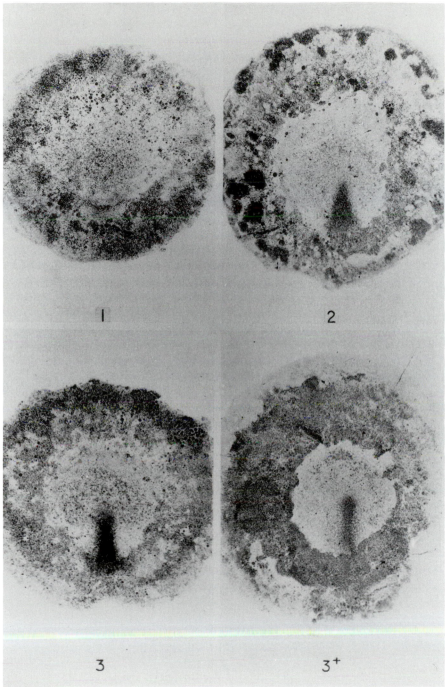

1

2

3

3⁺

Figure 1

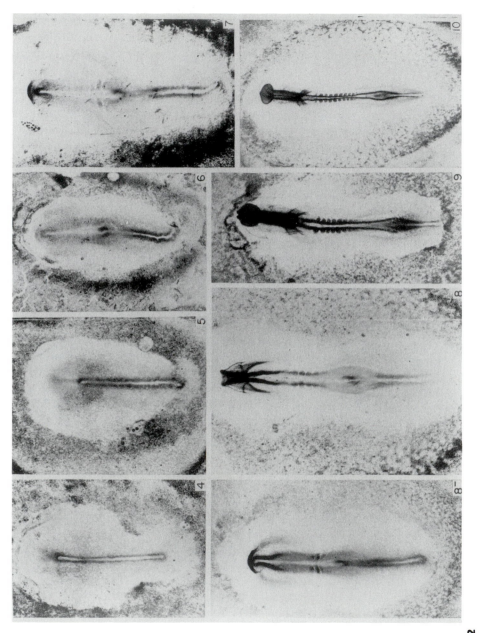

Figure 2

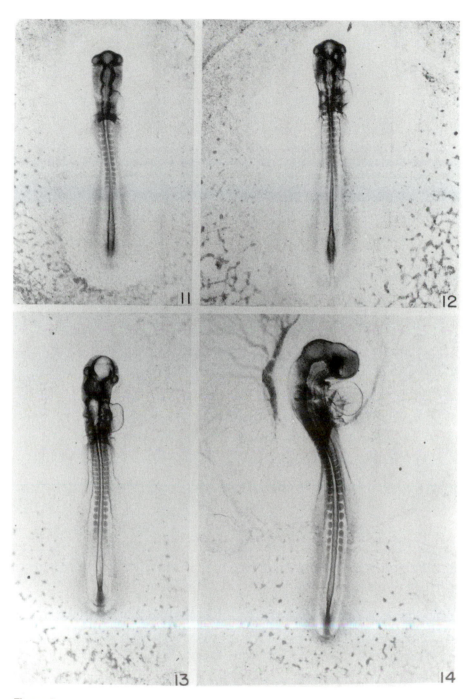

Figure 3

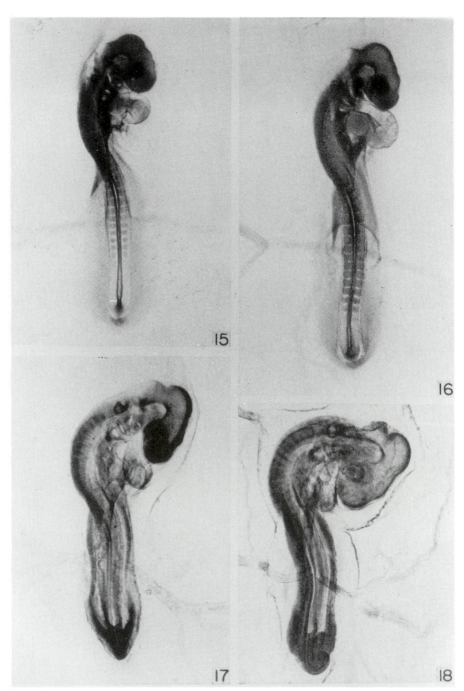

Figure 4

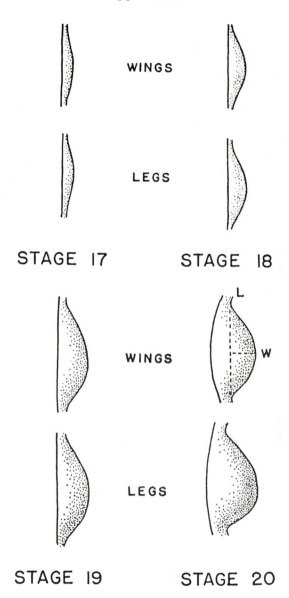

Figure 5

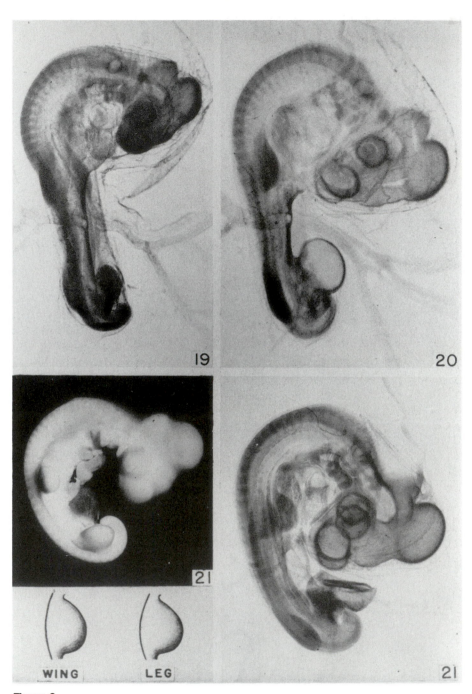

Figure 6

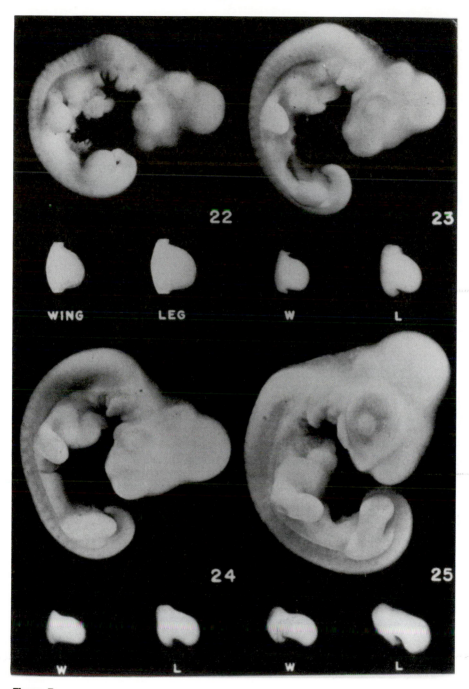

Figure 7

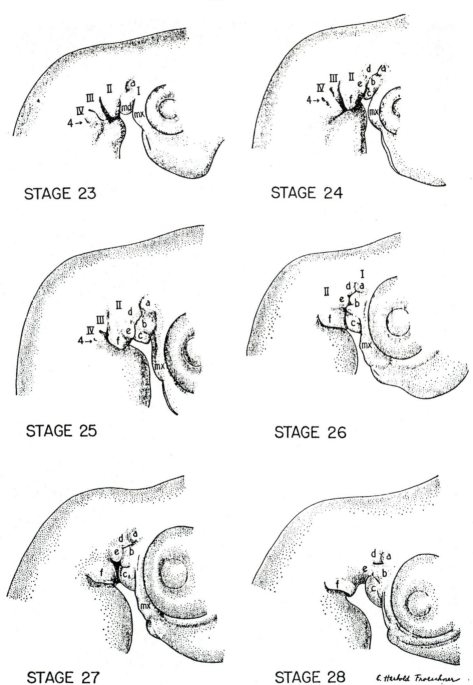

STAGE 23

STAGE 24

STAGE 25

STAGE 26

STAGE 27

STAGE 28

Figure 8

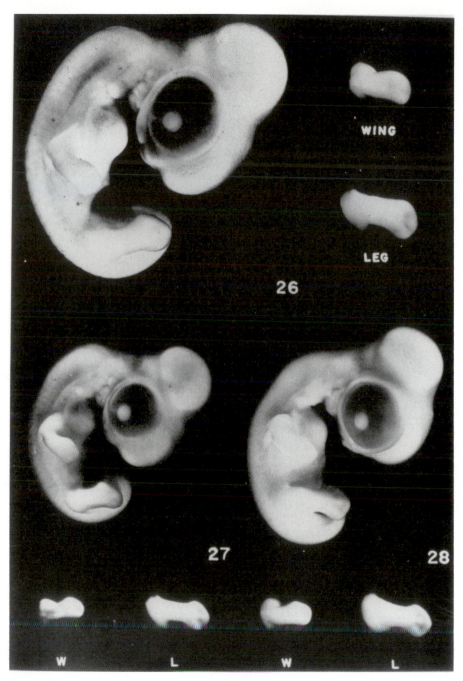

Figure 9

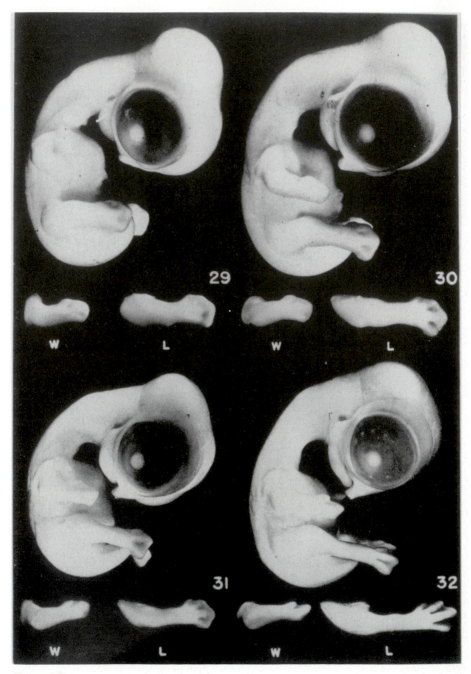

Figure 10

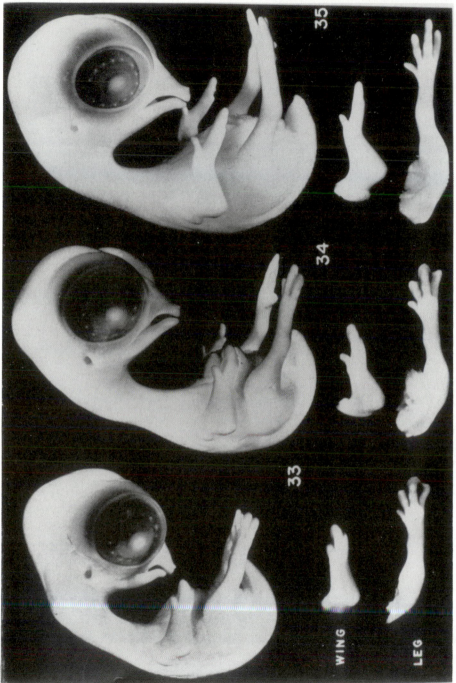

Figure 11

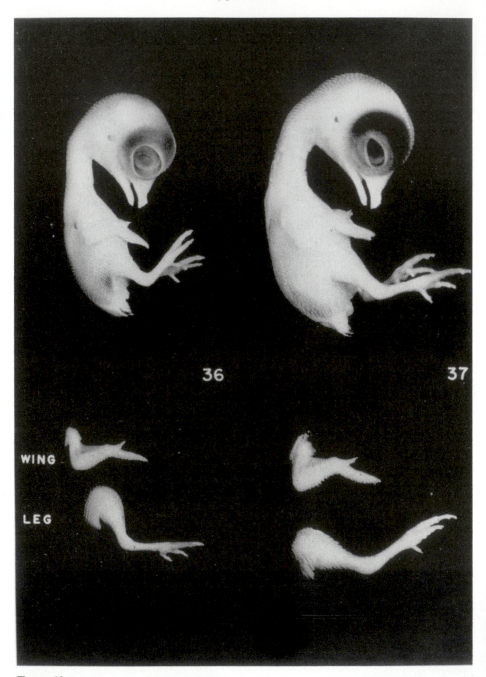

Figure 12

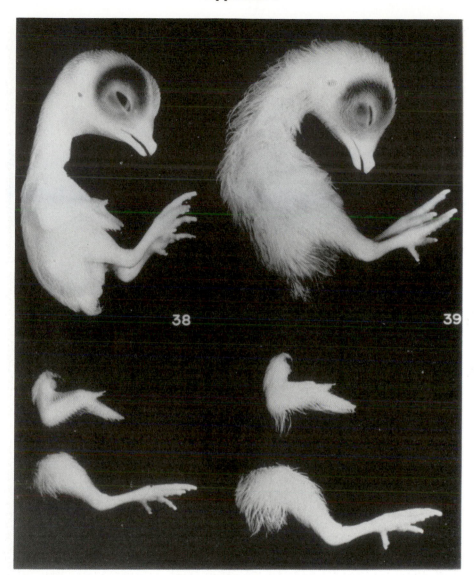

Figure 13

Figure 14

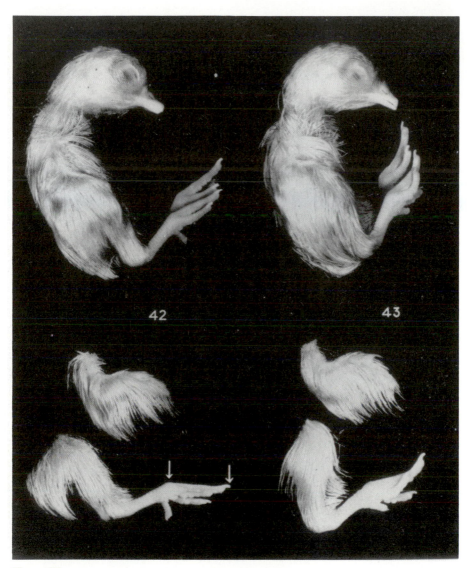

Figure 15

Figure 16

A2

Suppliers of specialist items

Aldrich Chemical Company, The Old Brickyard, New Road, Gillingham, SP8 4JL, UK.

Alpha Laboratories, 40 Parham Drive, Eastleigh, SO5 4NU, UK.

A-M Systems Inc., 11627A Airport Road, Everett, WA 98204, USA.

Atto, 2–3 Hongo 7-Chome, Bunkyo-ku, Tokyo, 113, Japan; distributed by Genetic Research Instrumentation.

American Type Culture Collection, ATCC, 12301 Parklawn Drive, Rockville, MD 20852, USA.

Amersham International plc, Lincoln Place, Green End, Aylesbury, HP20 2TP, UK.

Aquarium Systems Inc., 8141 Tyler Blvd, Mentor, OH 44060, USA.

BDH/Merck Ltd, Broom Road, Poole, BH12 4NN, UK.

Beckman Instruments, 2500 Harbor Blvd, Fullerton, CA 92634–3100, USA; Progress Road, Sands Industrial Estate, High Wycombe, HP12 4JL, UK.

Bio-Rad Laboratories, 3300 Regatta Boulevard, Richmond, CA 94804, USA; Maylands Avenue, Hemel Hempstead, HP2 7TD, UK.

Boehringer Mannheim UK Ltd, Bell Lane, Lewes, BN7 1LG, UK; Boehringer Mannheim Biochemicals, PO Box 50414, Indianapolis, IN 46250, USA.

Bright Instrument Company Ltd, St Margaret's Way, Huntingdon, PE18 6EB, UK.

Burroughs Wellcome, Langley Court, South Eden Park Road, Beckenham, BK3 3BS, UK; 3030 Cornwallis Road, Research Triangle Park, NC 27709–4498, USA.

Calbiochem, PO Box 12087, San Diego, CA 92112–4180, USA; Calbiochem Novabiochem Ltd, 3 Heathcoat Building, Highfields Science Park, Nottingham, NG7 2QJ, UK.

Candela Laser Corp., 530 Boston Post Rd., Wayland, MA 01778, USA.

Cel-Line Inc, PO Box 82, Newfield, NJ 08340, USA; distributed by Dynatech, Daux Road, Billingshurst RH14 9SJ, UK.

Clarke Electromedical, PO Box 8, Pangbourne, RG8 7AU, UK.

Collaborative Research Inc., Lexington, MA, USA; distributed by Stratech Scientific, 61–63 Dudley Street, Luton, LU2 0NP, UK.

Costar Corporation, 1 Alewife Center, Cambridge, MA 02140, USA; Costar UK Ltd, Victoria House, 28–38 Desborough Street, High Wycombe, HP11 2NF, UK.

Digitimer Ltd, 14 Tewin Court, Welwyn Garden City, AL7 1AF, UK.
Digital Instruments, 6780 Cortona Drive, Santa Barbara, CA 93117, USA.
Dow Corning Corp., Box 0994, Midland, MI 48686–0994, USA.
Drummond Scientific Co., Broomhall, PA, USA; supplied by Sigma.
Ethicon Ltd, PO Box 408, Bankhead Avenue, Edinburgh, EH11 4HE, UK; Ethicon Inc, NJ, USA.
Exciton Inc., PO Box 31126, Overlook Station, Drayton, OH 45431, USA.
Falcon, Available from Becton Dickinson, Between Towns Road, Cowley, Oxford, OX4 3LY; Becton Drive, Franklin Lakes, NJ 07417, USA.
Feather, distributed by **pfm GmbH,** Unterbuschweg 45, W-5000 Köln 50/ Sürth, Germany.
Fisher Scientific, 711 Forbes Avenue, Pittsburgh, PA 15219, USA.
Fisons plc, Bishop Meadow Road, Loughborough, LE11 0RG, UK; distributed by Curtin Matheson Scientific Inc, 9999 Veterans Memorial Drive, Houston, TX 77038, USA.
General Valve Corporation, 19 Gloria Lane, PO Box 1333, Fairfield, NJ 07006, USA.
Genetic Research Instrumentation, Gene House, Dunmow Road, Felsted, CM6 3LD, UK.
Gibco BRL, Life Technologies Ltd, PO Box 35, Trident House, Renfrew Road, Paisley, PA3 4EF, UK; PO Box 6009, 8717 Grovemont Circle, Gaitersburg, MD 20898, USA.
Gilson, distributed by Anachem Ltd, 20 Charles Street, LU2 0EB, UK.
Goodfellow Metals Ltd, Science Park, Cambridge, CB4 4DJ, UK.
Grass Instruments Co., 101 Colony Avenue, PO Box 516, Quincy, MA 021691, USA; Stag Instruments, 16 Monument Industrial Park, Chalgrove, OX9 7RW, UK.
Hanovia UV Ltd, 145 Farnham Road, Slough, SL1 4XB, UK.
Hitachi Scientific Instruments, Nisei Sangyo Co Ltd, Hogwood Industrial Estate, Finchampstead, Wokingham, RG11 4QQ, UK.
Invitrogen Corporation, 3985 B Sorrento Valley Blvd, San Diego, CA 92121, USA; distributed by British Bio-technology Ltd, 4–10 The Quadrant, Abingdon, OX14 3YS, UK.
Jackson Immunoresearch Labs Inc., 872 West Baltimore Pike, PO Box 9, West Grove, PA 19390, USA; distributed by Stratech Scientific, 61–63 Dudley Street, Luton, LU2 0NP, UK.
Jamarin Laboratories, Osaka, Japan.
Janssen Pharmaceutical Ltd, Grove, Wantage, OX12 0DQ, UK; 1125 Trenton-Harbourton Road, Titusville, NJ 08560, USA.
Kodak, Laboratory & Research Products Division, Eastman Kodak Company, Rochester, NY 14650, USA; Acornfield Road, Kirkby Industrial Estate, Liverpool, L33 7UR, UK.
Lab-Tek, PO Box 66617, Scotts Valley, CA 95066, USA.
Laser Scientific Inc., 75 Chapel Street, Newton, MA 02158, USA.

Leitz, Wild Leitz Ltd, CH-9435 Heerbrugg, Switzerland; Leica UK Ltd, Davy Avenue, Knowlhill, MK5 8LB, UK.

MedProbe, Postboks 2640 St Hanshaugen, N-0131 Oslo 1, Norway.

Miles Scientific, PO Box 37, Stoke Poges, Slough, SL2 4LY, UK.

Millipore, PO Box 255, Bedford, MA 01730, USA; The Boulevard, Blackmoor Lane, Watford, WD1 8YW, UK.

Mitsubishi Kasei Corp., 5–2 Marunouchi, 2-chome, Chiyodaku, 100 Japan.

Molecular Probes, PO Box 22010, Eugene, OR 97402–0414, USA; also distributed by Cambridge BioScience, 25 Signet Court, Cambridge, CB5 8LA, UK.

Narishige, 1 Plaza Road, Greenvale, NY 11548, USA; 166 Anerley Road, London, SE20 8BD, UK.

National Diagnostics Inc., 1013–1017 Kennedy Blvd, Manville, NJ 08835–2031, USA; Unit 3, Chamberlain Road, Aylesbury, HP19 3DY, UK.

Olympus Optical Co Ltd, 2–8 Honduras Street, London, EC1Y 0TX, UK; 4 Nevada Drive, Lake Success, NY 11042–1179, USA.

Oxoid, Wade Road, Basingstoke, RG24 0PW, UK; 9017 Red Branch Road, Columbia, MD 21045, USA.

Pacific Biomarine Laboratories, PO Box 1348, Venice, CA 92024, USA.

Perkin Elmer Corporation, 761 Main Avenue, Norwalk, CT 06859–0251, USA; Perkin Elmer Ltd, Maxwell Road, Beaconsfield, HP9 1OA, UK.

pfm GmbH, see **Feather.**

Pharmacia LKB Biotechnology, Davy Avenue, Knowlhill, MK5 8PH, UK; 800 Centennial Avenue, PO Box 1327, Piscataway, NJ 08855–1327, USA.

Polysciences Ltd, 24 Low Farm Place, Northampton, NN3 1HY, UK.

Produit Chimiques Egine-Kuhlmann, 25 Blvd L'Amiral Bruix, 75116 Paris, France.

Promega Corporation, 2800 Woods Hollow Road, Madison, WI 53711–5399, USA; Promega Ltd, Epsilon House, Enterprise Road, Chilworth Research Centre, Southampton, SO1 7NS, UK.

Quiagen Inc., PO Box 7401–737, Studio City, CA 91604, USA; distributed by Hybaid Ltd, 111–3 Waldegrave Road, Teddington, TW11 8LL, UK.

Roche Products Ltd, PO Box 8, Broadwater Road, Welwyn Garden City, UK; Hoffman-La Roche, 340 Kingsland Street, Nutley, NJ 07110, USA.

Sartorius GmbH, Postfach 3243, D-3400 Gottingen, Germany; Sartorius Ltd, Blenheim Road, Epsom, KT19 9QN, UK.

Schleicher and Schuell Inc., 543 Washington Street, Keene, NH 03431, USA; distributed by Anderman Ltd, 145 London Road, Kingston upon Thames, KT2 6NII, UK.

Sigma Chemical Company Ltd, Fancy Road, Poole, BH17 7BR, UK; PO Box 14508, St Louis, MO 63178, USA.

Singer Instruments Co Ltd, Treborough Lodge, Roadwater, Watchet, TA23 0QL, UK.

Stratagene, 11011 North Torrey Pines Road, La Jolla, CA 92037, USA;

Cambridge Innovation Centre, Science Park, Milton Road, Cambridge, CB4 4GF, UK.

Techne (Cambridge) Ltd, Duxford, Cambridge, CB2 4PZ, UK.

Tektronix, Fourth Avenue, Marlow, SL7 1YD, UK.

Thurlby Electronics, New Road, St Ives, PE17 4BG, UK.

Treff, distributed by Scotlab Ltd, Kirkshaws Road, Coatbridge, Strathclyde, ML5 8AD, UK.

Ultraviolet Products Inc, PO Box 1501, San Gabriel, CA 91778, USA; UVP Ltd, Science Park, Milton Road, Cambridge, CB4 4BN, UK.

United Electric Controls Company, PO Box 9143, Watertown, MA 02272–9143, USA.

United States Biochemical Corporation, PO Box 22400, Cleveland, OH 44122, USA; PO Box 317, Cambridge, CB1 2DQ, UK.

Universal Marine Industries, 1815 Williams Street, San Leandro, CA 94577, USA.

Video Scope International Ltd, Dulles International Airport, Washington DC 20041–1715, USA.

Whatman International Ltd, Springfield Mill, Maidstone, ME14 2LE, UK.

World Precision Instruments Inc, 375 Quinnipiac Ave, New Haven, CT 06513, USA.

Index

ORDER OTHER TITLES OF INTEREST TODAY

Price list for: UK, Europe, Rest of World (excluding US and Canada)

Forthcoming Titles

124. Human Genetic Disease Analysis Davies, K.E. (Ed)
...... Spiralbound hardback 0-19-963309-6 **£30.00**
...... Paperback 0-19-963308-8 **£18.50**
123. Protein Phosphorylation Hardie, G. (Ed)
...... Spiralbound hardback 0-19-963306-1 **£32.50**
...... Paperback 0-19-963305-3 **£22.50**
122. Immunocytochemistry Beesley, J. (Ed)
...... Spiralbound hardback 0-19-963270-7 **£32.50**
...... Paperback 0-19-963269-3 **£22.50**
121. Tumour Immunobiology Gallagher, G., Rees, R.C. & others (Eds)
...... Spiralbound hardback 0-19-963370-3 **£35.00**
...... Paperback 0-19-963369-X **£25.00**
120. Transcription Factors Latchman, D.S. (Ed)
...... Spiralbound hardback 0-19-963342-8 **£30.00**
...... Paperback 0-19-963341-X **£19.50**
119. Growth Factors McKay, I.A. & Leigh, I. (Eds)
...... Spiralbound hardback 0-19-963360-6 **£30.00**
...... Paperback 0-19-963359-2 **£19.50**
118. Histocompatibility Testing Dyer, P. & Middleton, D. (Eds)
...... Spiralbound hardback 0-19-963364-9 **£32.50**
...... Paperback 0-19-963363-0 **£22.50**
117. Gene Transcription Hames, D.B. & Higgins, S.J. (Eds)
...... Spiralbound hardback 0-19-963292-8 **£35.00**
...... Paperback 0-19-963291-X **£25.00**
116. Electrophysiology Wallis, D.I. (Ed)
...... Spiralbound hardback 0-19-963348-7 **£32.50**
...... Paperback 0-19-963347-9 **£22.50**
115. Biological Data Analysis Fry, J.C. (Ed)
...... Spiralbound hardback 0-19-963340-1 **£50.00**
...... Paperback 0-19-963339-8 **£27.50**
114. Experimental Neuroanatomy Bolam, J.P. (Ed)
...... Spiralbound hardback 0-19-963326-6 **£32.50**
...... Paperback 0-19-963325-8 **£22.50**
112. Lipid Analysis Hamilton, R.J. & Hamilton, S.J. (Eds)
...... Spiralbound hardback 0-19-963098-4 **£35.00**
...... Paperback 0-19-963099-2 **£25.00**
111. Haemopoiesis Testa, N.G. & Molineux, G. (Eds)
...... Spiralbound hardback 0-19-963366-5 **£32.50**
...... Paperback 0-19-963365-7 **£22.50**

Published Titles

113. Preparative Centrifugation Rickwood, D. (Ed)
...... Spiralbound hardback 0-19-963208-1 **£45.00**
...... Paperback 0-19-963211-1 **£25.00**
110. Pollination Ecology Dafni, A.
...... Spiralbound hardback 0-19-963299-5 **£32.50**
...... Paperback 0-19-963298-7 **£22.50**
109. In Situ Hybridization Wilkinson, D.G. (Ed)
...... Spiralbound hardback 0-19-963328-2 **£30.00**
...... Paperback 0-19-963327-4 **£18.50**
108. Protein Engineering Rees, A.R., Sternberg, M.J.E. & others (Eds)
...... Spiralbound hardback 0-19-963139-5 **£35.00**
...... Paperback 0-19-963138-7 **£25.00**

107. Cell-Cell Interactions Stevenson, B.R., Gallin, W.J. & others (Eds)
...... Spiralbound hardback 0-19-963319-3 **£32.50**
...... Paperback 0-19-963318-5 **£22.50**
106. Diagnostic Molecular Pathology: Volume I Herrington, C.S. & McGee, J. O'D. (Eds)
...... Spiralbound hardback 0-19-963237-5 **£30.00**
...... Paperback 0-19-963236-7 **£19.50**
105. Biomechanics-Materials Vincent, J.F.V. (Ed)
...... Spiralbound hardback 0-19-963223-5 **£35.00**
...... Paperback 0-19-963222-7 **£25.00**
104. Animal Cell Culture (2/e) Freshney, R.I. (Ed)
...... Spiralbound hardback 0-19-963212-X **£30.00**
...... Paperback 0-19-963213-8 **£19.50**
103. Molecular Plant Pathology: Volume II Gurr, S.J., McPherson, M.J. & others (Eds)
...... Spiralbound hardback 0-19-963352-5 **£32.50**
...... Paperback 0-19-963351-7 **£22.50**
101. Protein Targeting Magee, A.I. & Wileman, T. (Eds)
...... Spiralbound hardback 0-19-963206-5 **£32.50**
...... Paperback 0-19-963210-3 **£22.50**
100. Diagnostic Molecular Pathology: Volume II: Cell and Tissue Genotyping Herrington, C.S. & McGee, J.O'D. (Eds)
...... Spiralbound hardback 0-19-963239-1 **£30.00**
...... Paperback 0-19-963238-3 **£19.50**
99. Neuronal Cell Lines Wood, J.N. (Ed)
...... Spiralbound hardback 0-19-963346-0 **£32.50**
...... Paperback 0-19-963345-2 **£22.50**
98. Neural Transplantation Dunnett, S.B. & Björklund, A. (Eds)
...... Spiralbound hardback 0-19-963286-3 **£30.00**
...... Paperback 0-19-963285-5 **£19.50**
97. Human Cytogenetics: Volume II: Malignancy and Acquired Abnormalities (2/e) Rooney, D.E. & Czepulkowski, B.H. (Eds)
...... Spiralbound hardback 0-19-963290-1 **£30.00**
...... Paperback 0-19-963289-8 **£22.50**
96. Human Cytogenetics: Volume I: Constitutional Analysis (2/e) Rooney, D.E. & Czepulkowski, B.H. (Eds)
...... Spiralbound hardback 0-19-963288-X **£30.00**
...... Paperback 0-19-963287-1 **£22.50**
95. Lipid Modification of Proteins Hooper, N.M. & Turner, A.J. (Eds)
...... Spiralbound hardback 0-19-963274-X **£32.50**
...... Paperback 0-19-963273-1 **£22.50**
94. Biomechanics-Structures and Systems Biewener, A.A. (Ed)
...... Spiralbound hardback 0-19-963268-5 **£42.50**
...... Paperback 0-19-000007-7 **£25.00**
93. Lipoprotein Analysis Converse, C.A. & Skinner, E.R. (Eds)
...... Spiralbound hardback 0-19-963192-1 **£30.00**
...... Paperback 0-19-963231-6 **£19.50**
92. Receptor-Ligand Interactions Hulme, E.C. (Ed)
...... Spiralbound hardback 0-19-963090-9 **£35.00**
...... Paperback 0-19-963091-7 **£25.00**
91. Molecular Genetic Analysis of Populations Hoelzel, A.R. (Ed)
...... Spiralbound hardback 0-19-963278-2 **£32.50**
...... Paperback 0-19-963277-4 **£22.50**

90. **Enzyme Assays** Eisenthal, R. & Danson, M.J. (Eds)
...... Spiralbound hardback 0-19-963142-5 £35.00
...... Paperback 0-19-963143-3 £25.00

89. **Microcomputers in Biochemistry** Bryce, C.F.A. (Ed)
...... Spiralbound hardback 0-19-963253-7 £30.00
...... Paperback 0-19-963252-9 £19.50

88. **The Cytoskeleton** Carraway, K.L. & Carraway, C.A.C. (Eds)
...... Spiralbound hardback 0-19-963257-X £30.00
...... Paperback 0-19-963256-1 £19.50

87. **Monitoring Neuronal Activity** Stamford, J.A. (Ed)
...... Spiralbound hardback 0-19-963244-8 £30.00
...... Paperback 0-19-963243-X £19.50

86. **Crystallization of Nucleic Acids and Proteins** Ducruix, A. & Giegⱨ130⟩, R. (Eds)
...... Spiralbound hardback 0-19-963245-6 £35.00
...... Paperback 0-19-963246-4 £25.00

85. **Molecular Plant Pathology: Volume I** Gurr, S.J., McPherson, M.J. & others (Eds)
...... Spiralbound hardback 0-19-963103-4 £30.00
...... Paperback 0-19-963102-6 £19.50

84. **Anaerobic Microbiology** Levett, P.N. (Ed)
...... Spiralbound hardback 0-19-963204-9 £32.50
...... Paperback 0-19-963262-6 £22.50

83. **Oligonucleotides and Analogues** Eckstein, F. (Ed)
...... Spiralbound hardback 0-19-963280-4 £32.50
...... Paperback 0-19-963279-0 £22.50

82. **Electron Microscopy in Biology** Harris, R. (Ed)
...... Spiralbound hardback 0-19-963219-7 £32.50
...... Paperback 0-19-963215-4 £22.50

81. **Essential Molecular Biology: Volume II** Brown, T.A. (Ed)
...... Spiralbound hardback 0-19-963112-3 £32.50
...... Paperback 0-19-963113-1 £22.50

80. **Cellular Calcium** McCormack, J.G. & Cobbold, P.H. (Eds)
...... Spiralbound hardback 0-19-963131-X £35.00
...... Paperback 0-19-963130-1 £25.00

79. **Protein Architecture** Lesk, A.M.
...... Spiralbound hardback 0-19-963054-2 £32.50
...... Paperback 0-19-963055-0 £22.50

78. **Cellular Neurobiology** Chad, J. & Wheal, H. (Eds)
...... Spiralbound hardback 0-19-963106-9 £32.50
...... Paperback 0-19-963107-7 £22.50

77. **PCR** McPherson, M.J., Quirke, P. & others (Eds)
...... Spiralbound hardback 0-19-963226-X £30.00
...... Paperback 0-19-963196-4 £19.50

76. **Mammalian Cell Biotechnology** Butler, M. (Ed)
...... Spiralbound hardback 0-19-963207-3 £30.00
...... Paperback 0-19-963209-X £19.50

75. **Cytokines** Balkwill, F.R. (Ed)
...... Spiralbound hardback 0-19-963218-9 £35.00
...... Paperback 0-19-963214-6 £25.00

74. **Molecular Neurobiology** Chad, J. & Wheal, H. (Eds)
...... Spiralbound hardback 0-19-963108-5 £30.00
...... Paperback 0-19-963109-3 £19.50

73. **Directed Mutagenesis** McPherson, M.J. (Ed)
...... Spiralbound hardback 0-19-963141-7 £30.00
...... Paperback 0-19-963140-9 £19.50

72. **Essential Molecular Biology: Volume I** Brown, T.A. (Ed)
...... Spiralbound hardback 0-19-963110-7 £32.50
...... Paperback 0-19-963111-5 £22.50

71. **Peptide Hormone Action** Siddle, K. & Hutton, J.C.
...... Spiralbound hardback 0-19-963070-4 £32.50
...... Paperback 0-19-963071-2 £22.50

70. **Peptide Hormone Secretion** Hutton, J.C. & Siddle, K. (Eds)
...... Spiralbound hardback 0-19-963068-2 £35.00
...... Paperback 0-19-963069-0 £25.00

69. **Postimplantation Mammalian Embryos** Copp, A.J. & Cockroft, D.L. (Eds)
...... Spiralbound hardback 0-19-963088-7 £35.00
...... Paperback 0-19-963089-5 £25.00

68. **Receptor-Effector Coupling** Hulme, E.C. (Ed)
...... Spiralbound hardback 0-19-963094-1 £30.00
...... Paperback 0-19-963095-X £19.50

67. **Gel Electrophoresis of Proteins (2/e)** Hames, B.D. & Rickwood, D. (Eds)
...... Spiralbound hardback 0-19-963074-7 £35.00
...... Paperback 0-19-963075-5 £25.00

66. **Clinical Immunology** Gooi, H.C. & Chapel, H. (Eds)
...... Spiralbound hardback 0-19-963086-0 £32.50
...... Paperback 0-19-963087-9 £22.50

65. **Receptor Biochemistry** Hulme, E.C. (Ed)
...... Spiralbound hardback 0-19-963092-5 £35.00
...... Paperback 0-19-963093-3 £25.00

64. **Gel Electrophoresis of Nucleic Acids (2/e)** Rickwood, D. & Hames, B.D. (Eds)
...... Spiralbound hardback 0-19-963082-8 £32.50
...... Paperback 0-19-963083-6 £22.50

63. **Animal Virus Pathogenesis** Oldstone, M.B.A. (Ed)
...... Spiralbound hardback 0-19-963100-X £30.00
...... Paperback 0-19-963101-8 £18.50

62. **Flow Cytometry** Ormerod, M.G. (Ed)
...... Paperback 0-19-963053-4 £22.50

61. **Radioisotopes in Biology** Slater, R.J. (Ed)
...... Spiralbound hardback 0-19-963080-1 £32.50
...... Paperback 0-19-963081-X £22.50

60. **Biosensors** Cass, A.E.G. (Ed)
...... Spiralbound hardback 0-19-963046-1 £30.00
...... Paperback 0-19-963047-X £19.50

59. **Ribosomes and Protein Synthesis** Spedding, G. (Ed)
...... Spiralbound hardback 0-19-963104-2 £32.50
...... Paperback 0-19-963105-0 £22.50

58. **Liposomes** New, R.R.C. (Ed)
...... Spiralbound hardback 0-19-963076-3 £35.00
...... Paperback 0-19-963077-1 £22.50

57. **Fermentation** McNeil, B. & Harvey, L.M. (Eds)
...... Spiralbound hardback 0-19-963044-5 £30.00
...... Paperback 0-19-963045-3 £19.50

56. **Protein Purification Applications** Harris, E.L.V. & Angal, S. (Eds)
...... Spiralbound hardback 0-19-963022-4 £30.00
...... Paperback 0-19-963023-2 £18.50

55. **Nucleic Acids Sequencing** Howe, C.J. & Ward, E.S. (Eds)
...... Spiralbound hardback 0-19-963056-9 £30.00
...... Paperback 0-19-963057-7 £19.50

54. **Protein Purification Methods** Harris, E.L.V. & Angal, S. (Eds)
...... Spiralbound hardback 0-19-963002-X £30.00
...... Paperback 0-19-963003-8 £20.00

53. **Solid Phase Peptide Synthesis** Atherton, E. & Sheppard, R.C.
...... Spiralbound hardback 0-19-963066-6 £30.00
...... Paperback 0-19-963067-4 £18.50

52. **Medical Bacteriology** Hawkey, P.M. & Lewis, D.A. (Eds)
...... Spiralbound hardback 0-19-963008-9 £38.00
...... Paperback 0-19-963009-7 £25.00

51. **Proteolytic Enzymes** Beynon, R.J. & Bond, J.S. (Eds)
...... Spiralbound hardback 0-19-963058-5 £30.00
...... Paperback 0-19-963059-3 £19.50

50. **Medical Mycology** Evans, E.G.V. & Richardson, M.D. (Eds)
...... Spiralbound hardback 0-19-963010-0 £37.50
...... Paperback 0-19-963011-9 £25.00

49. **Computers in Microbiology** Bryant, T.N. & Wimpenny, J.W.T. (Eds)
...... Paperback 0-19-963015-1 £19.50

48. **Protein Sequencing** Findlay, J.B.C. & Geisow, M.J. (Eds)
...... Spiralbound hardback 0-19-963012-7 £30.00
...... Paperback 0-19-963013-5 £18.50

47. **Cell Growth and Division** Baserga, R. (Ed)
...... Spiralbound hardback 0-19-963026-7 £30.00
...... Paperback 0-19-963027-5 £18.50

46. **Protein Function** Creighton, T.E. (Ed)
...... Spiralbound hardback 0-19-963006-2 £32.50
...... Paperback 0-19-963007-0 £22.50

45. **Protein Structure** Creighton, T.E. (Ed)
...... Spiralbound hardback 0-19-963000-3 £32.50
...... Paperback 0-19-963001-1 £22.50

44. **Antibodies: Volume II** Catty, D. (Ed)
...... Spiralbound hardback 0-19-963018-6 £30.00
...... Paperback 0-19-963019-4 £19.50

ORDER FORM for UK, Europe and Rest of World

(Excluding USA and Canada)

Qty	ISBN	Author	Title	Amount
			P&P	
			TOTAL	

Please add postage and packing: £1.75 for UK orders under £20; £2.75 for UK orders over £20; overseas orders add 10% of total.

Name ....

Address ....

....

.... Post code ....

[ ] Please charge £ .... to my credit card

Access/VISA/Eurocard/AMEX/Diners Club (circle appropriate card)

Card No .... Expiry date ....

Signature ....

Credit card account address if different from above:

....

.... Postcode ....

[ ] I enclose a cheque for £......................

Please return this form to: OUP Distribution Services, Saxon Way West, Corby, Northants NN18 9ES

OR ORDER BY CREDIT CARD HOTLINE: Tel +44-(0)536-741519 or Fax +44-(0)536-746337

ORDER OTHER TITLES OF INTEREST TODAY

123. Protein Phosphorylation Hardie, G. (Ed)
...... Spiralbound hardback 0-19-963306-1 **$65.00**
...... Paperback 0-19-963305-3 **$45.00**

121. Tumour Immunobiology Gallagher, G., Rees,
R.C. & others (Eds)
...... Spiralbound hardback 0-19-963370-3 **$72.00**
...... Paperback 0-19-963369-X **$50.00**

117. Gene Transcription Hames, D.B. & Higgins,
S.J. (Eds)
...... Spiralbound hardback 0-19-963292-8 **$72.00**
...... Paperback 0-19-963291-X **$50.00**

116. Electrophysiology Wallis, D.I. (Ed)
...... Spiralbound hardback 0-19-963348-7 **$66.50**
...... Paperback 0-19-963347-9 **$45.95**

115. Biological Data Analysis Fry, J.C. (Ed)
...... Spiralbound hardback 0-19-963340-1 **$80.00**
...... Paperback 0-19-963339-8 **$60.00**

114. Experimental Neuroanatomy Bolam, J.P. (Ed)
...... Spiralbound hardback 0-19-963326-6 **$65.00**
...... Paperback 0-19-963325-8 **$40.00**

111. Haemopoiesis Testa, N.G. & Molineux, G.
(Eds)
...... Spiralbound hardback 0-19-963366-5 **$65.00**
...... Paperback 0-19-963365-7 **$45.00**

113. Preparative Centrifugation Rickwood, D. (Ed)
...... Spiralbound hardback 0-19-963208-1 **$90.00**
...... Paperback 0-19-963211-1 **$50.00**

110. Pollination Ecology Dafni, A.
...... Spiralbound hardback 0-19-963299-5 **$65.00**
...... Paperback 0-19-963298-7 **$45.00**

109. In Situ Hybridization Wilkinson, D.G. (Ed)
...... Spiralbound hardback 0-19-963328-2 **$58.00**
...... Paperback 0-19-963327-4 **$36.00**

108. Protein Engineering Rees, A.R., Sternberg,
M.J.E. & others (Eds)
...... Spiralbound hardback 0-19-963139-5 **$75.00**
...... Paperback 0-19-963138-7 **$50.00**

107. Cell-Cell Interactions Stevenson, B.R., Gallin,
W.J. & others (Eds)
...... Spiralbound hardback 0-19-963319-3 **$60.00**
...... Paperback 0-19-963318-5 **$40.00**

106. Diagnostic Molecular Pathology: Volume I
Herrington, C.S. & McGee, J.O'D. (Eds)
...... Spiralbound hardback 0-19-963237-5 **$58.00**
...... Paperback 0-19-963236-7 **$38.00**

105. Biomechanics-Materials Vincent, J.F.V. (Ed)
...... Spiralbound hardback 0-19-963223-5 **$70.00**
...... Paperback 0-19-963222-7 **$50.00**

104. Animal Cell Culture (2/e) Freshney, R.I. (Ed)
...... Spiralbound hardback 0-19-963212-X **$60.00**
...... Paperback 0-19-963213-8 **$40.00**

102. Molecular Plant Pathology: Volume II Gurr,
S.J., McPherson, M.J. & others (Eds)
...... Spiralbound hardback 0-19-963352-5 **$65.00**
...... Paperback 0-19-963351-7 **$45.00**

101. Protein Targeting Magee, A.I. & Wileman, T.
(Eds)
...... Spiralbound hardback 0-19-963206-5 **$75.00**
...... Paperback 0-19-963210-3 **$50.00**

**100. Diagnostic Molecular Pathology: Volume II:
Cell and Tissue Genotyping** Herrington, C.S. &
McGee, J.O'D. (Eds)
...... Spiralbound hardback 0-19-963239-1 **$60.00**
...... Paperback 0-19-963238-3 **$39.00**

99. Neuronal Cell Lines Wood, J.N. (Ed)
...... Spiralbound hardback 0-19-963346-0 **$68.00**
...... Paperback 0-19-963345-2 **$48.00**

98. Neural Transplantation Dunnett, S.B. &
Björklund, A. (Eds)
...... Spiralbound hardback 0-19-963286-3 **$69.00**
...... Paperback 0-19-963285-5 **$42.00**

**97. Human Cytogenetics: Volume II: Malignancy
and Acquired Abnormalities (2/e)** Rooney,
D.E. & Czepulkowski, B.H. (Eds)
...... Spiralbound hardback 0-19-963290-1 **$75.00**
...... Paperback 0-19-963289-8 **$50.00**

**96. Human Cytogenetics: Volume I: Constitutional
Analysis (2/e)** Rooney, D.E. & Czepulkowski,
B.H. (Eds)
...... Spiralbound hardback 0-19-963288-X **$75.00**
...... Paperback 0-19-963287-1 **$50.00**

95. Lipid Modification of Proteins Hooper, N.M. &
Turner, A.J. (Eds)
...... Spiralbound hardback 0-19-963274-X **$75.00**
...... Paperback 0-19-963273-1 **$50.00**

94. Biomechanics-Structures and Systems
Biewener, A.A. (Ed)
...... Spiralbound hardback 0-19-963268-5 **$85.00**
...... Paperback 0-19-963267-7 **$50.00**

93. Lipoprotein Analysis Converse, C.A. &
Skinner, E.R. (Eds)
...... Spiralbound hardback 0-19-963192-1 **$65.00**
...... Paperback 0-19-963231-6 **$42.00**

92. Receptor-Ligand Interactions Hulme, E.C. (Ed)
...... Spiralbound hardback 0-19-963090-9 **$75.00**
...... Paperback 0-19-963091-7 **$50.00**

91. Molecular Genetic Analysis of Populations
Hoelzel, A.R. (Ed)
...... Spiralbound hardback 0-19-963278-2 **$65.00**
...... Paperback 0-19-963277-4 **$45.00**

90. Enzyme Assays Eisenthal, R. & Danson, M.J.
(Eds)
...... Spiralbound hardback 0-19-963142-5 **$68.00**
...... Paperback 0-19-963143-3 **$48.00**

89. Microcomputers in Biochemistry Bryce, C.F.A.
(Ed)
...... Spiralbound hardback 0-19-963253-7 **$60.00**
...... Paperback 0-19-963252-9 **$40.00**

88. The Cytoskeleton Carraway, K.L. & Carraway,
C.A.C. (Eds)
...... Spiralbound hardback 0-19-963257-X **$60.00**
...... Paperback 0-19-963256-1 **$40.00**

87. Monitoring Neuronal Activity Stamford, J.A.
(Ed)
...... Spiralbound hardback 0-19-963244-8 **$60.00**
...... Paperback 0-19-963243-X **$40.00**

86. Crystallization of Nucleic Acids and Proteins
Ducruix, A. & Gieg‹130›, R. (Eds)
...... Spiralbound hardback 0-19-963245-6 **$60.00**
...... Paperback 0-19-963246-4 **$50.00**

85. Molecular Plant Pathology: Volume I Gurr,
S.J., McPherson, M.J. & others (Eds)
...... Spiralbound hardback 0-19-963103-4 **$60.00**
...... Paperback 0-19-963102-6 **$40.00**

84. Anaerobic Microbiology Levett, P.N. (Ed)
...... Spiralbound hardback 0-19-963204-9 **$75.00**
...... Paperback 0-19-963262-6 **$45.00**

No.	Title / Editor	Format	ISBN	Price
83.	Oligonucleotides and Analogues Eckstein, F. (Ed)	Spiralbound hardback	0-19-963280-4	$65.00
		Paperback	0-19-963279-0	$45.00
82.	Electron Microscopy in Biology Harris, R. (Ed)	Spiralbound hardback	0-19-963219-7	$65.00
		Paperback	0-19-963215-4	$45.00
81.	Essential Molecular Biology: Volume II Brown, T.A. (Ed)	Spiralbound hardback	0-19-963112-3	$65.00
		Paperback	0-19-963113-1	$45.00
80.	Cellular Calcium McCormack, J.G. & Cobbold, P.H. (Eds)	Spiralbound hardback	0-19-963131-X	$75.00
		Paperback	0-19-963130-1	$50.00
79.	Protein Architecture Lesk, A.M.	Spiralbound hardback	0-19-963054-2	$65.00
		Paperback	0-19-963055-0	$45.00
78.	Cellular Neurobiology Chad, J. & Wheal, H. (Eds)	Spiralbound hardback	0-19-963106-9	$73.00
		Paperback	0-19-963107-7	$43.00
77.	PCR McPherson, M.J., Quirke, P. & others (Eds)	Spiralbound hardback	0-19-963226-X	$55.00
		Paperback	0-19-963196-4	$40.00
76.	Mammalian Cell Biotechnology Butler, M. (Ed)	Spiralbound hardback	0-19-963207-3	$60.00
		Paperback	0-19-963209-X	$40.00
75.	Cytokines Balkwill, F.R. (Ed)	Spiralbound hardback	0-19-963218-9	$64.00
		Paperback	0-19-963214-6	$44.00
74.	Molecular Neurobiology Chad, J. & Wheal, H. (Eds)	Spiralbound hardback	0-19-963108-5	$56.00
		Paperback	0-19-963109-3	$36.00
73.	Directed Mutagenesis McPherson, M.J. (Ed)	Spiralbound hardback	0-19-963141-7	$55.00
		Paperback	0-19-963140-9	$35.00
72.	Essential Molecular Biology: Volume I Brown, T.A. (Ed)	Spiralbound hardback	0-19-963110-7	$65.00
		Paperback	0-19-963111-5	$45.00
71.	Peptide Hormone Action Siddle, K. & Hutton, J.C.	Spiralbound hardback	0-19-963070-4	$70.00
		Paperback	0-19-963071-2	$50.00
70.	Peptide Hormone Secretion Hutton, J.C. & Siddle, K. (Eds)	Spiralbound hardback	0-19-963068-2	$70.00
		Paperback	0-19-963069-0	$50.00
69.	Postimplantation Mammalian Embryos Copp, A.J. & Cockroft, D.L. (Eds)	Spiralbound hardback	0-19-963088-7	$70.00
		Paperback	0-19-963089-5	$50.00
68.	Receptor-Effector Coupling Hulme, E.C. (Ed)	Spiralbound hardback	0-19-963094-1	$70.00
		Paperback	0-19-963095-X	$45.00
67.	Gel Electrophoresis of Proteins (2/e) Hames, B.D. & Rickwood, D. (Eds)	Spiralbound hardback	0-19-963074-7	$75.00
		Paperback	0-19-963075-5	$50.00
66.	Clinical Immunology Gooi, H.C. & Chapel, H. (Eds)	Spiralbound hardback	0-19-963086-0	$69.95
		Paperback	0-19-963087-9	$50.00
65.	Receptor Biochemistry Hulme, E.C. (Ed)	Spiralbound hardback	0-19-963092-5	$70.00
		Paperback	0-19-963093-3	$50.00
64.	Gel Electrophoresis of Nucleic Acids (2/e) Rickwood, D. & Hames, B.D. (Eds)	Spiralbound hardback	0-19-963082-8	$75.00
		Paperback	0-19-963083-6	$50.00
63.	Animal Virus Pathogenesis Oldstone, M.B.A. (Ed)	Spiralbound hardback	0-19-963100-X	$68.00
		Paperback	0-19-963101-8	$40.00
62.	Flow Cytometry Ormerod, M.G. (Ed)	Paperback	0-19-963053-4	$50.00
61.	Radioisotopes in Biology Slater, R.J. (Ed)	Spiralbound hardback	0-19-963080-1	$75.00
		Paperback	0-19-963081-X	$45.00
60.	Biosensors Cass, A.E.G. (Ed)	Spiralbound hardback	0-19-963046-1	$65.00
		Paperback	0-19-963047-X	$43.00
59.	Ribosomes and Protein Synthesis Spedding, G. (Ed)	Spiralbound hardback	0-19-963104-2	$75.00
		Paperback	0-19-963105-0	$45.00
58.	Liposomes New, R.R.C. (Ed)	Spiralbound hardback	0-19-963076-3	$70.00
		Paperback	0-19-963077-1	$45.00
57.	Fermentation McNeil, B. & Harvey, L.M. (Eds)	Spiralbound hardback	0-19-963044-5	$65.00
		Paperback	0-19-963045-3	$39.00
56.	Protein Purification Applications Harris, E.L.V. & Angal, S. (Eds)	Spiralbound hardback	0-19-963022-4	$54.00
		Paperback	0-19-963023-2	$36.00
55.	Nucleic Acids Sequencing Howe, C.J. & Ward, E.S. (Eds)	Spiralbound hardback	0-19-963056-9	$59.00
		Paperback	0-19-963057-7	$38.00
54.	Protein Purification Methods Harris, E.L.V. & Angal, S. (Eds)	Spiralbound hardback	0-19-963002-X	$60.00
		Paperback	0-19-963003-8	$40.00
53.	Solid Phase Peptide Synthesis Atherton, E. & Sheppard, R.C.	Spiralbound hardback	0-19-963066-6	$58.00
		Paperback	0-19-963067-4	$39.95
52.	Medical Bacteriology Hawkey, P.M. & Lewis, D.A. (Eds)	Spiralbound hardback	0-19-963008-9	$69.95
		Paperback	0-19-963009-7	$50.00
51.	Proteolytic Enzymes Beynon, R.J. & Bond, J.S. (Eds)	Spiralbound hardback	0-19-963058-5	$60.00
		Paperback	0-19-963059-3	$39.00
50.	Medical Mycology Evans, E.G.V. & Richardson, M.D. (Eds)	Spiralbound hardback	0-19-963010-0	$69.95
		Paperback	0-19-963011-9	$50.00
49.	Computers in Microbiology Bryant, T.N. & Wimpenny, J.W.T. (Eds)	Paperback	0-19-963015-1	$40.00
48.	Protein Sequencing Findlay, J.B.C. & Geisow, M.J. (Eds)	Spiralbound hardback	0-19-963012-7	$56.00
		Paperback	0-19-963013-5	$38.00
47.	Cell Growth and Division Baserga, R. (Ed)	Spiralbound hardback	0-19-963026-7	$62.00
		Paperback	0-19-963027-5	$38.00
46.	Protein Function Creighton, T.E. (Ed)	Spiralbound hardback	0-19-963006-2	$65.00
		Paperback	0-19-963007-0	$45.00
45.	Protein Structure Creighton, T.E. (Ed)	Spiralbound hardback	0-19-963000-3	$65.00
		Paperback	0-19-963001-1	$45.00
44.	Antibodies: Volume II Catty, D. (Ed)	Spiralbound hardback	0-19-963018-6	$58.00
		Paperback	0-19-963019-4	$39.00
43.	HPLC of Macromolecules Oliver, R.W.A. (Ed)	Spiralbound hardback	0-19-963020-8	$54.00
		Paperback	0-19-963021-6	$45.00
42.	Light Microscopy in Biology Lacey, A.J. (Ed)	Spiralbound hardback	0-19-963036-4	$62.00
		Paperback	0-19-963037-2	$38.00
41.	Plant Molecular Biology Shaw, C.H. (Ed)	Paperback	1-85221-056-7	$38.00
40.	Microcomputers in Physiology Fraser, P.J. (Ed)	Spiralbound hardback	1-85221-129-6	$54.00
		Paperback	1-85221-130-X	$36.00
39.	Genome Analysis Davies, K.E. (Ed)	Spiralbound hardback	1-85221-109-1	$54.00
		Paperback	1-85221-110-5	$36.00
38.	Antibodies: Volume I Catty, D. (Ed)	Paperback	0-947946-85-3	$38.00
37.	Yeast Campbell, I. & Duffus, J.H. (Eds)	Paperback	0-947946-79-9	$36.00
36.	Mammalian Development Monk, M. (Ed)	Hardback	1-85221-030-3	$60.00
		Paperback	1-85221-029-X	$45.00
35.	Lymphocytes Klaus, G.G.B. (Ed)	Hardback	1-85221-018-4	$54.00
34.	Lymphokines and Interferons Clemens, M.J., Morris, A.G. & others (Eds)	Paperback	1-85221-035-4	$44.00
33.	Mitochondria Darley-Usmar, V.M., Rickwood, D. & others (Eds)	Hardback	1-85221-034-6	$65.00
		Paperback	1-85221-033-8	$45.00

32. **Prostaglandins and Related Substances** Benedetto, C., McDonald-Gibson, R.G. & others (Eds)
.... Hardback 1-85221-032-X **$58.00**
.... Paperback 1-85221-031-1 **$38.00**
31. **DNA Cloning: Volume III** Glover, D.M. (Ed)
.... Hardback 1-85221-049-4 **$56.00**
.... Paperback 1-85221-048-6 **$36.00**
30. **Steroid Hormones** Green, B. & Leake, R.E. (Eds)
.... Paperback 0-947946-53-5 **$40.00**
29. **Neurochemistry** Turner, A.J. & Bachelard, H.S. (Eds)
.... Hardback 1-85221-028-1 **$56.00**
.... Paperback 1-85221-027-3 **$36.00**
28. **Biological Membranes** Findlay, J.B.C. & Evans, W.H. (Eds)
.... Hardback 0-947946-84-5 **$54.00**
.... Paperback 0-947946-83-7 **$36.00**
27. **Nucleic Acid and Protein Sequence Analysis** Bishop, M.J. & Rawlings, C.J. (Eds)
.... Hardback 1-85221-007-9 **$66.00**
.... Paperback 1-85221-006-0 **$44.00**
26. **Electron Microscopy in Molecular Biology** Sommerville, J. & Scheer, U. (Eds)
.... Hardback 0-947946-64-0 **$54.00**
.... Paperback 0-947946-54-3 **$40.00**
25. **Teratocarcinomas and Embryonic Stem Cells** Robertson, E.J. (Ed)
.... Hardback 1-85221-005-2 **$62.00**
.... Paperback 1-85221-004-4 **$0.00**
24. **Spectrophotometry and Spectrofluorimetry** Harris, D.A. & Bashford, C.L. (Eds)
.... Hardback 0-947946-69-1 **$56.00**
.... Paperback 0-947946-46-2 **$39.95**
23. **Plasmids** Hardy, K.G. (Ed)
.... Paperback 0-947946-81-0 **$36.00**
22. **Biochemical Toxicology** Snell, K. & Mullock, B. (Eds)
.... Paperback 0-947946-52-7 **$40.00**
19. **Drosophila** Roberts, D.B. (Ed)
.... Hardback 0-947946-66-7 **$67.50**
.... Paperback 0-947946-45-4 **$46.00**
17. **Photosynthesis: Energy Transduction** Hipkins, M.F. & Baker, N.R. (Eds)
.... Hardback 0-947946-63-2 **$54.00**
.... Paperback 0-947946-51-9 **$36.00**
16. **Human Genetic Diseases** Davies, K.E. (Ed)
.... Hardback 0-947946-76-4 **$60.00**
.... Paperback 0-947946-75-6 **$34.00**
14. **Nucleic Acid Hybridisation** Hames, B.D. & Higgins, S.J. (Eds)
.... Hardback 0-947946-61-6 **$60.00**
.... Paperback 0-947946-23-3 **$36.00**
13. **Immobilised Cells and Enzymes** Woodward, J. (Ed)
.... Hardback 0-947946-60-8 **$0.00**
12. **Plant Cell Culture** Dixon, R.A. (Ed)
.... Paperback 0-947946-22-5 **$36.00**
11a. **DNA Cloning: Volume I** Glover, D.M. (Ed)
.... Paperback 0-947946-18-7 **$36.00**
11b. **DNA Cloning: Volume II** Glover, D.M. (Ed)
.... Paperback 0-947946-19-5 **$36.00**
10. **Virology** Mahy, B.W.J. (Ed)
.... Paperback 0-904147-78-9 **$40.00**

9. **Affinity Chromatography** Dean, P.D.G., Johnson, W.S. & others (Eds)
.... Paperback 0-904147-71-1 **$36.00**
7. **Microcomputers in Biology** Ireland, C.R. & Long, S.P. (Eds)
.... Paperback 0-904147-57-6 **$36.00**
6. **Oligonucleotide Synthesis** Gait, M.J. (Ed)
.... Paperback 0-904147-74-6 **$38.00**
5. **Transcription and Translation** Hames, B.D. & Higgins, S.J. (Eds)
.... Paperback 0-904147-52-5 **$38.00**
3. **Iodinated Density Gradient Media** Rickwood, D. (Ed)
.... Paperback 0-904147-51-7 **$36.00**

Sets

Essential Molecular Biology: Volumes I and II as a set Brown, T.A. (Ed)
.... Spiralbound hardback 0-19-963114-X **$118.00**
.... Paperback 0-19-963115-8 **$78.00**
Antibodies: Volumes I and II as a set Catty, D. (Ed)
.... Paperback 0-19-963063-1 **$70.00**
Cellular and Molecular Neurobiology Chad, J. & Wheal, H. (Eds)
.... Spiralbound hardback 0-19-963255-3 **$133.00**
.... Paperback 0-19-963254-5 **$79.00**
Protein Structure and Protein Function: Two-volume set Creighton, T.E. (Ed)
.... Spiralbound hardback 0-19-963064-X **$114.00**
.... Paperback 0-19-963065-8 **$80.00**
DNA Cloning: Volumes I, II, III as a set Glover, D.M. (Ed)
.... Paperback 1-85221-069-9 **$92.00**
Molecular Plant Pathology: Volumes I and II as a set Gurr, S.J., McPherson, M.J. & others (Eds)
.... Spiralbound hardback 0-19-963354-1 **$0.00**
.... Paperback 0-19-963353-3 **$0.00**
Protein Purification Methods, and Protein Purification Applications, two-volume set Harris, E.L.V. & Angal, S. (Eds)
.... Spiralbound hardback 0-19-963048-8 **$98.00**
.... Paperback 0-19-963049-6 **$68.00**
Diagnostic Molecular Pathology: Volumes I and II as a set Herrington, C.S. & McGee, J. O'D. (Eds)
.... Spiralbound hardback 0-19-963241-3 **$0.00**
.... Paperback 0-19-963240-5 **$0.00**
Receptor Biochemistry; Receptor-Effector Coupling; Receptor-Ligand Interactions Hulme, E.C. (Ed)
.... Spiralbound hardback 0-19-963096-8 **$193.00**
.... Paperback 0-19-963097-6 **$125.00**
Signal Transduction Milligan, G. (Ed)
.... Spiralbound hardback 0-19-963296-0 **$60.00**
.... Paperback 0-19-963295-2 **$38.00**
Human Cytogenetics: Volumes I and II as a set (2/e) Rooney, D.E. & Czepulkowski, B.H. (Eds)
.... Hardback 0-19-963314-2 **$130.00**
.... Paperback 0-19-963313-4 **$90.00**
Peptide Hormone Secretion/Peptide Hormone Action Siddle, K. & Hutton, J.C. (Eds)
.... Spiralbound hardback 0-19-963072-0 **$135.00**
.... Paperback 0-19-963073-9 **$90.00**

ORDER FORM for USA and Canada

Qty	ISBN	Author	Title	Amount
			S&H	
CA and NC residents add appropriate sales tax				
			TOTAL	

Please add shipping and handling: $2.50 for first book, ($1.00 each book thereafter)

Name ....

Address ....

....

.... Zip ....

[ ] Please charge $ .... to my credit card
Mastercard/VISA/American Express (circle appropriate card)

Acct. .... Expiry date ....

Signature ....

Credit card account address if different from above:

....

.... Zip ....

[ ] I enclose a cheque for $............

Mail orders to: Order Dept. Oxford University Press, 2001 Evans Road, Cary, NC 27513